The Structure of Small Molecules and Ions

The Structure of Small Molecules and Ions

Edited by

**Ron Naaman and
Zeev Vager**
Weizmann Institute of Science
Rehovot, Israel

Plenum Press • New York and London

Library of Congress Cataloging in Publication Data

International Workshop on the Structure of Small Molecules and Ions (1987: Neve Ilan, Israel)
  The structure of small molecules and ions / edited by Ron Naaman and Zeev Vager.
     p.   cm.
  "Proceedings of an International Workshop on the Structure of Small Molecules and Ions, held December 13-18, 1987, in Neve Ilan, Israel."
  Bibliography: p.
  Includes index.

  ISBN-13: 978-1-4684-7426-8   e-ISBN-13: 978-1-4684-7424-4
  DOI: 10.1007/978-1-4684-7424-4

  1. Molecular structure—Congresses. 2. Ionic structure—Congresses. I. Naaman, Ron. II. Vager, Zeev. III. Title.
QD461.I65   1987   88-25333
539′.6—dc19   CIP

Proceedings of an International Workshop on the Structure of
Small Molecules and Ions, held December 13-18, 1987, in
Neve Ilan, Israel

© 1988 Plenum Press, New York
Softcover reprint of the hardcover 1st edition 1988
A Division of Plenum Publishing Corporation
233 Spring Street, New York, N.Y. 10013

INTERNATIONAL WORKSHOP ON
"THE STRUCTURE OF SMALL MOLECULES AND IONS"
Held in the memory of Prof. Itzhak Plesser
Neve Ilan Guesthouse, Jerusalem, Israel
December 13 - 18, 1987

The Workshop is organized by the Department of Nuclear Physics, the Weizmann Institute of Science, Rehovot.

SPONSORED BY:

The Weizmann Institute of Science

ORGANIZING COMMITTEE:

Z. Vager, Rehovot - Chairman

R.S. Berry, Chicago

D.S. Gemmell, Argonne

U. Kaldor, Tel-Aviv

C. Lifshitz, Jerusalem

N. Moiseyev, Haifa

R. Naaman, Rehovot

The Organizing Committee wishes to express its thanks for the financial support to:

The Albert Einstein Center for Theoretical Physics at the Weizmann Institute of Science

The Israel Academy of Sciences and Humanities

Isramex; Representatives of Spectra-Physics

The Maurice & Gabriela Goldschleger Conference Foundation at the Weizmann Institute of Science

The Ministry of Science and Development

New Tec, RK; Representatives of Coherent, Quantel, and Lambda Physik

PREFACE

The workshop on "The structure of small molecules and ions" was held at the Neve-Ilan guest house, near Jerusalem, Israel on December 13 to 18 in memory of the late Professor Itzhak Plesser.

Professor Plesser played a central role in the research done both at the Weizmann Institute and at Argonne National Laboratories on the "Coulomb explosion" method. His friends honored his memory by organizing a meeting in which subjects related to Plesser's interests would be discussed. Just a week before the conference started we were struck by another tragedy - the death of our graduate student Ms. Hana Kovner, who participated in many of the Coulomb explosion experiments at the Weizmann Institute. We would like to dedicate these proceedings to her memory as well.

The goal of the workshop was to bring together chemists and physicists working on different aspects of the structural problems of small molecular entities. The time seemed appropriate for discussing experimental and theoretical concepts, since in recent years new methods have been introduced, and a large amount of information has been accumulated on systems not studied before, like unstable molecules, ions, van der Waals molecules and clusters. The program of the workshop reflects, we believe, these new developments.

The meeting was characterized by intensive discussions in which the weaknesses and strengths of new and of well established concepts were revealed. We hope that it measured up to the high standards Itzhak Plesser maintained all through his scientific life.

We would like to thank the organizing committee:C. Lifshitz, N. Mosayev, U. Kaldor, and special thanks to S. Berry who inspired us through all the organization. Many thanks to the conference technician Mr. Leo Sapir and to Ms. Ana Weksler, our secretary, who has to be admired for her patience and ability which helped greatly in making this meeting possible.

Ron Naaman and Zeev Vager

CONTENTS

COULOMB EXPLOSION

PHOTOIONIZATION AND PHOTODETECHMENT

ELECTRONIC STRUCTURE-THEORY

SMALL MOLECULES AND IONS:

A THING OR TWO WE STILL DON'T UNDERSTAND

R. Stephen Berry

Department of Chemistry
The University of Chicago
Chicago, Illinois 60637

INTRODUCTION

This essay is intended as a thorn to start off the discussions
of the Workshop on the Structure of Small Molecules and Ions in memory
of Professor Itzhak Plesser, by reviewing some of the aspects of
small molecules which we do not yet understand. In many of the
situations described here, we at least think we know how to express
relevant questions; in some cases, we clearly do not yet know how to
ask questions that we can address in a systematic way. And in some
cases, perhaps the tools for answering the questions are only now
emerging, some perhaps in the context of this Workshop.

The first group of subjects concern structure; these topics
lead us directly into a group of questions concerned with nonrigidity.
Next we will look at some problems of dynamics and some questions
regarding how we distinguish equilibrium properties from transient and
dynamic phenomena at the scale of the species on which this Workshop
focuses. At that point, I would like to change my perspective and ask
what we mean by "understanding" small molecules and ions. What do we
expect of models and what do we mean by "validating" or "testing"
them? What is the role of theory in studying small molecules? This
viewpoint forces us to look critically at what we mean by the *salient*
questions we should address to add to our understanding of small
molecules.

QUESTIONS CONCERNING STRUCTURE

We are comfortable with the concept of molecular structure.
That is, we accept with essentially no question[1,2] the idea that
molecules can be described by well-defined geometric structures
corresponding to the stable, minimum-energy forms that the nuclei
would assume if they were infinitely massive, around which the real
nuclei vibrate. Moreover these geometric structures permit us to
define Eckart frames of coordinates[3] which rotate with the molecule--
because the molecule does rotate more or less as a rigid body. Most
of us can disregard the paradox of identical-- i.e. indistinguishable-
-nuclei occupying inequivalent sites, whose inequivalence permits us
to distinguish those nuclei[1,4]. The reason we can overlook this
paradox is the long time required to establish particle identity by
any kind of exchange process. However the nuclear resonance
spectroscopist has dealt for many years with systems exhibiting both
inequivalence and equivalence, with temperature or medium determining

which aspect one observes[5]. And we now know some systems such as the Li3 molecule which can pass from one equilibrium geometry to another simply within the zero-point motions of their ground states[6]. Now, with our capacity to use multistep excitation with a single quantum or to carry our elaborate multiquantum transitions, we can bring small molecules up to all sorts of identifiable excited vibrational states, even very close to dissociation limits. Such states (I revert to classical language for convenience) may well exhibit large-amplitude oscillations or complex motions hardly classifiable even as multidimensional Lissajous figures. There is even some experimental reason to suspect that molecular rearrangements can occur in excited electronic states long supposed to differ at most from ground states in details of bond lengths and bond angles[7].

If a molecule is too nonrigid to be described in terms of an Eckart frame, it poses a difficult dilemma indeed, because, in our traditional bag of tricks, we have essentially no way to start to describe its nuclear motions. Even the validity of the Born-Oppenheimer approximation is questionable in such cases[8], although its basis of validity is more general than the original derivation indicated[9]. One approach has now appeared[10] which is efficient and powerful for describing the vibrational states of a triatomic molecule, thus far only for states with no total angular momentum, even states with as much as 3eV of vibrational energy and therefore possibly quite nonrigid. The application of this method, a description in which stretching modes are represented by distributed Gaussians and bending by a discrete numerical basis, to states of triatomics with J>0 and to larger molecules remains to be done. Still another group of approaches is based on separation of the modes and the equations that govern them. Those based on analogies with scattering problems are illustrated by, for example, the artificial channel method of Shapiro and collaborators[11], and by the more recent work on O3 by Atabek et al.[12] Separations based on analogy with the Born-Oppenheimer approach are illustrated by work of our own group[13].

One way to look at the problem of loss of rigidity is in terms of the magnitude of the effects that spoil the rigid rotor--small-amplitude oscillator model. These can be centrifugal or Coriolis interactions, or internal motions such as low-frequence bending modes, especially degenerate bending modes, or internal rotations, either of which carry angular momentum. If the magnitudes of the couplings or the spacings of the internal modes are at all comparable with the spacings of relevant rotational levels of the molecule as a whole, the rigid-rotor, small-amplitude oscillator model must fail. We ourselves have been pursuing one approach to this problem in the context of triatomic molecules, by treating the bending and rotations together, as slow modes in the effective average field of the faster stretching modes[13]. However the approach has not yet been brought to the stage of introducing the couplings between slow and fast modes so such important effects as the interaction between the state of H_2O with two quanta in the v_2 (bend) mode and that with one quantum of symmetric stretch, v_1, have not yet been accounted for; furthermore, as with the powerful approach of Bacic and Light, no application has yet been made to anything larger than a triatomic.

All the methods mentioned thus far and most of the others that allow us to treat molecular structures, dynamics and kinetics at the microscopic level require that we assume the Born-Oppenheimer approximation and develop a potential surface. We have already mentioned the question of the Born-Oppenheimer approximation and need only point out that it was originally derived to be valid just in the vicinity of stable minima on the potential surface, although it may be valid elsewhere as well, particularly on grounds different from those invoked by Born and Oppenheimer. But even then, we now have to check the validity of that approximation for each case individually, if we check it at all; it would be very helpful indeed if there were some general test that did not require solving the full electronic-vibrational problem in order to test the soundness of the separability of the two kinds of modes. Once we assume the approximation, we have to deal with the problem of constructing a potential surface. While we have the technology to determine points on this surface and even to use spectroscopic constants to determine some aspects of the shape of a surface around a minimum, we do not yet really know how to construct a potential surface in the sense of knowing how many points are required or where they should be chosen in the configuration space of the molecule if we are to be able to treat some chosen class of problems to some desired level of accuracy. Some thinking has of course been done about this issue[14] but in essence, the problem stands as a major barrier to our progress in understanding molecular dynamics.

NONRIGIDITY

The problem of nonrigidity plagues not only the interpretation of molecular behavior; it is even more pervasive in the description of electrons in atoms and of nucleons in nuclei, for which we usually start with an independent-particle picture, rather than a rotor-vibrator model. With such tools as single-photon multistep spectroscopy and multiple-resonance many-photon spectroscopy, we can now observe assignable rotation-vibration levels even up virtually to the limits of dissociation. Many of such high levels of many molecules, and even the lowest-lying levels of a few systems, can be expected to be associated with large-amplitude motions (here I again lapse into the language of classical description) and therefore with nonrigid behavior.

One well-established kind of breakdown of the collective model of molecular structure is local-mode excitation[15]. The transition from collective, normal, harmonic modes of the O-H stretching motions in H_2O to local, anharmonic (but of course symmetric or antisymmetric) modes is well understood[16,17]. In general this behavior occurs when one class of mode such as the hydrogen stretching mode is not strongly coupled with the other modes of the molecule, *and when the effects of anharmonicity are stronger than the harmonic couplings* among the modes of the class of interest. If anharmonicity is that important, normal modes lose their meaning. Hence localization becomes more important with increasing energy. Not all localization of vibrational modes is restricted to oscillations of individual atoms; in some of the octahedral hexafluoride molecules, the normal, three-fold degenerate (vector-like) stretching modes transform with increasing energy from vibration of two axial fluorines against four equatorial fluorines into just the antisymmetric stretching mode of the two axial fluorines, but so far as we know there is no evidence that the two fluorines decouple from each other[17]. However there is no evidence yet for bound states in which individual atoms are excited enough to have their own *angular momentum* quantum numbers, which would be one signature of extreme nonrigidity. The closest thing to such states are orbiting resonances in low-energy collisions. One way to search for nonrigidity in highly excited vibrational states would be to look for a transition in a pattern of levels, e.g. in a triatomic molecule such as H_2O, from bending and overall rotation to independent-particle rotation, a change with a relatively unambiguous signature.

DYNAMICS

One way to approach an understanding of the dynamics of small molecules is to ask what happens to energy deposited in such a molecule. This is not a precise enough question, as it stands, to be useful, but it sets a direction. One direction that has been particularly popular in recent years has been to ask how separable are the vibrational modes. The significance of this is, in a diffuse sense, to inquire to what extent do all the atoms of the molecule share the energy; in a more precise way, it asks whether there are approximate constants of the motion, in addition to the total energy and total angular momentum, which characterize excited rotation-vibration states. The approximate constants of the motion in question would be at least the energies of subsystems of the entire molecule. The lowest states of most molecules are of course characterized by the assignment of quantum numbers to individual normal modes, and thus by constants of the motion in addition to the total energy. Higher on the energy scale, as the previous section's discussion indicated, some states are still characterized by modal quantum numbers but the modes take on the meaning of local rather than normal modes. Whether these are representative of most of the highly excited states, or are an unusual subset is the question. One interpretation holds that the states with extra approximate constants of motion constitute a very small set among a much larger set characterizable only by their total energy. Another holds that a large fraction of the excited states, even up to very high energies, can be characterized by assignment to particular modes, at least in an approximate fashion.

In one sense, this question can be settled at least on a case-by-case basis by a detailed search for suitable approximate modes, in a manner such as that proposed by Kellman and Lynch[18]. However it may well be--my view is that it is inevitable-- that we will have to face up to resolving what we mean by "approximate constants of the motion" or, more precisely put, *quantifying* the degree to which a quantity fails to be constant. The way to do this is clear[2] but has not been done. Approximate separability implies that we can write the full Hamiltonian $\mathcal{H}$ in terms of two parts,

$$\mathcal{H} = \mathcal{H}_o + \mathcal{H}_1$$

in which the first term corresponds to the part presumed approximately separable and the second contains the rest of the Hamiltonian including the parts that spoil the separability of $\mathcal{H}_O$. The time evolution of any operator is given by its commutator with the full Hamiltonian. If an observable is a constant of the motion, its operator commutes with the Hamiltonian. The deviation of that observable from constancy is therefore given by the expectation value of the commutator of the corresponding operator with the Hamiltonian:

$$\langle d\mathcal{H}_o/dt \rangle = (i\hbar)^{-1} \langle [\mathcal{H}, \mathcal{H}_o] \rangle$$

Hence one effective way to characterize how much an observable in a particular state deviates from being a constant of the motion is to specify the time required for the energy of that state to deviate halfway to the next quantum state, that is, to specify the time required for the state to lose the identity we would like to attribute to it. The same kind of criterion can be used with other operators such as the angular momentum of some subset such as an internal rotor, some l_z; we can ask how long a time is required for its expectation value to wander from its original value halfway to the next eigenvalue. The difficulty in carrying out such calculations is somewhat like the difficulty in computing variational upper bounds: one must evaluate the rather troublesome operator, the square of the Hamiltonian. The expectation value of a commutator is, in general, more difficult to evaluate than the expectation value of most of the operators with which we usually deal. However such evaluations are within our present capacities, at least for simple systems. This then may be one possible way to approach the quantification of non-constancy.

Another approach to inquiring what happens to the energy in an excited molecule is to invoke the Correspondence Principle and use classical mechanics to give us answers. We can in that way ask how an excited molecule fills or otherwise occupies its phase space. This method will surely be used and is already being applied to triatomic molecules by Sumpter and Ezra[19] and by ourselves[20]. Here, the same problem of quantifying a suitable measure arises: how can we quantify the degree of ergodicity of a molecule? A related but different and easier question is: how can we quantify the degree of chaos of a system? The question of ergodicity is related to a specific trajectory and therefore to a characteristic that can be related to a quantized trajectory and thence to a quantum state. It is therefore likely to be useful to quantify the ergodicity of a system. Whether it will be useful to quantify the degree of chaos of a classical counterpart of a real system is much more questionable and we will not enter here into that controversy. The degree of chaos of a classical system can be quantified by its Liapunov spectrum and its Kolmogorov entropy, which measure the degree to which trajectories diverge from one another. Whether a property relating various classical trajectories to each other has a useful or meaningful quantum counterpart is one way of putting that question which we now drop, as we return to the question of ergodicity.

Quantum ergodicity is definable[21]; whether it is calculable and useful is still open. Classical ergodicity makes its appearance in at least two ways, which require different measures. One is the specification of the volume of the available phase space swept out by classical trajectories. The other is the specification of the dimensionality of the manifold on which the phase point moves. Neither of these alone tells us how well a classical trajectory fills its phase space; together they tell us a great deal about how that phase space is filled. The dimensionality can be calculated, albeit inefficiently, with the algorithm of Grassberger and Procaccia[22]. It will be helpful to have more efficient ways to do those calculations because there is little hope of computing the dimensionality of any system of more than six or eight particles, at least with computers now available. Then we will need to quantify the degree to which the trajectory fills the manifold whose dimensionality we calculate.

SALIENT QUESTIONS

This essay concludes with a little discussion of what it is we want to know about atoms and molecules, that is, of what constitute salient questions and what it is that we consider improvement in our "understanding" of these things. I will particularly scrutinize the role of theory, but will have to ask a bit about what constitutes a salient experiment as well.

Much of the effort in theoretical studies related to molecular structure in recent years has exploited our vastly increased power to compute. The variational principle guarantees that if we have the correct equation and sufficient computing power, we can achieve as accurate a set of wave functions as we wish, and therefore can reproduce or predict the outcome of an experimental measurement as closely as we wish. Practicalities frequently change that situation into one of achieving a replication or prediction as accurate as we can afford, but the principle remains.

New insights, from this style of approach, come in two ways. One is inventive realization of new phenomena consistent with accurate equations we already know, as, for example, with the prediction of many of the effects of intense laser fields, which are all derivable from the Schroedinger equation but with the radiation field treated much more carefully than in traditional one-photon spectroscopy[22]. The other is the realization by comparison of theory and experiments that the equation cannot be the right one. But we then ask whether there is any reason to use anything other than the "correct" equation, at least for atomic and molecular phenomena. We all know the answer to that: there are many, many situations in which we have no expectation of using the "exact" equation, and make do with some approximation. But for the present, let us stick with situations in which we use equations we know are likely to be accurate, not equations constructed deliberately to represent simplified models of very complicated real situations.

In the best of cases, we can find the ranges of validity and stay within them when we use such approximate but "good" equations. Sometimes we are happily surprised by finding that the range of validity of an approximation is broader than we expected; the Born-Oppenheimer approximation is one, as we discussed previously. Sometimes we can be less happily surprised, for example, if we try to describe valence electrons of elements in the middle or bottom of the Periodic Table with effective core potentials based on nonrelativistic treatments of their cores. And sometimes we are trapped between having tractable equations we know are insufficiently accurate, on one hand, and on the other, intractable but less approximate equations which are at best likely to be accurate enough for current needs. This is a situation illustrated now by the state of affairs for developing potential surfaces extensive enough to model dynamics of clusters of metal or semiconductor atoms. In this illustration, the idea of developing dynamics, classical, semiclassical or quantum, for the behavior of atoms on a potential surface is already rooted in the Born-Oppenheimer approximation, an approximation likely to break down at least in selected regions of such potential surfaces. So the point of this paragraph is that, apart from ingenious new phenomena we might discover, the most important function of atomic and molecular theory-- used in the mode we are now discussing, i.e. of comparing experimental results with the most accurate results we can obtain with a given equation--is to validate the equation or, in the case of a well validated equation, to raise the question of the interpretation or accuracy of the measurement.

If we are using an equation whose range of validity is rather well established, as it is for the nonrelativistic Schroedinger equation, unless we find a very surprising discrepancy, we add little to our stock of new knowledge or understanding. "Proving the Schroedinger equation again" is a term of opprobrium used to describe experiments whose outcome is so predictable that one wonders what, other than elegance, justifies them. An analogous situation occurs with computations that add nothing but higher accuracy and more terms in a variational series. This is not the case, of course, when one shows that a new kind of series is advantageous, for example in the use of logarithmic terms or the equivalent in the variational calculation of the ground state of the helium atom. But the unimaginative application of known methods to the replication of high-quality experimental data is not a useful way to deepen our understanding.

There are, I believe, other ways in which we can use theory now to deepen our understanding of atomic and molecular phenomena. One involves the use of approximate Hamiltonians in the manner of deliberate simplifications of reality; another requires that we use accurate Hamiltonians and compute accurate solutions to the equations, but then analyze the results in terms of simple models. Both approaches are directed toward finding very good or best approximations which we can then interpret in terms of simple, comprehensible physical pictures. This viewpoint is a direct challenge to any claim that possessing a 1000-term variational function with high accuracy is *in itself* equivalent to having understanding. This may be a straw man that I challenge, but I think the point should be made in a time when we have the power to generate those accurate, 1000-term functions. Having such a function is like having a key to the library; being able to open the door is not quite equivalent to having read the books.

The first of our two paths is one way to find what approximate model, and particularly what approximate constants of motion and quantum numbers are suitable for representing a real system. Model Hamiltonians are vital in the early stages of a subject, as in the use of the Huckel Hamiltonian and the π-electrons alone to represent the significant properties of conjugated hydrocarbons. Such uses are best treated as stepping stones, to be left behind as soon as a firmer foundation can be given to the subject. And in the case of the conjugated hydrocarbons, we were able to do so. But sometimes it can be useful to return to systems we can already describe accurately and try to use new, simple first-step models because they may lead us to new ways of thinking about the problems or to new insights into the constants of the motion or the separability or ergodicity of the system. An example of this is the use of algebraic Hamiltonians to describe the rotations and vibrations of small molecules[18,24]. The potential of this approach has not yet been fully explored by any means but some new insights are emerging which the traditional approaches such as the Dunham expansion have not revealed to us.

The other direction, which we have thus far been pursuing in the context of electronic structure of atoms rather than the atomic structure of molecules, is to analyze the results of accurate calculations in terms of simple models. In particular, one can *interpret* the very complicated but accurate solutions to accurate equations by projecting those solutions onto the accurate solutions of equations for simpler systems. When we examine the coefficients of a configurational expansion, and particularly when we pick out the largest coefficient, we are carrying out such a projection--in this case, onto the solutions of an effective, independent-particle equation based on a mean-field approximation. The physics we learn is a measure of how good the independent-particle quantum numbers are, that is, how nearly constant the single-particle constants of motion are. But a well-converged variational wave function can be examined by projection onto other basis sets than that in which it was developed. By carrying out projections of well-converged configurational expansions onto optimized rotor-vibrator functions, Hunter and I found[25] that the ground and most but not all of the low-lying excited states of the alkaline earth atoms are extremely well represented by rotor-vibrator functions. The implications are tantalizing but as yet unexplored, that perhaps a) the accurate representation of the valence electrons of the alkaline earth atoms can be accomplished significantly more efficiently by expansion in rotor- vibrator functions than by a configurational expansion, and b) the valence electrons of the Group 3, Group 4 and perhaps other elements may be more accurately described in terms of collective rotor-vibrator quantum numbers than in terms of one-electron quantum numbers.

The point of describing this second path here is that we are learning to compute accurate wave functions to describe the behavior of atoms within molecules and therefore should prepare ourselves to go beyond the replication of numbers extracted from molecular spectra. We should plan to examine those accurate functions to interpret what they mean, by comparing them with the wave functions that correspond to plausible simplified models of the complex systems. Understanding, in these terms, comes from seeing how closely a complicated thing can be made to resemble something we can picture and describe in detail-- in this case, by comparing the accurate description of the real object with accurate descriptions of several, perhaps many simple objects. I conjecture that this approach will be a useful means to interpret the motions of nonrigid molecules, in which there are too many dimensions for us to visualize the information contained in an accurate wave function. We will, however, be able to construct models for different kinds of motion that we can interpret, solve the corresponding model equations and project their solutions onto the accurate solution. The results will give us a heirarchy of validity of the models, on which we can then base our physical interpretation.

ACKNOWLEDGMENT

    That research described here which was done by our group at the
University of Chicago was supported by a Grant from the National
Science Foundation.

REFERENCES

1. c.f. R. G. Woolley, J. Am. Chem. Soc. **100**, 1073 (1978) and R. G.
Woolley, in *Molecules in Physics, Chemistry and Biology* , Proc. Conf.
dedicated to Prof. R. Daudel, Paris, June, 1986 (to be published), for
concerns about this point.
2. R. S. Berry, Rev. Mod. Phys.**32**, 447 (1960).
3. J. D. Louck and H. W. Galbraith, Rev. Mod. Phys. **48**, 69 (1976).
4. R. S. Berry, Rev. Mod. Phys. **32**, 447 (1960).
5. c.f. J.-F. Labarre, *Structure and Bonding* **35**, 1 (1978).
6. J. L. Martins, R. Car and J. Buttet, J. Chem. Phys. **78**, 5646 (1983)
and refs. therein; for an excited state of Na_3, see G. Delacretaz, E.
R. Grant, R. L. Whetten, L. Wöste and J. Zwanziger, Phys. Rev. Lett.
56, 2598 (1985).
7. c.f. I. Ohmine, J. Chem. Phys. **83**, 2348 (1985).
8. M. Born and J. R. Oppenheimer, Ann. Phys. **84**, 457 (1927).
9. H. Essén, Int. J. Quantum Chem. **12**, 721 (1977).
10. Z. Bacic and J. C. Light, J. Chem. Phys. (in press); I. P.
Hamilton and J. C. Light, J. Chem. Phys. **84**, 306 (1986); J. V. Lill,
G. A. Parker and J. C. Light, J. Chem. Phys. **85**, 900 (1986).
11. M. Shapiro and G. G. Balint-Kurti, J. Chem. Phys. **71**, 1461 (1979);
E. Segev and M. Shapiro, J. Chem. Phys. **77**, 5604 (1982).
12. O. Atabek, S. Miret-Artes and M. Jacon, J. Chem. Phys. **83**, 1769
(1985).
13. G. A. Natanson, G. S. Ezra, G. Delgado-Barrio and R. S. Berry, J.
Chem. Phys. **81**, 3400 (1984); ibid. **84**, 2035 (1986).
14. c.f. I. Csizmadia, in *Symmetries and Properties of Non-Rigid
Molecules,* J. Maruani and J. Serre, eds. (Elsevier, New York, 1983),
p.315;D. Liotard, ibid., p.323.
15. B. R. Henry, Acc. Chem. Res. **10**, 207 (1977).; Vib. Spectra
Structure **10**, 269 (1982).
16. R. T. Lawton and M. S. Child, Mol. Phys. **37**, 1799 (1979); M. S.
Child and R. T. Lawton, Chem. Phys. Lett. **87**, 217 (1982).
17. L. Halonen and M. S. Child, J. Chem. Phys. **79**, 559 (1983); M. S.
Child,Accts. Chem. Res. **18**, 45 (1985).
18. M. E. Kellman and E. D. Lynch, J. Chem. Phys. **85**, 5855 (1986).
19. B. G. Sumpter and G. S. Ezra, Chem. Phys. Lett. (submitted); G.
S. Ezra, C. C. Martens and L. E. Fried, J. Phys. Chem. **91**, 3721
(1987).
20. T. L. Beck, D. M. Leitner and R. S. Berry (submitted for
publication); D. M. Leitner, T. L. Beck and R. S. Berry (submitted for
publication).
21. E. Stechel, J. Chem. Phys. **82**, 364 (1985).
22. P. Grassberger and I. Procaccia, Phys. Rev. Lett. **50**, 346 (1983);
ibid., Physica **9**D, 189 (1983); ibid., Phys. Rev. A **28**, 2591 (1983).
23. c.f. R. Loudon, *The quantum theory of light* , Second Edition
(Oxford University Press, Oxford, 1983).
24. F. Iachello and R. Levine, J. Chem. Phys. **77**, 3046 (1982); O. S.
van Roosmalen, F. Iachello, R. D. Levine and A. E. L. Dieperink, J.
Chem. Phys. **79**, 2515 (1983); I. Benjamin, R. D. Levine and J. L.
Kinsey, J. Phys. Chem. **87**, 727 (1983); O. S. van Roosmalen, I.
Benjamin and R. D. Levine, J. Chem. Phys. **81**, 5986 (1984); C. E.
Wulfman and R. D. Levine, Chem. Phys. Lett. **104**, 9 (1984).
25. J. E. Hunter III and R. S. Berry, Phys. Rev. A **36**, 3042 (1987).

APPROXIMATE SEPARABILITY AND CHOICE OF COORDINATES FOR EXCITED VIBRATIONS OF POLYATOMIC MOLECULES AND CLUSTERS

R.B. Gerber and T.R. Horn

Department of Physical Chemistry and
The Fritz Haber Research Center for Molecular Dynamics
The Hebrew University, Jerusalem 91904, Israel

and

M.A. Ratner

Department of Chemistry
Northwestern University
Evanston, Illinois 60201, USA

ABSTRACT

To simplify the dynamics of coupled anharmonic vibrations in polyatomic systems, methods such as the self-consistent field (SCF) approximation and the adiabatic approximation assume that motions in different modes are separable. Success of such methods thus depends considerably on a good choice of the coordinates that are being mutually separated.

This study examines the physical considerations that can be used in making adequate choices of coordinate systems in SCF calculations of small molecules and van der Waals clusters. Both Cartesian and curvilinear coordinate systems are discussed. Part of this article reviews results of previous studies on this topic, e.g. the optimization of Cartesian coordinates by coordinate rotation, and the introduction of ellipsoidal coordinates for the bending-stretching spectrum of HCN. New results are presented for $Xe(He)_2$, a prototype of the "Three Balls" problem; and for I_2He, a prototype of the "Stick and Ball" systems. Comparative SCF calculations using hyperspherical, ellipsoidal and Jacobi coordinates are made, in a full 3D framework. Hyperspherical modes are found optimal for the "Three Balls", ellipsoidal coordinates prove optimal for the "Stick and Ball". Physical explanation for these findings is offered, and related insight is obtained into the vibrational motions involved. Suggestions are made of possible extensions of some of the above themes for other systems.

I. INTRODUCTION

The energy spectra and the dynamics of motion of coupled anharmonic oscillators have been the focus of extensive theoretical research in the last few years. Undoubtedly, much of this activity is motivated by remarkable progress in two pertinent experimental areas: The study of highly excited vibrational states of (chemically bound) small polyatomic molecules, such as HCN, C_2H_2, CH_2O, etc.[1], and the spectroscopic investigation of van der Waals clusters, such as I_2He, ArHF, CO_2HF, etc.[2] The theoretical interpretation of experiments in these areas requires quantitative treatment of several vibrational degrees of freedom in the anharmonic regime, where significant coupling between the modes is involved. For the van der Waals clusters this situation is often pertinent even to the ground state and to the lowest excited states, since anharmonicity in such systems may be substantial even for these levels. Both with regard to the van der Waals species and to the highly excited states of (chemically bound) molecules, an important aspect of the challenge to theory is that one must typically account in these cases for large amplitude motions. These create difficulties for many of the available techniques for energy level calculations. For instance, consider expansion into a basis of eigenfunctions of some zeroth-order Hamiltonian in which the modes are uncoupled: Since the true eigenfunctions of the coupled-mode system are spread across a wide region of configuration space, many basis states are required to represent each delocalized eigenfunction. (A ratio of 10-50 basis functions per calculated eigenstate is often necessary). Progress in recent years in dealing with coupled anharmonic modes, and in particular with large amplitude motion in such systems, has led to the development of new, promising "numerically exact" methods[3], and also to the emergence of powerful approximation methods. As an example of a successful and elegant "numerically exact" algorithm, we mention the DVR-DGB (Discrete Variable Representation - Distributed Gaussian Basis) technique of Light and Bacic: This method efficiently distributes a basis set across the region of the large-amplitude motion, avoiding pitfalls of some of the conventional basis-expansion procedures. Even this method could thus far be applied only to triatomic systems. For all but the smallest molecules, approximation methods therefore appear indispensable.

Among the most fruitful approximations suggested for this problem appear to be those which assume separability between motion in different modes, in one sense or another. The self-consistent field (SCF) approximation[4-6], and the adiabatic approximation[7-9] are both of this type. Separability of modes in the sense introduced in the SCF method, or in that employed in the adiabatic approximation, is far more refined a notion than the mere neglect of coupling between the modes. Clearly, as in all approximations, one trades in these methods some loss of accuracy for a gain in simplicity of computation and insight. The quantitative results of these methods are, however, often of striking accuracy[4-6,10]. It is recognized now that success of the SCF method (and indeed also of the adiabatic approximation) in any given application depends on the choice of coordinates that are treated as mutually separated[6,11]. In fact, the optimal choice of coordinates for SCF (or adiabatic) separation may depend in general on the state and energy of the system[11]. A general and practical algorithm for making the optimal choice of coordinate is not yet available. However, from physical intuition and from experience in the application of the method to various systems, insight is beginning to emerge as to which types of coordinates are suitable for certain classes of molecules (characterized by mass ratios, properties of the potential surface, etc.).

In this article we examine the choice of "good coordinates" for SCF in several cases of molecules and clusters. Part of the material reviews previous results on this topic including the variational method for coordinate optimization. New results presented here focus on determining the best coordinates for two types of van der Waals clusters: The "Three Balls Problem" - three weakly bound atoms; and the "Ball and Stick Problem" - an atom weakly bound to a linear molecule. This part reports calculations on realistic models of $Xe(He)_2$, I_2He, and it becomes apparent how successful the SCF method can be when carried out in suitable coordinates, and how important a good coordinate choice is for improving the results.

II. THE SCF APPROXIMATION AND COORDINATE OPTIMIZATION

The vibrational Self-Consistent Field approximation was developed rather recently[12], although the corresponding method for electrons in atoms and molecules dates back to the earliest stages of quantum theory. Several reviews of the vibrational SCF were published in the last few years[4-6], and the emphasis in Ref. 6 is particularly close to the topic of the present discussion. We therefore restrict ourselves to a very brief summary of the method.

Consider a molecule having N vibrational modes. Its Hamiltonian in a set of given Cartesian coordinates $q_1, ..., q_N$ is given by:

$$H = - \sum_{i=1}^{N} \frac{\hbar^2}{2m_i} \frac{\partial^2}{\partial q_i^2} + V(q_1, ..., q_N) \tag{1}$$

where m_i is the mass associated with the mode i. The SCF method approximates each eigenfunction of (1) by a product of single-mode wavefunctions:

$$\Psi(q_1, ..., q_N) = \prod_{i=1}^{N} \phi^{(i)}(q_i). \tag{2}$$

Using (2) to evaluate the expectation value of (1), then employing the variational principle to determine the best single-mode wavefunctions leads to the SCF equations:

$$h_i^{SCF} \phi_n^{(i)}(q_i) = \varepsilon_n^{(i)} \phi_n^{(i)}(q_i) \tag{3}$$

where

$$h_i^{SCF}(q_i) = - \frac{\hbar^2}{2m} \frac{\partial^2}{\partial q_i^2} + \bar{U}_i(q_i) \tag{4}$$

$$\bar{U}_i(q_i) = < \prod_{j \neq i} \phi^{(j)}(q_j) | V(q_1, ..., q_N) | \prod_{j \neq i} \phi^{(j)}(q_j) > . \tag{5}$$

The SCF approximation for the total energy of the system is given by:

$$E_{n(1),...,n(N)} = \sum_{j=1}^{N} \varepsilon_{n(j)}^{(j)} + (1-N) < \prod_{j=1}^{N} \phi_{n(j)}^{(j)}(q_j) \,|\, V \,|\, \prod_{j=1}^{N} \phi_{n(j)}^{(j)}(q_j) > , \qquad (6)$$

where $n(i)$ is the quantum number of the i mode. There is no reason to expect good results from SCF in an arbitrary coordinate system. The above equations are variationally the best way to separate the motions in $q_1,...,q_N$ but motions in these coordinates may be strongly coupled in a given problem. Separability in the SCF sense is clearly an improvement over straightforward neglect of the coupling of the modes, since by Eqs. (3)-(5), each mode in the SCF field has the effect of all the other modes through an effective averaged field (the $\bar{U}_i(q_i)$). Still, this separability, as any other approximate separability, may be better in some coordinates than in others. Consider, for example, a linear transformation on the fixed coordinates $q_1, \cdots, q_N$:

$$Q_i = \sum_{j=1}^{N} \alpha_{ij} q_j \quad \text{for} \quad i = 1,...,N \ . \qquad (7)$$

The SCF approximation can also be applied with respect to the Q_i, that is:

$$\Psi(Q_1,...,Q_N) = \prod_{i=1}^{N} \tilde{\phi}^{(i)}(Q_i) \qquad (8)$$

where the $\sim$ sign indicates that the wavefunctions in (8) are not necessarily the same as in (2). SCF equations of the form (3)-(7) hold for the best $\tilde{\phi}^{(i)}(Q_i)$, given the coordinates Q_i. However, we can now ask for the "best" coordinates Q_i which yield the best SCF. This is translated variationally into the condition

$$\frac{\delta E_{n(1),...n(N)}}{\delta \alpha_{ij}} = 0 \qquad (9)$$

for an independent subset of the α_{ij}. (The α_{ij} are restricted by the orthogonality condition).

Actually, the SCF equation need not be restricted to Cartesian modes. Likewise, the search for an optimal coordinate systems should not necessarily be confined to the framework of linear transformations, Eq. (7). Indeed, it will subsequently be seen in several realistic examples that good coordinates for SCF are often curvilinear[11], and the search for the optimal one involves nonlinear transformations. The search for good curvilinear coordinates is not a trivial one, and general algorithms are not yet available. Successful applications of the method have relied so far on a physically motivated guess. If the proposed coordinate system depends also upon one or several continuous parameters (which implies that a family of coordinate systems is available), then the variational principle can be used to determine the optimal modes, as in Eq. (9).

While a generic procedure for the construction of suitable coordinates in a given system is not yet available, there are several obvious guidelines one can employ in making a reasonable guess: (i) The modes chosen should be such that the effect of short-range hard-wall repulsions between any pair of atoms can be accommodated within a single mode description: The coordinates of the various atoms in a molecule cannot be used in an SCF calculation, because this is tantamount to ignoring a correlation consisting of hard-wall repulsion between the atoms, which correspond to very large errors. This serious problem does not arise in bond modes (local modes) or in normal modes, where the repulsive wall part of the interatomic potentials can be incorporated within the single-mode description. Another way of stating the above conclusion is that one should not use in SCF calculations coordinates that can produce an artificial overlap of the repulsive cores of the constituent atoms in the molecule[13]. While this excludes single-atom positions and several other related choices as SCF modes, this restriction still leaves an extremely wide range of coordinates. (ii) One should choose coordinates that are as far as possible from yielding a zero-order resonance between the modes: Degeneracy clearly breaks the SCF treatment, and amplifies the effects of correlations[13]. (iii) The shape of the minimum-energy path associated with a large amplitude motion is often a useful guide in choosing the optimal modes: The modes should be such as to naturally follow the minimum energy path. The study of Bacic *et al.*[11] of excited bending states of the HCN ⇌ HNC system illustrates that clearly. In this case the minimum energy path is roughly ellipsoidal in shape, and ellipsoidal coordinates proved very successful in the SCF calculations. The excited bending states of the I_2He cluster, results on which are shown in the present study, offer another example. (iv) In studying a new molecular system it may be possible to draw on previous experience from a similar or related system. The pertinent similarity may be in the mass ratios or in the potential energy surface. For instance, the mass ratios in the HCN molecule suggest the usefulness of ellipsoidal coordinates for treating the exciting bending motions in that system. Qualitatively similar mass ratios are found in I_2He, and indeed ellipsoidal coordinates prove optimal for describing the bending states in this case as well.

We now proceed to examine the choice of good coordinates in several applications of SCF theory, focussing on the physical considerations involved in these cases.

III. EXAMPLES AND ANALYSIS OF GOOD COORDINATE CHOICES

1. *Coordinate rotation and optimal Cartesian modes*[14-18]: The set of all Cartesian coordinates for a given molecule can conveniently be generated by applying all linear, orthogonal transformations to any given Cartesian coordinate system for that molecule. This yields the convenient parametrization Eq. (7) of the Cartesian modes. The transformations involved are, of course, coordinate relations. In a two-mode system one can write:

$$Q_1 = q_1 \cos\alpha + q_2 \sin\alpha: \quad Q_2 = -q_1 \sin\alpha + q_2 \cos\alpha \qquad (10)$$

which represents any Cartesian system in terms of the fixed system $\{q_1,q_2\}$ and the transformation parameter α. The consequences of such coordinate rotations in the context of SCF calculations were pursued by Truhlar and coworkers[15,16], by Lefebvre[17] and by Moiseyev[18]. Thompson and Truhlar[16] used coordinate rotation combined with the variational condition (9) in SCF calculations of low-lying vibrational excited states of H_2O. The energies obtained with the optimal Cartesian modes gave rms errors that were only about 40% of the rms errors in the SCF calculations using normal modes. This indicates that mode optimization even in the Cartesian framework above can significantly improve the SCF approximation. An important advantage in using Cartesian coordinates is the mathematical simplicity. Physically, the choice of such coordinates may be motivated in some systems by the fact that both local modes and normal modes are of this type. *Cases which are expected to be transitional between local and normal mode behaviour seem therefore obvious candidates for an SCF/optimal Cartesian coordinates treatment.* A comparison of SCF calculations with normal and with local modes was reported by Roth *et al.*[14] for overtone stretching excitations in HDO and DTO[19]. The Hamiltonian used in this study included only momentum coupling between the two bond (local) modes. If SCF in local modes is used, the SCF average of the momentum coupling vanishes, hence the mean field method in this case offers no improvement beyond the straightforward neglect of the coupling between the modes. In normal modes there is no momentum coupling, and the coupling between the modes for the same Hamiltonian enters in the potential function. The SCF correction for this interaction using normal mode separation does not vanish, so the mean field method in these modes offers more than straightforward neglect of the coupling. It was thus found that the results for the excited states of HDO and DTO were much better when SCF in normal, rather than local modes is used. Only for H_2O itself, where the Wilson momentum coupling between the two O-H modes is very weak does SCF in local modes (as in the decoupled local modes model) emerge as better than SCF in normal modes. This further supports the conclusion mentioned previously: That coordinate optimization of SCF in the Cartesian framework (by coordinate rotation) should prove very useful in systems and states that correspond to the transition range between local and normal mode behavior, such as the first 5 or 6 overtones in HDO, DTO.

2. *Hyperspherical versus Cartesian modes for stretching vibrations of triatomic molecules*: Unlike the case of Cartesian coordinates, no method is available for optimization of SCF over the entire set of curvilinear coordinate systems, or over a large subset of this set. Several familiar curvilinear coordinate systems were however tested in SCF calculations. Gibson *et al.*[19] compared SCF in hyperspherical modes with SCF in both local and in normal modes for the stretching vibration in H_2O and $C^{18}O^{16}O$. The calculations were carried out in a framework that includes only the stretching modes, neglecting both the bending vibrations and molecular rotations. For a linear, nonbending triatomic molecule ABC the hyperspherical coordinates used r and ϕ were defined by[19]:

$$r\cos\phi = [r_{AB} + \mu_2 r_{BC}]/\mu_1 \tag{11a}$$

$$r\sin\phi = r_{BC} \tag{11b}$$

and the mass ratios μ_1, μ_2 are given by:

$$\mu_1^2 = \frac{m_B m_C (m_A + m_B + m_C)}{m_A (m_B + m_C)^2} \;\; ; \;\; \mu_2 = \frac{m_B}{m_B + m_C} \;\; . \tag{12}$$

The Hamiltonian in these coordinates is:

$$H = -\frac{\hbar^2}{2\mu_{BC}}[\frac{1}{r}\frac{\partial}{\partial r} r \frac{\partial}{\partial r} + \frac{1}{r^2}\frac{\partial^2}{\partial\phi^2}] + V(r,\phi) . \tag{13}$$

where μ_{BC} is the reduced mass of B and C. The SCF factorization for the wavefunction is then

$$\psi_{SCF} = \Phi(\phi)R(r) . \tag{14}$$

The SCF equations in these modes are:

$$\left[<R\,|-\frac{\hbar^2}{2\mu_{BC}r^2}\,|R>\frac{d^2}{d\phi^2} + <R\,|V|R> -\varepsilon_\phi \right] \Phi(\phi) = 0 \tag{15}$$

$$\left[-\frac{\hbar^2}{2\mu_{BC}r}\frac{d}{dr}\left[r\frac{d}{dr} \right] + <\Phi|V|\Phi> -\frac{\hbar^2}{2\mu_{BC}r^2}<\Phi\,|\frac{d^2}{d\phi^2}\,|\Phi> -\varepsilon_r \right] R(r) = 0 , \tag{16}$$

The total energy in this case is just ε_r. Table I lists the vibrational energies computed by Gibson *et al.* for the low-mode model of $C^{18}O^{16}O$. The energies from SCF in hyperspherical modes, E_{hs}^{SCF}, are compared with these from SCF in normal modes, E_{nor}^{SCF}, and with the numerically exact energies for this model.

Table I. Computed vibrational energies (eV) for a two-mode model of $C^{18}O^{16}O$.
The results are from Ref. 19.

State	E_{hs}^{SCF}	E_{exact}	E_{nor}^{SCF}
0,0	.2303	0.2309	0.2309
1,0	0.3849	0.3849	0.3862
0,1	0.5305	0.5306	0.5316
2,0	0.5390	0.5377	0.5404
1,1	0.6824	0.6808	0.6858
3,0	0.6932	0.6894	0.6934
0,2	0.8259	0.8248	0.8264

The SCF result is lower than the exact one for the (0,0) level because of a minor approximation mode in the evaluation of the potential in hyperspherical coordinates (for computational convenience). The hyperspherical SCF is of superb accuracy (giving an average deviation of 0.2% only from the exact results for the states in Table I!), and it offers a significant improvement over SCF in normal (and also local) coordinates in this case. The same is found in the case of H_2O, but there the improvement made by using hyperspherical modes is not as large as in the case of CO_2.

The physical basis for the success of hyperspherical coordinates in the above example are twofold: First, the hyperspherical modes have a highly delocalized character, and the displacements they describe are spread over all the atoms involved. Such delocalized "waves" involve less correlations than highly localized motions. In particular, spurious "hard collisions" between the atoms, in which they penetrate into the repulsion core of the mutual interaction between them are avoided. The hyperspherical modes seem to be more delocalized for the systems studied than the normal modes, which are also collective. A second reason for the success of the hyperspherical is that the very different nature of the r and the ϕ motions assure a good separation of frequencies between these two modes - the situation is even further from a resonance condition in hypersphericals than in normal modes. We expect this to be quite typical for coupled stretching vibrations, *hence hyperspherical modes are expected quite generally to yield better* SCF *results than any Cartesian coordinates for the stretching motions.*

3. *Good curvilinear coordinates for large amplitude bending-stretching - ellipsoidal modes for* HCN: The energy levels corresponding to the coupled bending and stretching H-C vibrations in HCN were studied by Bacic *et al.*[11] (The CN distance was kept frozen at equilibrium in these calculations). The focus was on high bending excitations, which, as the energy increases, lead to the HCN   HNC isomerization. In the case of a light particle moving in the field of two, much heavier centers the motion is expected to approximate an ellipsoidal path (or part thereof). Indeed, the minimum energy path computed for the potential energy surface of this system does resemble an ellipse to first approximation. This suggests using ellipsoidal (also referred to as spheroidal) coordinates for this system, which are defined by:

$$\xi = \frac{r_1 + r_2}{2a} \quad ; \quad \eta = \frac{r_1 - r_2}{2a} \tag{17}$$

where r_1, r_2 are the distances between the particle (H) and the two focii of the ellipse, and $2a$ is the distance between the focii. Although one expects the focii to correspond at least approximately to the C and N atoms, there is in fact no reason for making such a correspondance quantitatively. In fact, since the minimum energy path for the H motion is by no means an exact ellipse, it seems reasonable to retain a as a parameter: The low-lying bending states that correspond to nearly linear geometries of HCN, may be better described by one value of the ellipsoidal parameter, the high-lying bending motions may be better represented by another a value. The SCF equations in the ellipsoidal coordinates[11]:

$$h_1^{SCF}(\xi, a)\chi_m(\xi) = \varepsilon_m^{(1)}(n)\chi_m(\xi) \tag{18a}$$

$$h_2^{SCF}(\eta, a)\phi_n(\eta) = \varepsilon_n^{(2)}(m)\phi_n(\eta) \tag{18b}$$

depend parametrically upon a. The optimal coordinates condition is

$$\frac{\delta\eta_m^{(n)}}{\delta a} = 0 \ , \tag{19}$$

(with the definitions used in (11), $\varepsilon_m k^{(1)}(n) = \varepsilon_n^{(2)}(m) = E_{mn}$, the total energy). Condition (19) determines which parameter a, hence which specific scaling of the ellipsoidal coordinates, will be best for each stretching-bending state (m,n). Fig. 1 from Ref. 11 shows the dependence of the stretching-bending energy levels E_{mn} upon the parameter a for the levels $(m,n) = (0,2)$ and $(m,n) = (0,4)$. The SCF energy for the same level in polar (hyperspherical) modes is also shown in each of the two cases, and also the results of bare (i.e., decoupled) mode calculations are also given. The bare mode results depend as well on the value of a used.

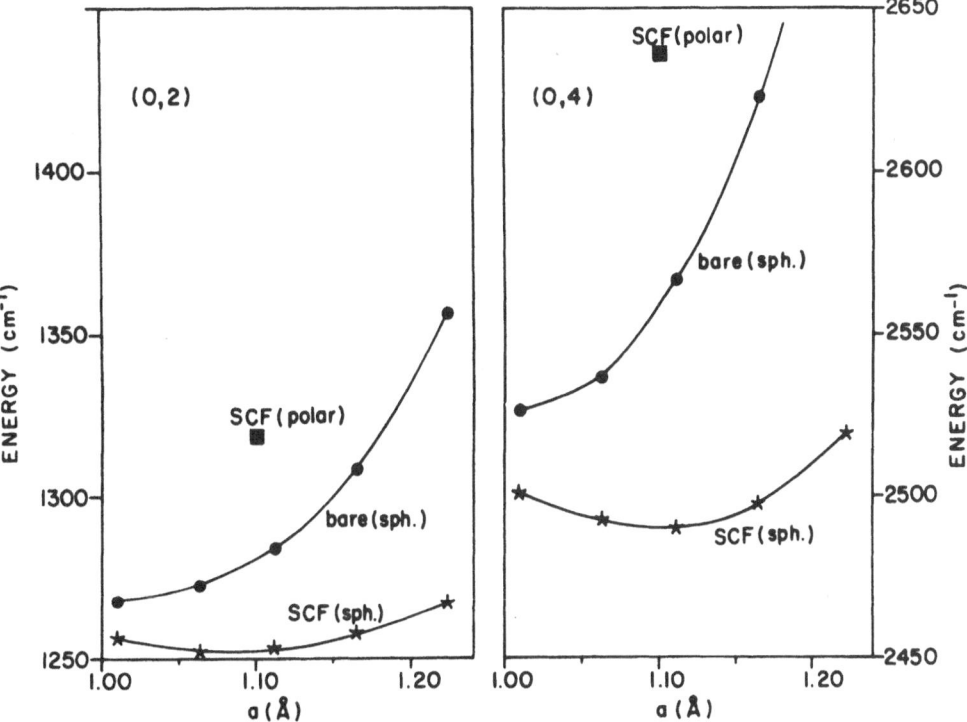

Fig. 1. Optimal coordinate behavior of the (0,2) and (0,4) states. The ellipsoidal SCF energies are plotted vs. a, the parameter of the coordinate system. The bare (uncoupled) mode results vs. a are also shown and the hyperspherical SCF energy is given also for comparison. The results are from Ref. 11.

The results show that: (i) The choice of coordinates is extremely important. The ellipsoidal modes give much better results than the hyperspherical ones. Indeed, the calculation improves roughly as much by using good coordinates within SCF, as by having the SCF correction to the decoupled mode treatment!

(ii) For these low-lying states, choice of the coordinate type, is more important than the additional improvement gained by the variational optimization of *a*.

Consider now the corresponding results for the levels $(m,n) = (0,16)$, $(m,n) = (0,18)$, shown in Fig. 2. The SCF energies in polar modes are poor for these states and are off the scale of Fig. 2. The same is true of the uncoupled mode result in the case of (0,18). It is evident that the differences between the results of the good spheroidal modes and those of the less satisfactory hyperspherical modes are much larger for higher excitation energies. Also, within the framework of the "good" type of coordinates, the importance of coordinate optimization is much greater for the higher levels. Finally, we note by comparison of Fig. 1 and Fig. 2, that the coordinates (or *a* value) which are optimal can vary considerably with the excitation energy.

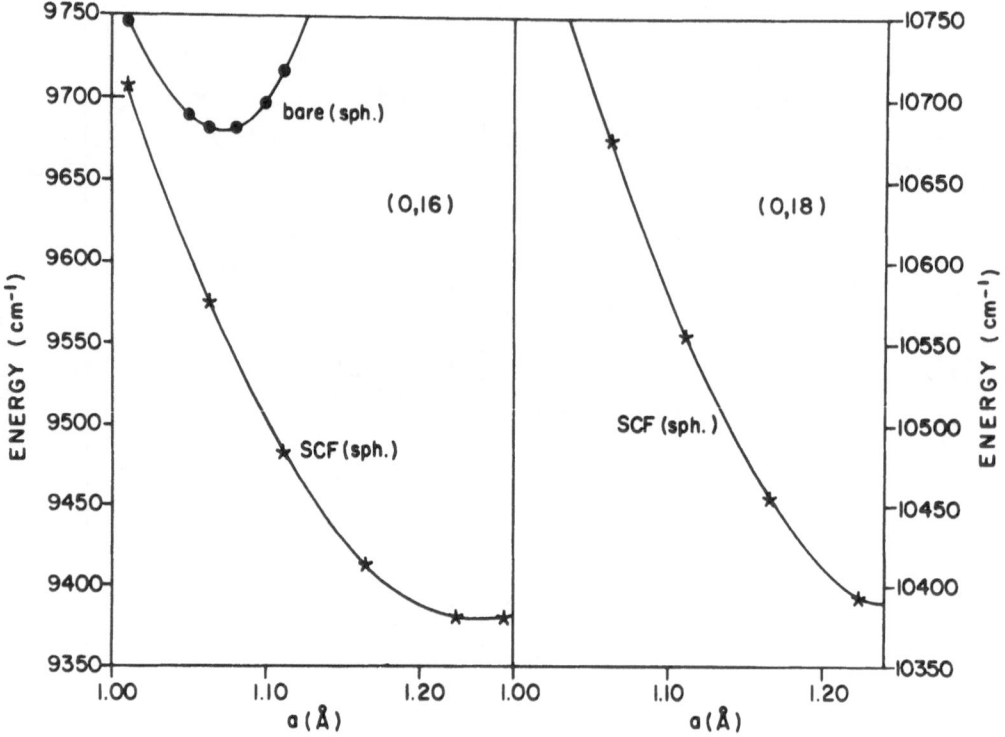

Fig. 2. Optimal coordinate behavior of the (0,16) and (0,18) states. The ellipsoidal SCF energies are plotted *vs. a*, the coordinate system parameter. For the (0,16) state, also the uncoupled-mode energies are shown. The results are from Ref. 11.

We can sum up the implications of these results as follows: The shape of the minimum energy path can be a useful guide in estimating suitable coordinates for a system where large amplitude is involved. When dealing more specifically with excited states that involve the motion of a light atom across a linear molecular frame, ellipsoidal coordinates should prove suitable. The latter case represents a frequently encountered type of large amplitude motion, e.g. in isomerization process. It will be seen later that it is also relevant to an important class of clusters, the "Stick and Ball" systems. We expect therefore that ellipsoidal coordinates should arise among the most useful coordinate systems in optimal-mode SCF studies.

4. *Good modes for the "Three Balls" problem, hyperspherical coordinates*: We use here the terminology of the "Three Balls" system in reference to a cluster of three weakly interacting atoms. Very recently, the question of good SCF coordinates for this important class of systems was explored by Horn et al.[20]. The specific system considered in that study is $Xe(He)_2$. A fascinating aspect of this, and similar clusters, is the extremely "fluxional" nature of the system. As we shall see later on, even at the zero point energy the amplitudes of motion are very large, and it is quite meaningless to discuss the system in terms of a static structure. Indeed, one may think of such clusters as microscopic, quantum liquid drops. It is of considerable interest to determine the signatures of such a state in the energy level structure.

The He-Xe bonds are the stiffest ones in this cluster. Thus, the excitations that increase the mean radius of the cluster are expected to be the most costly in energy for this system. In addition, the He-He distance is expected to be an important coordinate, since the lowest states (at least) must involve avoidance of close contacts or "collisions" between the repulsive cores of the atoms. All this suggests intuitively that one of the good modes for this cluster may be a "hyper-radius" defines as some weighted combination of r_{12}, r_{13} and r_{23} (where r_{12}, r_{13} are the two Xe-He distances, and r_{23} the He-He distance), that is:

$$\rho^2 = a\, r_{12}^2 + a\, r_{13}^2 + b\, r_{23}^2 \,, \tag{20}$$

where a,b are numerical coefficients. ρ is expected to be the "stiffest" mode, while the lowest excitations should pertain to the "angular" motions of the atoms while staying on the hyper-radius. A realization of such coordinates is the familiar hyperspherical coordinate system, given for this case by:

$$\rho^2 = C^2 r_{23}^2 + C^{-2} R^2 \,, \tag{21}$$

where R is the distance between Xe and the c.m. of the two He atoms. C^2 is given by:

$$C^2 = [\frac{(2m+M)}{4M}]^{1/2} \tag{22}$$

where m,M are the masses of He, Xe, respectively. The other modes are:

$$k = 2\arctan[C^2\frac{r_{23}}{R}] \tag{23}$$

$$\chi = \arccos[\frac{\vec{r}_{23}\cdot\vec{R}}{r_{23}R}] \tag{24}$$

$$\Psi(\rho,k,\chi) = \sin^{-1}k\sin^{-1/2}\chi\rho^{-5/2}\Phi_n^{(1)}(\rho)\Phi_m^{(2)}(k)\Phi_l^{(3)}(\chi) \,. \tag{25}$$

The SCF equations have the form:

$$[\frac{\partial^2}{\partial\rho^2} + \frac{2\mu}{\hbar^2}(\varepsilon_n^{(1)} - <ml\,|V|\,ml>_{k,\chi} - \frac{\hbar^2}{2\mu}(\frac{15}{4\rho^2}) \tag{26}$$

$$+ \frac{\hbar^2}{2\mu}(\frac{4}{\rho^2})<ml\,|\frac{1}{\sin^2 k}\frac{\partial}{\partial k}\sin^2 k\frac{\partial}{\partial k} + \frac{1}{\sin^2 k\sin\chi}\frac{\partial}{\partial\chi}\sin\chi\frac{\partial}{\partial\chi}\,|\,ml>_{k,\chi}]\times\Phi_n^{(1)}(\rho) = 0$$

$$[-\frac{\hbar^2}{2\mu_k}(\frac{\partial^2}{\partial k^2} + 1 + <l\,|\frac{1}{\sin\chi}\frac{\partial}{\partial\chi}\sin\chi\frac{\partial}{\partial\chi}\,|\,l>_\chi\sin^{-2}k) \tag{27}$$

$$+ <nl\,|V|\,nl>_{\rho\chi}]\Phi_m^{(2)} = \varepsilon_m^{(2)}\Phi_m^{(2)}$$

where

$$[\frac{\partial^2}{\partial\chi^2} + \frac{2\mu_\chi}{\hbar^2}(\varepsilon_l^{(3)} - <nm\,|V|\,nm>_{\rho,k} \tag{28}$$

$$+ \frac{\hbar^2}{2\mu_\chi}(\frac{1}{4}\cot^2\chi + \frac{1}{2}))]\Phi_l^{(3)} = 0$$

$$\mu_k = \mu[<n\,|\frac{4}{\rho^2}\,|\,n>_\rho]^{-1} \tag{29}$$

and

$$\mu_\chi = \mu[<n\,|\frac{4}{\rho^2}\,|\,n>_\rho]^{-1}[<m\,|\frac{1}{\sin^2 k}\,|\,m>_k]^{-1}. \tag{30}$$

Results of SCF calculations in hyperspherical coordinates for XeHe$_2$ are shown in Table II. The energies are compared with those obtained from SCF in Jacobi coordinates (in which case the modes are r - the He-He distance, R - the distance between Xe and the c.m. of the two He atoms, θ - the angle between $\vec{R}$ and $\vec{r}$). These are full 3D calculations. Realistic pairwise potentials are used, and the details are found in Ref. 20.

The quantum numbers n,m,l correspond respectively to the modes ρ,k,χ. The dissociation energy of Xe(He)$_2$ with the potentials used is 36 cm^{-1}, so all states of higher energy in Table II are metastable states. The energies in Table II are measured from the classical minimum of the potential surface. We stress that (i) The hyperspherical SCF clearly gives better (lower) energies than the Jacobi SCF. Indeed, the hyperspherical SCF predicts two bound excited states, while in the Jacobi SCF the energy of the second excited state already lies in the continuum. (ii) All the lowest excitations in the system correspond to the k mode, representing essentially an angular motion on the hypersphere. The lowest excitation in the hyper-radius mode ρ is already predicted to be embedded in the continuum, although it is not far from the energetic threshold of being a true bound state.

The SCF wavefunctions offer a convenient, single-mode description of the vibrational states of the cluster. The single mode (h.s.) SCF potentials and wavefunctions for the ground state $m = n = l, = 0$ of Xe ^{4}He$_2$ are shown in Figs. 3,4.

An important feature of the results is the large amplitude, considerably delocalized nature of the ground-state motions, especially with regard to the k-mode. The potential function for this system has two local minima, as is evident from Fig. 3. However, the zero point energy in the k-mode is already above the barrier which separates the two minima along this coordinate. $\Phi_0^{(2)}(k)$ is therefore not peaked at either of

Table II. Vibrational energy levels of $XeHe_2$
in Hyperspherical (h.s.) and in Jacobi SCF

quantum numbers of state (in hyperspherical SCF)	energy (cm^{-1}) of h.s. - SCF	energy of corresponding level in Jacobi SCF
$n = m = l = 0$	27.5	29.1
$n = 0, \ m = 1, \ l = 0$	29.5	35.4
$n = 0, \ m = 2, \ l = 0$	33.3	41.2
$n = 0, \ m = 3, \ l = 0$	37.9	-
$n = 1, \ m = 0, \ l = 0$	39.1	-

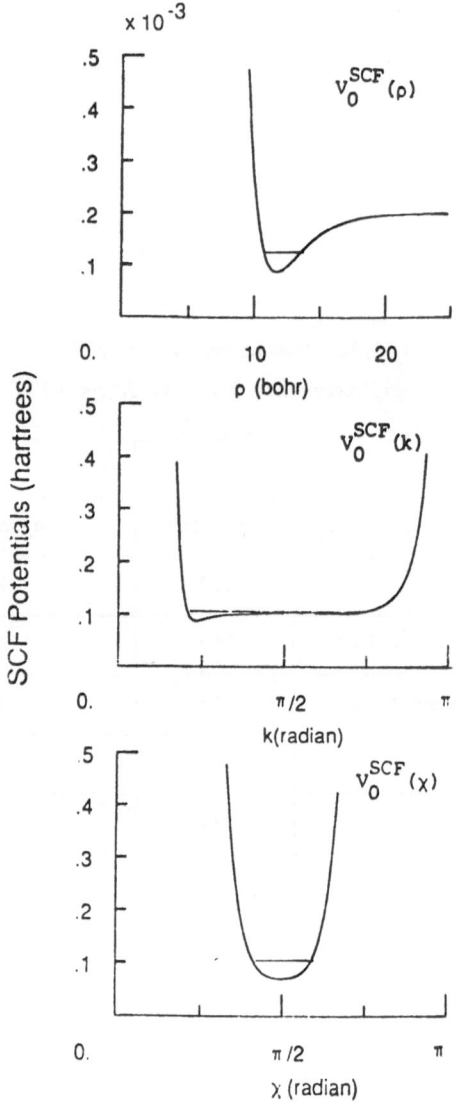

Fig. 3. The single-mode SCF potentials for the ground state of Xe ^{4}He$_2$. The zero point energy for each hyperspherical mode is also shown. ρ is in bohr, the V^{SCF} are in Hartree.

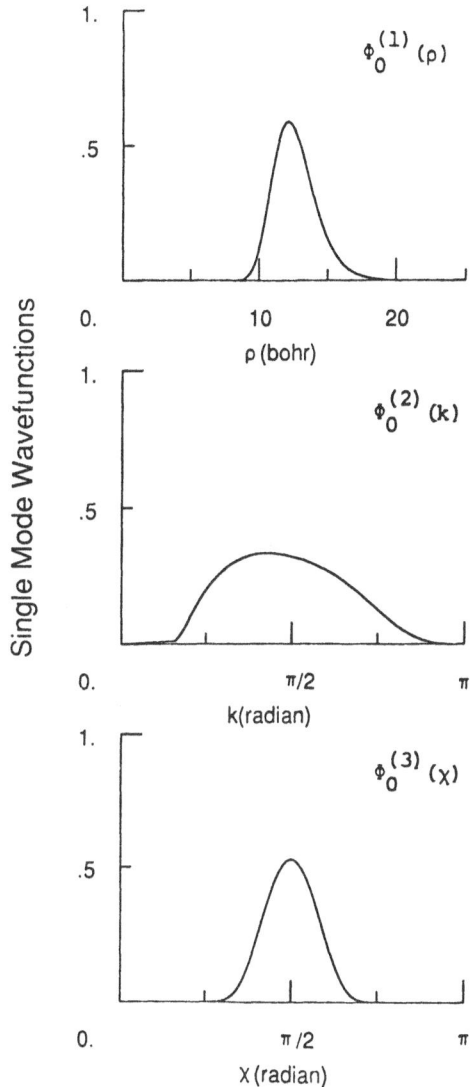

Fig. 4. Single-mode hyperspherical wavefunctions for the ground state of Xe ^{4}He$_2$.

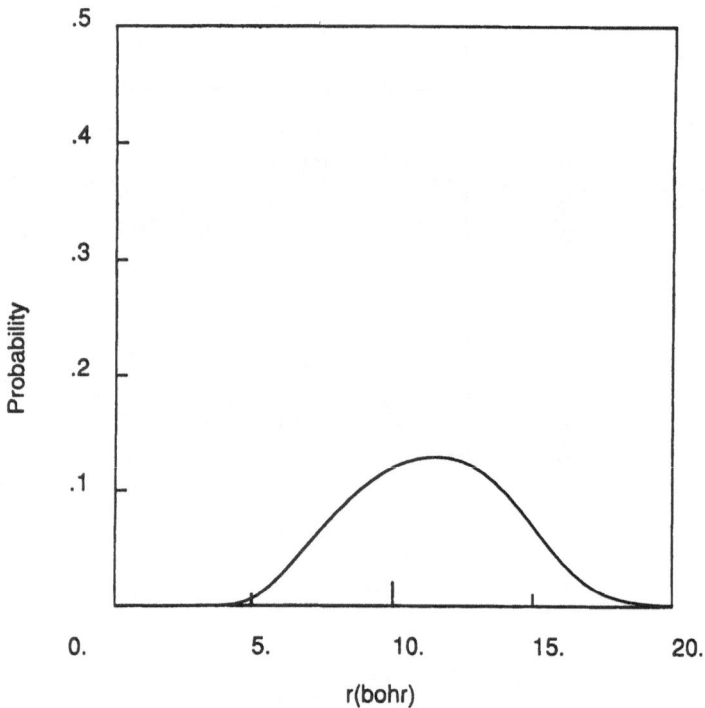

Fig. 5. The probability distribution $P(r)$ for finding the He atoms at a mutual distance r.

the minima, but is broad-spread across most of the allowed domain of the k-variable. The state of this cluster resembles to some extent that of a liquid, where all the local barriers can be surmounted and all the local minima of the potential energy are accessible. Here, however, this is realized to be the ground state.

Fig. 5 reveals the delocalized nature of the ground state from another point of view.

This figure gives the probability distribution for finding the He atoms at various mutual distances. The distribution is very broad covering an effective range of distances of approximately 10 bohr. This is an extraordinarily delocalized behavior for small molecule vibrations in the ground state.

5. *Good modes for the "Stick and Ball" problem - ellipsoidal coordinates for* I_2He [20]: The terminology of "Ball and Stick" refers to a weak complex between an atom and a linear molecule. I_2He will be used here as an example. The large amplitude motions of interest here correspond to the motion of the light He atom with respect to the two Iodine "centers". This strongly suggests that ellipsoidal (spheroidal) coordinates should be very suitable as for the case discussed in part (3) above. Horn *et al.*[20] have recently carried out SCF calculations for this system in ellipsoidal, Jacobi and hyperspherical coordinates. Further, using the SCF states as a basis, a numerically-exact (converged) configuration interaction calculated was carried out to assess the errors of the "optimal" SCF energies. Table III gives the results of the ellipsoidal SCF, the Jacobi SCF and the converged CI calculations. The energies obtained with the hyperspherical SCF are very poor in this case, and are therefore not listed. Although the hyperspherical modes were earlier found to be the right ones for three atoms interacting through roughly similar pairwise potentials, these are inadequate as SCF modes when one pair of atoms is stiffly bound.

Table III: Vibrational energies of I_2He from
ellipsoidal SCF, Jacobi SCF and from converged CI calculations

Quantum Numbers	E^{SCF} (ellipsoidal)	E^{SCF} (Jacobi)	E^{CI}
(n_1,n_2)	(cm^{-1})	(cm^{-1})	(cm^{-1})
(0,0)	14.8	15.9	13.4
(0,1)	20.1	23.0	18.3
(0,2)	25.3	25.3	22.1
(0,3)	26.0	-	25.8
(1,0)	26.6	26.3	25.7

The results of Table II are for calculations in which the vibrational stretching mode of I_2 was frozen (calculations on the full three-mode system will be reported in a future article[20]). The quantum numbers (n_1,n_2) correspond respectively to the coordinates ξ,η in the ellipsoidal (spherical) case, as given in Eq. (17), and to the coordinates R,θ (I_2-He distance and polar angle) in the Jacobi case. Only the true bound states of the complex are given: The energy of the (0,3) given by the Jacobi SCF lies in the dissociative continuum.

The results of Table III clearly show the ellipsoidal SCF to be, for this system, superior to the Jacobi SCF. Indeed, the ellipsoidal SCF energies are within about 1 cm^{-1} of the true value, which may be considered to be of very good accuracy. Many systems of the "Stick and Ball" type were studied experimentally, so we expect the ellipsoidal SCF to prove a very useful tool in the interpretation of spectroscopic data.

IV. CONCLUDING REMARKS

In this article we examined the importance of choosing the appropriate coordinates in applying the approximate Self-Consistent Field separation to the vibrational dynamics of polyatomic systems. We found that the choice of coordinates is often of critical impact for such success of the SCF method. By survey of previous and of new examples it was demonstrated that intuitive physical considerations can be a very useful guide to a successful choice of "good modes" in specific applications. Such considerations may involve factors such as mass ratios, or may focus on the shape of the minimum energy path on the potential surface. One pragmatic approach to developing guidelines for the choice of good coordinates is to find such modes for prototypical examples which represent classes of systems. In that respect, the demonstration here of successful choices of good SCF modes in the cases of $XeHe_2$ and I_2He may prove important. The "Three Balls" and the "Stick and Ball" systems correspond to two large classes of van der Waals clusters. Our expectation is that good coordinates can be found by similar arguments for many other types of clusters. At the same time there should be strong motivation for pursuing also a search for systematic algorithms to determine good modes, algorithms that would be free of the need to make intuitive guesses. In the Cartesian framework, a variational principle offers a solution for this problem. There is clearly a need to develop such methods for more general types of coordinates.

Finally, the question of choosing the best coordinates in approximations based on separability is not confined to stationary states. The same issue arises in the Time-Dependent Self-Consistent Field (TDSCF) approximation, which has been applied recently to a wide range of problems in intramolecular vibrational energy flow[5,6]. So far, only the effect of Cartesian coordinate rotations in the framework of the TDSCF approximation was studied[21]. Efforts at coordinate optimization for TDSCF in both the Cartesian and the curvilinear framework show promise direction in developing improved approximations for time-dependent phenomena.

Acknowledgment: The Fritz Haber Research Center is supported by the Minerva Gesellschaft für die Forschung, mbH, Munich, BRD.

REFERENCES

1. C.E. Hamilton, J.L. Kinsey and R.W. Field, Ann. Rev. Phys Chem. **37**, 493 (1986).

2. R.E. Miller, J. Phys. Chem. **93**, 301 (1986).

3. See, for instance: Z. Bacic and J.C. Light, J. Chem. Phys. **85**, 4594 (1986); **86**, 3065 (1987).

4. J.M. Bowman, Accts. Chem. Res. **19**, 202 (1986).

5.   M.A. Ratner and R.B. Gerber, J. Phys. Chem. **90**, 20 (1986).

6.   R.B. Gerber and M.A. Ratner, Adv. Chem. Phys. **70** (Part 1) 97 (1988).

7.   M. Shapiro and M.S. Child, J. Chem. Phys. **76**, 6176 (1982).

8.   G. Hose, H.S. Taylor and Y.Y. Bai, J. Chem. Phys. **80** 4313 (1984).

9.   P.R. Certain and N. Moiseyev, J. Chem. Phys. **86**, 2146 (1987).

10. G.C. Schatz, M.A. Ratner and R.B. Gerber, J. Chem. Phys. (in press).

11. Z. Bacic, R.B. Gerber and M.A. Ratner, J. Phys. Chem. **90**, 3606 (1986).

12. See, for instance: (a) G.D. Carney, L.I. Sprandel and C.W. Kern, Adv. Chem. Phys. **37**, 305 (1978); (b) J.M. Bowman, K. Christoffel and F. Tobin, J. Phys. Chem. **83**, 905 (1979); (c) R.B. Gerber and M.A. Ratner, Chem. Phys. Lett. **68**, 195 (1979).

13. M.A. Ratner, R.B. Gerber and V. Buch in "Stochasticity and Intramolecular Redistribution of Energy", edited by R. Lefebvre and S. Mukamel (Reidel, Dordrecht, Holland 1986) p. 57.

14. R.M. Roth, M.A. Ratner and R.B. Gerber, J. Phys. Chem. **87**, 2376 (1983).

15. B.C. Garrett and D.G. Truhlar, Chem. Phys. Lett. **92**, 64 (1982).

16. T.C. Thompson and D.G. Truhlar, J. Chem. Phys. **77**, 3031 (1982).

17. R. Lefebvre, Int. J. Quant. Chem. **23**, 543 (1983).

18. N. Moiseyev, Chem. Phys. Lett. **98**, 223 (1983).

19. L.L. Gibson, R.M. Roth, M.A. Ratner and R.B. Gerber, J. Chem. Phys. **85**, 3425 (1986).

20. T.R. Horn, R.B. Gerber and M.A. Ratner, to be published.

21. J. Kucar, H.-D. Meyer and L.S. Cederbaum, Chem. Phys. Lett. **140**, 525 (1987).

THE APPLICATION OF SPECTRAL MEASURES

TO THE STIMULATED EMISSION PUMPING OF ACETYLENE

J.L. Kinsey[*]
Department of Chemistry and George Harrison Spectroscopy Lab.
Massachusetts Institute of Technology
Cambridge, Massachusetts 02139, USA

and

R.D. Levine
The Fritz Haber Research Center for Molecular Dynamics
The Hebrew University, Jerusalem 91904, Israel

INTRODUCTION

The analysis of spectroscopic data for the signature of classically chaotic motion is of current interest[1-10]. Nuclear spectroscopy[11,12] and computational studies[13,14] have tended to emphasize the level spacing statistics involving primarily the short range level correlations. Here we discuss a complementary aspect, that of the statistics of the intensity distribution. This requires sufficient resolution to distinguish adjacent independent lines and a well established base line so that weak transitions can be discerned. So far, these conditions could only be met for computationally generated spectra. Stimulated Emission Pumping[1] (SEP) is a double resonance technique which eliminates rotational congestion, yet for a given intermediate state one can access into very many highly vibrationally excited final states. Since such states are strongly mixed, it might appear that the transition intensities will vary in a smooth manner. In particular, such may be the case when a "bright" zero order state (i.e., a wave function carrying oscillator strength with respect to the intermediate level) is distributed amongst many final states. On the other hand, when good quantum numbers can be assigned, neighboring states can have quite different character and the spectral intensity can rapidly vary as a function of energy[2]. The extent of fluctuation of intensities (with respect to their smooth envelope) can thus serve as a diagnostic for the onset of extensive state mixing and the resulting "intensity-sharing" amongst many states. One can argue however that even in the chaotic regime the fluctuations do not die out but settle to a universal limit[3]. The resulting distribution of intensities is then the Porter-Thomas[4] distribution which can also be derived from random matrix theory. The approach to that limit has been discussed both theoretically[3] and for computational results[5,15].

We consider the results of the distribution of intensities in light of earlier findings[1,8,16] on the distribution of level spacings and of long range level correlations in the SEP spectrum of acetylene. An interpretation of the results in terms of broad gateway states[17] is proposed.

---
[*]Present address: Department of Chemistry and Rice Quantum Institute, Rice University, Houston, Texas 77001, USA

The SEP spectrum of $\tilde{X}\,^1\Sigma_g^+$ acetylene[8] at ca. 26500 cm^{-1} of vibrational excitation provides particularly interesting data for studying intensity fluctuations. At lower resolution, the spectrum consists of broad "clumps" with a density of ca. 0.6/cm^{-1}. These were interpreted[1,8,16] as the bright states which carry oscillator strength. (There was however evidence that these bright states are themselves mixed, and we shall have more to say on this point). At higher resolution each clump has a fine structure of purely vibrational nature[16], with a density which increases with energy. The structure within the clump can be interpreted as the mixing of a single bright level with numerous dark vibrational levels. The density of spectral lines as compared with estimates of the total density of vibrational states (estimated using a Dunham expansion for the energies) indicates that there are few missing lines.

THE GAUSSIAN ENSEMBLES

We begin by a short review of the Gaussian ensembles[11,12,18,19] which have been extensively discussed as means for characterization of the spectra for random Hamiltonians.

The distribution of intensity fluctuations in acetylene (as discussed in detail below) is found to be peaked at a finite intensity and about four times narrower than a Thomas-Porter distribution. Independently, previous work[8,9,16] has shown that the longer range spectral rigidity (of line positions) as measured by the two level correlation function[18,19] spans roughly one quarter of the range expected for a Gaussian orthogonal ensemble (this is also further discussed below). The two results taken together imply that the distribution of matrix elements is that characteristic not of the Gaussian *orthogonal* ensemble (GOE), but of the Gaussian *symplectic* ensemble (GSE)[11,19]. As far as we know, this is the first experimental example of the GSE.

The joint distributions of eigenvalues of the three Gaussian ensembles can be characterized by the value of the level repulsion parameter, β. $\beta = 1, 2$ and 4 for the orthogonal, unitary and symplectic ensembles respectively[11,19]. One can also compute the distribution of amplitudes x for these ensembles[11,20] with the result

$$<(y-<y>)^2> = (2/\beta)<y>^2 \qquad (1)$$

where $y = x^2$ is the intensity. A complementary approach[3,10] is to use information theory to determine the distribution of amplitudes. The relevant parameter is the number, ν, of degrees of freedom that occur when y is expressed as a quadratic form $y = x^\top \rho x$, with ν being the rank of the matrix ρ. In other words, the quantity y is comprised of ν independently fluctuating components. The distribution of y is of the χ^2 family with ν degrees of freedom. This also leads to eq. (1) with $\beta = \nu$. We find (see below) that for either the high resolution of low resolution spectra $\nu \sim 4$. This suggests the existence of four (giant) gateways amongst which the (downward) optical transition to the final state is distributed[17]. In a time-dependent picture of spectroscopy[21], one can say that the SEP of acetylene dumps into a wavepacket which bifurcates to four uncorrelated components whose temporal evolution determines the spectrum.

The two level position correlation function at energies of ca. 26500 cm^{-1} and 27900 cm^{-1} was previously determined[8,9] using the Fourier transform technique[9]. The value of β was determined using the result (theorems 9,1, 9.2 and eq. (5.84) of ref. 19) that the spacing distribution in the symplectic ensemble is essentially equivalent to superposing four energy level sequences from the GOE and selecting every fourth level. The cluster correlation functions as computed[8,9] from the data correspond to selecting one of every six levels of the GOE at $\sim$ 26500 cm^{-1} and one of every three levels at $\sim$ 27900 cm^{-1}. Both estimates of β (6 and 3, respectively) are uncertain due to the uncertainty in extracting the correlation function from the averaged squared Fourier transform of the spectra and to the uncertainly in the value of the mean level spacing S.

30

The long range spectra rigidity leads to $\beta \approx 3$ to 6 and the intensity fluctuations yield $\nu \sim 4$, conforming to our proposal that $\beta = \nu$. The determination of ν is directly from experimental intensities[10] (see below). This determination is subject to the uncertainty in the location of the base line, to the finite experimental resolutions (which is typically[16] below 1/3 of the mean spacing), and to uncertainties in the intensities themselves. The value of β is determined from the time dependence of the cluster function $b_2(St)$ where S is the mean level spacing. We have used for S the value obtained by counting observed lines. This value is lower than the theoretical one. (Computed as the inverse density of states). However, the theoretical computation relies on an extrapolation of the Dunham level equation far outside its tested range of validity. The experimental value of S is subject to uncertainty due to weak and/or over-

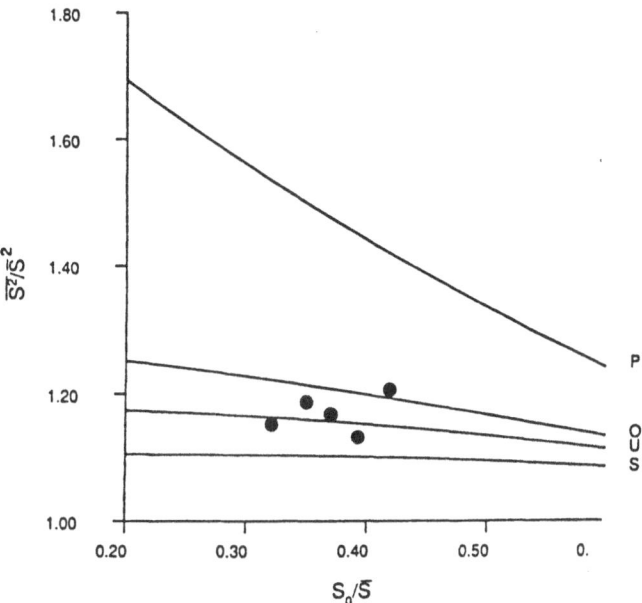

Figure 1. The second moment $\bar{S}^2$ of the near neighbor spacing distribution computed for a Poisson function ($\beta = 0$) and for the three Gaussian ensembles. S_o is the lowest spacing that can be experimentally resolved. The five experimental points are those shown in fig. 3 of ref. 16.

lapping lines. Hence, we regard the determination of $\nu \sim 4$ as more secure than the value for β which is found to be about 6 for the lower energy range (ca. 27000 cm^{-1}) and 4 for the upper energy range (ca. 28000 cm^{-1}. Much of the current literature focuses on the distribution of near neighbor level spacings. This approach readily distinguishes the Poisson distribution from those distributions for which $\beta \geq 1$. We are, however, unable to use this procedure to distinguish among the different Gaussian ensembles, because the low moments of the near-neighbor spacing distribution are only weakly β-dependent, cf. Figure 1. The distributions all have Gaussian tails and differ significantly only near the origin, Figure 2. This is precisely the region where the experiment cannot probe since the finite width, S_o, of the laser is a finite fraction of the mean spacing S for the highly excited molecules of interest. It appears, therefore, that determining the statistics of the intensity fluctuations is the method of choice for dense experimental spectra.

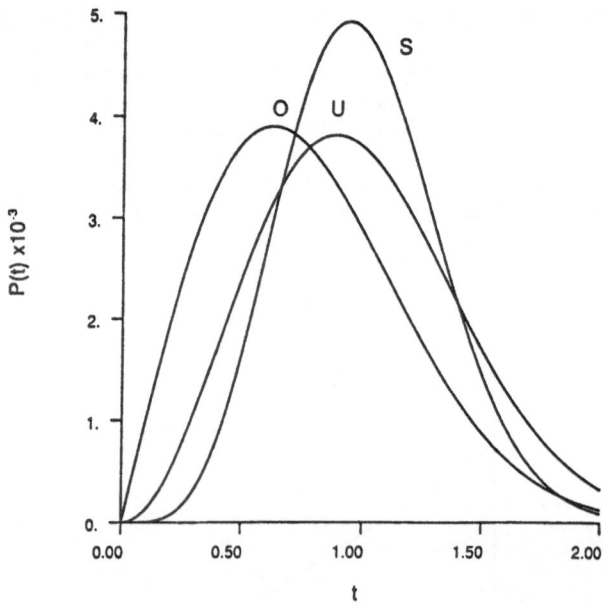

Figure 2. The distribution of near neighbor spacings in the three Gaussian ensembles.

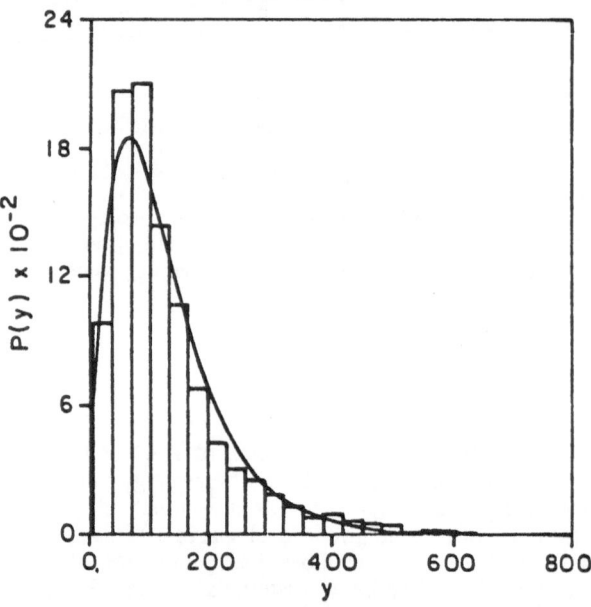

Figure 3. A histogramic representation of the distribution of intensities (abitrary units) and its fit by a χ^2 distribution of ν degrees of freedom for a high (ca. $0.5\ cm^{-1}$) resolution spectrum, using $N = 10162$ lines. $\nu = 4.2$.

INTENSITY FLUCTUATIONS

The distribution of intensity fluctuations was determined directly from the observed signal strength using two methods. The first converts the observed results into a histogram which is then fitted with a χ^2 distribution for ν degrees of freedom, Figure 3. Since the χ^2 distribution has been derived on information theoretic grounds, the fit is carried out using the quality of fit procedure[22] treating ν as a Lagrange multiplier[10]. For all the results reported here the minimal value of the quality of fit function H (which equals the amount of information in the data which is not accounted for by the χ^2 distribution) was below 0.03. This is to be compared[22] with the mean squared fractional error in the intensities, $s^2 = <\delta y/y)^2>$, $s^2 \sim .025$ and .006 for the low and high resolution spectra, respectively. For a good fit, $minH < s^2$. There is thus no persuasive evidence in the data for systematic deviations. The main contribution to s^2 is the uncertainty in the intensities due to the variation of the base line. The value of ν for the high resolution data is much less sensitive to variation in the base line.

It is important to emphasize that we measure the fluctuations with respect to the local envelope of the spectra. This was generated using a Gaussian window with a width σ equal to the inverse recurrence time. However, extensive variations in this width have a negligible effect on the value of ν (which changes by 3% for an order of magnitude change in σ^2). Figure 3 shows the results for both the high and low resolution spectra in the energy range 25850-27250 cm^{-1}.

The second numerical procedure operates directly on the raw intensities and determines ν from the implicit equation[4,10]

$$\frac{1}{N} \sum_{n=1}^{N} \ln(y_i/<y>) = \Psi(\frac{\nu}{2}) - \ln(\frac{\nu}{2}) . \qquad (2)$$

Here N is the number of data points and y_i is the ratio of the intensity of the i'th line to the value of the spectral envelope at the frequency. $<y> = N^{-1}\Sigma_i y_i$. Ψ is the digamma function. The values of ν determined from (2) are within $\pm 5\%$ of those obtained by minimizing H. The advantage of the second procedure is that it provides[10] error limits on ν expressed in terms of s^2. The results are $\nu = 4.0 \pm .3$ for the low resolution spectra and $\nu = 4.2 \pm .6$ for the high resolution data.

DISCUSSION OF SPACING DISTRIBUTIONS

For the three Gaussian ensembles the distribution of intensities can be computed (proceeding as in computation[19,20] of the distribution of widths) to be

$$P(y) \propto \exp(-x^\dagger \rho x) \qquad (3)$$

where x has a Gaussian distribution of zero mean and ρ is a $\beta \times \beta$ Hermitian matrix. The information theoretic result for this distribution has also the form (3) with ν being the *rank* of the matrix ρ. This provides the identification $\beta = \nu$. It is important to note, however, that the level repulsion parameter β, introduced via random matrix theory, is so far restricted to the values $\beta = 1,2$ and 4. Specifically, β is the parameter appearing in the joint distribution of N eigenvalues. (theorem 3.2 of ref. 18). E_0 is the location parameter and a is the scale parameter)

$$P_\beta(E_1, E_2, \ldots, E_n) \propto \exp\left[-\frac{1}{4a^2} \sum_{j-1}^{N}(E_j - E_0)^2\right] \prod_{1 \le k < j < N} |E_k - E_j|^\beta . \qquad (4)$$

The near-neighbor spacing distribution implied by the joint distribution (4) can also be computed for the three Gaussian ensembles, Figure 2. Approximately,[11]

$$P_\beta(S/D) \propto (S/D)^\beta \exp\left[-\frac{1}{2}(S/2a)^2\right] \qquad (5)$$

Here S is the spacing and $D \equiv <S> = a2^{3/2}\Gamma(1+\beta/2)/\Gamma/(2+\beta)/2)$ is the mean. The higher moments of (S/D) are readily computed. As is evident from (4), $<(S/D)^n> \to 1$ as β increases. The most probable (S/D) is $(2a\beta/D)^{1/2}$, with the values $4\pi^{-1/2}$, $2\pi^{1/2}$ and $3\pi^{1/2}$ for $\beta = 1,2$ and 4, respectively. At the high density of states of interest in molecular spectroscopy in particular and in studies of chaos in general, one is unlikely to be easily able to distinguish the three distributions (5). As is evident from figure 1, at a finite laser bandwidth, the measured lower spacing moments for acetylene cannot resolve the precise value of β except that $\beta \geq 1$. it remains to be seen how well a functional form of the type (4) performs during the onset of chaos.

THE NUMBER ν OF DEGREES OF FREEDOM

One can argue on both theoretical and computational grounds that, in general, for a one channel case, ν increases towards 1 as the classical dynamics becomes increasingly chaotic. In terms of our suggestion that $\beta \equiv \nu$, note that the limit of (4) as $\beta \to 0$ is a Poisson distribution as expected for the regular limit. However, the approximation (5) breaks down as $\beta \to 0$. An accurate approximation for (4) at low β's remains to be worked out, but, it is clear that (4) extrapolates correctly to the $\beta \to 0$ limit and hence that the identity $\beta \equiv \nu$ can be maintained also for $\beta < 1$. Our considerations suggest therefore that the number ν of degrees of freedom of the intensity fluctuations can be regarded as a fractal index. In the regular regime, $\nu \to 0$ (very wide fluctuations). As the fraction of classical phase space which is chaotic increases, so does ν. $\nu = 1$ in the fully chaotic limit in the absence of any constraints (apart from the sum rule[3]) on the transition intensities. If phase space is fragmented into a number r of regions (or into r channels for a decay process or into r gateways for a prepation process) then $\nu \to 1$ within each region. The observed value of ν, when the regions are not resolved, will thus tend to r. Hence ν is bounded from above by the number r of distinct regions of phase space.

GATEWAY STATES

How can one interpret the value $\nu = 4$? It was already argued[8,16] that the Hamiltonian H of acetylene, appears, from the spectra, to have the form $H = H_0 + V' + V''$ (ref. 8, eq. (7)). The bright states (clumps in the spectrum) diagonalize $H_0 + V'$. The weaker perturbation V'' is responsible for the fine structure seen, at high resolution, within each clump. The result $\nu = 4$ suggests that a suitable basis for H_0 at about 27000 cm^1 of vibrational energy is four giant gateway states. Since V' is a strong perturbation (with a range exceeding[8] 60 cm^{-1}), the broad gateways are completely split into bright levels by V'. The correlation time for the bright states can be determined from the spectrum to be $\sim 1/2$ psec. Hence, the initial wavepacket formed upon dumping from the higher (intermediate) state will bifurcate in less than 1/2 psec into its ν components.

To exhibit the spectrum of H_0 we have computed an averaged spectrum, Figure 4, as

$$S^o(\omega) = \sum_j y_j A(\omega-\omega_j). \qquad (6)$$

Here y_j is the recorded intensity at the frequency ω_j and $A(\omega-\omega_j)$ is a normalized window function centered at ω_j. The averaged spectra from a lower resolution recording using a Gaussian window function (FWHM = 60 cm^{-1}) is shown in Figure 4. For windows wider than FWHM ~ 60 cm^{-1} the structure seen in Figure 4 survives but the widths of the broad features are no longer intrinsic but rather those of the window.

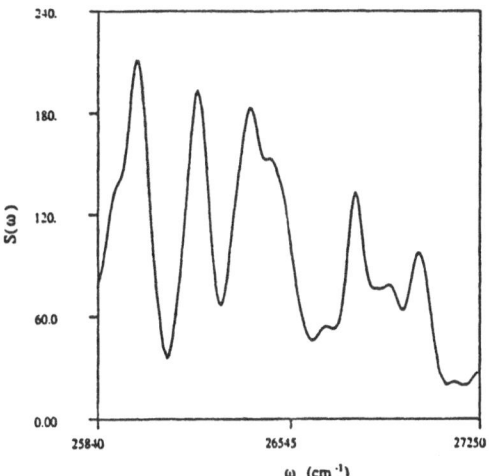

Figure 4: Averaged SEP spectrum of acetylene for downward transitions from a specific level $\tilde{A} \leftarrow \tilde{X}V_0^2, Q(1), K_a$ of the upper electronic state. Intensity (arbitrary units) vs. the vibrational energy in the ground electronic state (in cm^{-1}) in the range 25840 cm^{-1} to 27250 cm^{-1}. Averaged spectrum computed as in (6) from a broadband (resolution ~ 0.3 cm^{-1}) experiment using a Gaussian window function FWHM = 60 cm^{-1}

There is therefore a three tier hierarchical structure in the spectrum: (i) Broad states with an interaction range* (FWHM) $\omega_r \sim 70$ cm^{-1}. These correspond to the initial breakup of the wavepacket into four uncorrelated components. (ii) Clumps with $\omega_r \sim 7$ cm^{-1} which probably correspond to a mixture of ν_4 (trans-bend), ν_2 (C-C stretch) and ν_1 (symmetric C-H stretch) states, and (iii) fine structure with $\omega_r \sim 3$ cm^{-1} and whose density is assumed to be approximately equal to the total symmetry allowed density of states that can be accessed by the down SEP transition. The structure of the energy levels suggests the following scheme for the intramolecular dynamics: The initial wavepacket on the ground electronic state S_0 splits in about ~ 0.5 ps into four 'gateway' components. In the energy range shown in Figure 4 about four such components are accessed from the upper S_1 state whose equilibrium geometry is trans-bent and C-C extended with respect to S_0. The coupling between the ν_4, ν_2 and ν_1 modes is responsible for the splitting of the gateway into the clumps at a timescale of ca. 5 ps ($\equiv 7$ cm^{-1}). This is further supported by the density of clumps being of the same order of magnitude as the calculated density of vibrational states based only on the ν_4, ν_2 and ν_1 modes[16]. The clumps are dispersed over the entire range of vibrational states (i.e., including the dark ones which do not have highly excited trans-bend character) in about 10 ps ($\equiv 3$ cm^{-1}). What is now required is dynamic computations or C-C stretch that would answer the question of what kinds of motion cause this sequence of wavepacket splittings, such as was recently accomplished for the high CH overtones of benzene[24]. Experiments on isotopically substituted molecules[25], such as HCCD would then serve as a check on the assumed potential energy function.

---

*The interaction range ω_r is related to the interaction time by $\omega_r = (ct_r)^{-1}$. See in particular ref. 8 for results of interaction times in the SEP spectrum of acetylene in the same vibrational energy range as Figure 4.

CONCLUSIONS

To conclude, we have presented a consistent picture of the intramolecular dynamics of acetylene at ~ 27000 cm^{-1} vibrational excitation. Using both the value $\nu \sim 4$ for the fluctuation in spectral intensities and $\beta \approx 4$ from the long range correlations in level positions we determine that a symplectic ensemble is representative. This suggests the interpretation that $\beta = \nu$ even when the level repulsion parameter β is not an integer. The value $\beta = 4$ indicates that four broad (≥ 60 cm^{-1}) gateways steer the excitation strength. The two characteristic time scales in the spectrum reflect the breadth of the broad and of the bright ("clump") states respectively. An interpretation of these two time scales in terms of the intramolecular dynamics of highly vibrationally excited acetylene is clearly called for.

ACKNOWLEDGMENT

We thank our coworkers, E. Abramson, Y.M. Engel, R.W. Field and J.-P. Pique, whose work was discussed here. This work was supported by the Air Force Office of Scientific Research under grant AFOSR 86-0011. The experimental work on which it is based was supported by the Department of Energy Contract No. DE-FG02-87ER13671. The Fritz Haber Research Center is supported by the Minerva Gesellschaft für die Forschung, mbH, München, BRD.

REFERENCES

1. C.E. Hamilton, J.L. Kinsey and R.W. Field, *Ann. Rev. Phys. Chem.* 37:493 (1986).
2. E.J. Heller, *J. Chem. Phys.* 72:1337 (1980); E. b. Stechel and E.J. Heller, *Ann. Rev. Phys. Chem.* 35:563 (1984).
3. Y. Alhassid and R.D. Levine, *Phys. Rev. Lett.* 57:2879 (1986).
4. C.E. Porter and R.G. Thomas, *Phys. Rev.* 104:483 (1956).
5. Y.M. Engel, J. Brickmann and R.D. Levine, *Chem. Phys. Lett.* 137:441 (1987).
6. E.J. Heller and R.L. Sundberg, *in* "Chaotic Behavior in Quantum Systems", G. Casati, ed., Plenum, N.Y. (1985).
7. P. Pechukas, *Phys. Rev. Lett.* 51:943 (1983); O. Bohigas, M.J. Gionnoni and C. Schmit, *Phys. Rev. Lett.* 52:1 (1984).
8. J.-P. Pique, Y. Chen, R.W. Field and J.L. Kinsey, *Phys. Rev. Lett.* 58:475 (1987).
9. L. Leviander, M. Lombardi, R. Jost and J.-P. Pique, *Phys. Rev. Lett.* 56:2449 (1986).
10. R.D. Levine, *Adv. Chem. Phys.* 70:53 (1987).
11. C.E. Porter, "Statistical Theory of Spectra: Fluctuations", Academic Press, N.Y. (1965).
12. T.A. Brody, J. Flores, J.B. French, D.A. Mello, A. Pandey and S.S. Wong, *Rev. Mod. Phys.* 53:385 (1981).
13. E. Haller, H. Köppel and L.S. Cederbaum, *Phys. Rev. Lett.* 52:1665 (1984); T.H. Seligman, J.J.M. Verbaarschot and M.R. Zirnbauer, *Phys. Rev. Lett.* 53:215 (1984).
14. M.V. Berry and M.J. Robnik, *J. Phys.* A17:2413 (1984); See also M.V. Berry, *Proc. Roy. Soc. (London)* A400:229 (1985).
15. J.G. Leopold, R.D. Levine and D. Richards, submitted.
16. R.L. Sundberg, E. Abramson, J.L. Kinsey and R.W. Field, *J. Chem. Phys.* 83:466 (1985).
17. J.-P. Pique, Y.M. Engel, R.D. Levine, Y. Chen, R.W. Field and J.L. Kinsey, *J. Chem. Phys.*, in press.

18. F.J. Dyson, *J. Math. Phys.* 3:166 (1966).

19. M.L. Mehta, "Random Matrices", Academic Press, N.Y. (1967).

20. N. Ullah, *Phys. Lett.* 7:153 (1963); *J. Math. Phys.* 4:1279 (1963).

21. E.J. Heller, *Acc. Chem. Res.* 14:368 (1981).

22. R.D. Levine and J.L. Kinsey, *in* "Atom-Molecule Collision Theory", R.B. Bernstein, ed., Plenum, N.Y. (1979).

23. J.L. Kinsey and R.D. Levine, *Chem. Phys. Lett.* 65:413 (1979).

24. E.L. Sibert III, W.P. Reinhardt and J.T. Hynes, *J. Chem. Phys.* 81:1115,1135 (1984).

25. Y. Chen, D.M. Jonas, C.E. Hamilton, P.G. Green, J.L. Kinsey and R.W. Field, *Ber. Bunsenges. Phys. Chem.,* in press.

DYNAMICAL EFFECTS IN THE "PHASE CHANGE" BEHAVIOR OF SMALL CLUSTERS*

Julius Jellinek and Paul G. Jasien

Chemistry Division
Argonne National Laboratory
Argonne, IL 60439

ABSTRACT

Results of a detailed dynamical simulation study of a 13-particle Lennard-Jones cluster are presented. They point to the fact that the pattern of the time-evolution of the cluster, and thus its "phase" and "phase change" properties, depend not only on the total energy of the system but also on its initial state. It is found that the maximal Liapunov number can adequately label the different "phases" of a cluster system.

I. INTRODUCTION

In our inquiries into structural and dynamical properties of matter on the atomic and molecular level, ever-growing attention is being paid to small clusters. Historically, clusters have been (and continue to be) viewed as "laboratories" for studies of the surface and bulk properties of the condensed phase on the microscopic level. But the sharply increased interest in clusters in the past few years, which resulted in the emergence of cluster physics and chemistry as a field in its own right, can be attributed first of all to the recognition that clusters (especially those of small size) form a new state of matter. As intermediates between single atoms and molecules on the one end and the condensed phase on the other, clusters possess properties which often are distinctly different from those of either gases or solids/liquids. Understanding of these properties is of immense importance for establishing the relation between the "microworld" of atoms and molecules and the "macroworld" of the condensed phase and for unraveling the mechanisms governing the transitions between the two. It is becoming widely recognized that cluster studies are of extreme importance also from the practical-technological point of view. Processes of catalysis and corrosion, the insulator-metal-superconductor transitions, thin films and microelectronics technologies, materials design and fabrication all are dependent on and largely defined by clusters and the clustering process.

One of the intriguing aspects of cluster behavior relates to the observation that even very small clusters (probably as small as three

*Work supported by the U. S. Department of Energy, Office of Basic Energy Sciences, Division of Chemical Sciences, Contract W-31-109-Eng-38.

particles) may exhibit different "phases" which resemble the solid-like state and the liquid-like state, respectively. These "phases" and the transitions between them were observed both in molecular dynamics[1-4] (MD) and Monte Carlo[5-7] numerical simulations. They were also predicted in the framework of an analytical statistical-mechanical model.[8] For a recent review and extensive list of citations we refer the reader to Ref. 9.

A pronounced feature of the "phase change" in small clusters, which makes it distinctly different from the bulk phase transition, is that this change takes place over a finite energy (or temperature) range. Over this range the different forms of the cluster can coexist in equilibrium. A detailed isoergic MD study[2] on a 13-particle Lennard-Jones system (with parameters corresponding to Ar) produced a caloric curve

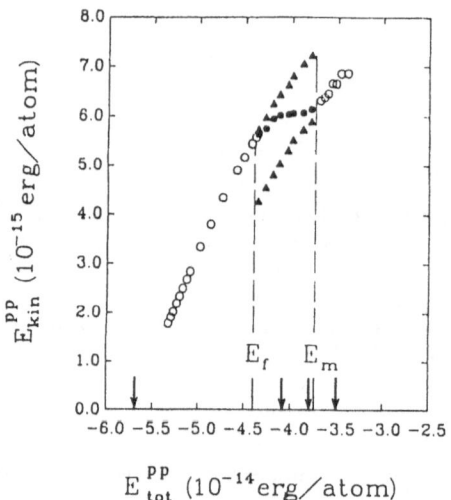

$$E_{tot}^{pp} \; (10^{-14} erg/atom)$$

Fig. 1. The caloric curve (time-averaged kinetic energy per particle vs total energy per particle) for a 13-particle Ar cluster. The open circles correspond to the solid-like "phase" (for $E_{tot}^{pp} < E_f$) and the liquid-like "phase" (for $E_{tot}^{pp} > E_m$). The triangles correspond to the two "phases" in coexistence, while the full circles represent the equilibrium averages over the two "phases". E_f stands for the "freezing" energy, E_m is the "melting" energy. The superscript "pp" means "per particle". (The figure is adopted from Ref. 2.) The arrows show the four energies used in the present study.

(time-averaged kinetic energy per particle vs total energy per particle) shown in Fig. 1. The main feature to note is that the curve consists of two single-valued portions representing the solid-like "phase" (for $E_{tot}^{pp} < E_f$) and the liquid-like "phase" (for $E_{tot}^{pp} > E_m$), respectively, and a two-valued portion (in the range $E_f < E_{tot}^{pp} < E_m$) representing the two "phases" in an equilibrated coexistence (for more details, see Ref. 2). The conclusion deduced from this graph was that the "phase" of the cluster is defined essentially by its total energy.

The goal of this study is to show that the "phase" behavior of small clusters is defined in fact not only by the energetics but also by the dynamics of these systems. It turns out that purely dynamical effects can and actually will lead to different "phases" and "phase changes" at the same total energy. In section II we present the methodology and the results. A discussion is given in section III. We conclude with a brief summary.

II. METHODOLOGY AND RESULTS

By fixing the number of particles in the cluster and its total energy we essentially define conditions corresponding to a microcanonical ensemble. The time-evolution of systems in such an ensemble takes place on an energy surface in the phase space (the word phase in this latter context should not be confused with the thermodynamic notion of "phase"). The customary way to actually mimic the classical time-evolution of an N-body system on a given energy surface is to solve its Hamilton (or Newtonian) equations of motion. The assumption made explicitly or implicitly in such an approach is that the generated dynamics is ergodic, i.e., that the trajectory visits the vicinity of (almost) each point on the energy surface with equal probability. But this assumption is invalid already at its inception. The classical equations of motion conserve not only the total energy but also the total linear and the total angular momenta of the system. Thus points on an energy surface corresponding to different values of the total linear and/or angular momentum cannot lie on the same trajectory. The way to explore all of a given energy surface is to start with different initial conditions (of course, if the emphasis is on the word "all" the task becomes impractical). The caloric curve of Fig. 1, adopted from Ref. 2, was obtained for the particular case of total linear and angular momenta equal to zero. In order to find out how the choice of initial conditions affects the time-evolution of a cluster system as well as its "phases" and "phase changes", we carried out MD simulations for a 13-particle Ar cluster at four different total energies shown by arrows in Fig. 1. We shall refer to these energies as number 1, 2, 3, and 4, labeling them in the ascending order of their values. The choice of the Ar_{13} system was made to facilitate direct comparison with existing results. The technical details of the simulation are similar to those described in Ref. 2, except that a more accurate Hamming's fourth-order predictor-corrector procedure was used to propagate the trajectories and the calculations were carried out on a CRAY-2 machine. The calculations were repeated at each of the four energies using different distributions of the initial momenta between the atoms. Since we were not interested in the overall translational motion of the cluster, these distributions were chosen so as to give zero total linear momentum. The restriction of zero total angular momentum, however, was removed. The initial configurations of the cluster for different runs at energies 2, 3, and 4 gave total potential energies in the range $-(0.632 \div 0.653) \times 10^{-12}$ erg. At energy 1 ($E_{tot}^{pp} = -5.68 \times 10^{-14}$ erg/atom, initial potential energy -0.739×10^{-12} erg) which placed the cluster almost at the bottom of its lowest potential well, corresponding to the icosahedral structure, the different distributions of the initial momenta did not have any effect, as it was expected. The short-time averages of the kinetic energy (over 500 steps with the step size of 10^{-14} sec -- approximately 3 "breathing" vibrations of the cluster) remained almost constant as a function of time and assumed the same value irrespective of the particular distribution of the initial momenta. The cluster maintained its icosahedral geometry and could be viewed as almost absolutely rigid.

At energy 2 ($E_{tot}^{pp} = -4.09 \times 10^{-14}$ erg/atom) different initial distributions of momenta produced distinctly different evolution patterns. Samples of these patterns are presented in Fig. 2. While on panel A the short-time average E_{kin}^{pp} of the kinetic energy fluctuates around a single

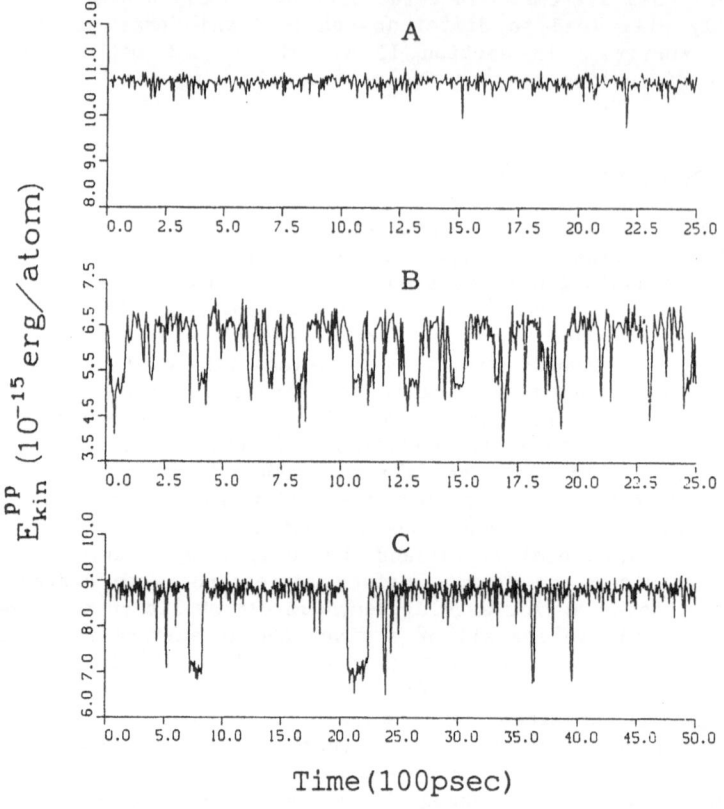

Fig. 2. Different patterns of time-evolution of the short-time
average of the kinetic energy at total energy 2 (see
text).

well-defined value, on panel B it oscillates around at least two such
values jumping frequently from one to the other. These two values essentially coincide with those denoted by triangles corresponding to E_{tot}^{pp} =
-4.09×10^{-14} erg/atom on Fig. 1. The graph on panel A indicates a very
limited sampling of the configuration space which confines the cluster to
only one of its isomeric forms. The pattern on panel B corresponds to a
more extensive sampling of the configuration space involving at least two
different isomers and frequent transitions between them. Note that the
two isomers of panel B are considerably colder than the isomer of panel A
which means that the average geometric structures corresponding to panel B
are of higher potential energy than that corresponding to panel A. In
order to test the ability of the cluster to show behavior intermediate between those presented on panels A and B and to answer the question whether
the two patterns can at all be connected in any continuous fashion, we
repeated the calculations by mixing the initial conditions which led to
patterns A and B. The two initial distributions of momenta were mixed in
different proportions and then rescaled to give the same total energy of
the cluster. The proportion in which the initial momenta which led to the
pattern of panel A (to be labeled as "R" for "rigid") were mixed with
those which gave the pattern of panel B (to be labeled as "NR" for "nonrigid") can be denoted as "NR"/"R". The first value of the ratio
"NR"/"R", taken in increasing order, for which an isomerization transition
has been observed on the time scale of the run is 0.4/0.6; it occurred
after 6700 psec. The cluster visited its higher potential energy strucure
for about 125 psec and then returned to its more stable form. As the

Table 1. Summary of the "mixing experiment" for energy 2 (see text). E_{kin}^{pp} here is the long-time (over an entire run) average of the kinetic energy per particle. The last column displays the maximal Liapunov number.

INITIAL COND. 'NR'/'R'	LENGTH OF RUN (psec)	E_{kin}^{pp} (10^{-15}erg/atom)	LIAPUNOV NUMBER (psec^{-1})
0.0/1.0	8000	10.72	0.52
0.1/0.9	20500	10.64	0.52
0.2/0.8	500	10.43	0.54
0.3/0.7	500	10.08	0.58
0.4/0.6	8000	9.48	0.61
0.5/0.5	24155	8.70	0.67
0.55/0.45	2500	8.41	0.70
0.6/0.4	3000	7.99	0.72
0.7/0.3	500	7.25	0.77
0.8/0.2	3000	6.58	0.80
0.9/0.1	500	6.48	0.80
1.0/0.0	5500	6.11	0.83

value of the ratio "NR"/"R" was being increased the isomerization transitions were occurring more readily and with increasing frequency. The pattern of the short-time average of the kinetic energy per particle for "NR"/"R" = 0.5/0.5 is shown in Fig. 2, panel C. The two characteristic values of E_{kin}^{pp} on this panel are lower than the characteristic values of E_{kin}^{pp} on panel A, but higher than the respective values on panel B. This is in accord with the general observation that the characteristic values of E_{kin}^{pp} are decreasing functions of the "NR"/"R" ratio. As a result, the long-time average of the kinetic energy per particle over an entire run (which in the case of sampling more than one isomeric form of the cluster is an average over the different isomers) also decreases as "NR"/"R" increases. The values of this long-time average as a function of "NR"/"R" are presented in Table 1 which summarizes the "mixing experiment". The last column of this table displays the values of the so-called maximal Liapunov number. The Liapunov numbers[10] characterize the rate of exponential divergence (or convergence) of two trajectories started in close proximity of each other. The maximal Liapunov number gives this rate in the direction of fastest divergence. The sum of the positive Liapunov numbers can be related[11] to the Kolmogorov (or metric) entropy and can be used to indicate a chaotic vs regular character of a dynamics. The zero value of this sum is a signature of regular (quasiperiodic) motion while a positive value measures the degree of chaoticity of a dynamical system. The numerical procedure[11,12] for calculating the maximal Liapunov number expresses this number as a long-time average of the local rates of divergence along the trajectory representing the time-evolution of the system of interest. We applied this procedure to the phase space trajectory of our 13-particle cluster. The procedure substantially increases the computational extent of the problem since it requires simultaneous propagation of both the primary and the accompanying trajectory. This forced us to limit the calculations to the maximal Liapunov number only. As seen in Table 1 this number is an increasing function of the "NR"/"R" ratio.

At energies 3 and 4 we observed again strong dependence on the initial conditions. Two different sets of initial momenta were used. At energy 3 (E_{tot}^{pp} = −3.79 × 10^{-14} erg/atom) we obtained the following two patterns for the short-time average E_{kin}^{pp} considered as a function of time. The first pattern shows at least five distinct values of this average and thus corresponds to the situation when the cluster samples at least five of its different isomeric forms. In each of these forms the cluster spends well-identifiable finite time-intervals. The long-time average (over the entire trajectory) of the kinetic energy per particle and the maximal Liapunov number corresponding to this pattern are 11.84 × 10^{-15} erg/atom and 0.55 psec^{-1}, respectively. The second pattern is characterized by strongly oscillating values of the short-time average E_{kin}^{pp}; hardly any particular value of E_{kin}^{pp} persists for a continuous time-interval long enough to allow for identification of a particular isomer. The cluster is constantly wandering between its different forms and vigorously explores a considerable part of its configuration space. The long-time average of the kinetic energy per particle and the maximal Liapunov number corresponding to the second pattern are 6.34 × 10^{-15} erg/atom and 0.93 psec^{-1}, respectively. Note that this value of the kinetic energy essentially coincides with that denoted by a full circle for E_{tot}^{pp} = −3.79 × 10^{-14} erg/atom on Fig. 1. Summarizing, we can say that the two patterns at energy 3 point to a "NR"-cluster, which samples at least five of its well-identifiable isomeric forms, and to a "L"- or almost "L"-cluster (where "L" stands for "liquid") in which separate isomers do not survive long enough to allow for their identification.

The different behaviors of the cluster corresponding to different initial conditions at energy 4 (E_{tot}^{pp} = −3.50 × 10^{-14} erg/atom) are shown in Fig. 3. While panel A clearly displays a pattern characteristic of a nonrigid cluster (at least four isomers can be identified with certainty, up to six isomers were seen in longer runs), the pattern of panel B unmistakenly points to a liquid-like cluster. The long-time average of the kinetic energy per particle and the maximal Liapunov number corresponding to the nonrigid cluster are 11.72 × 10^{-15} erg/atom and 0.61 psec^{-1}, respectively. These quantities for the liquid-like cluster are 6.77 × 10^{-15} erg/atom and 0.98 psec^{-1}. And again the value of the kinetic energy for the liquid-like cluster is the same as that corresponding to E_{tot}^{pp} = −3.50 × 10^{-14} erg/atom on Fig. 1.

III. DISCUSSION

As the results presented in the previous section indicate, the picture of "phases" and "phase changes" in small clusters is richer than that suggested by Fig. 1. The number and the magnitude of the different characteristic values of the short-time average of the kinetic energy, and thus the number and the types of the different isomeric forms actually sampled by the cluster in the course of its time-evolution, are not uniquely defined by its total energy. They are also strongly dependent on the initial state in which the cluster is prepared. More specifically, the initial distribution of momenta between the different degrees of freedom of the cluster defines its "phase" and "phase change" properties. The different "phases", usually related to the structural characteristics of the cluster, can be associated with different parts of its phase and configuration spaces. The "phase change" is then the result of communication

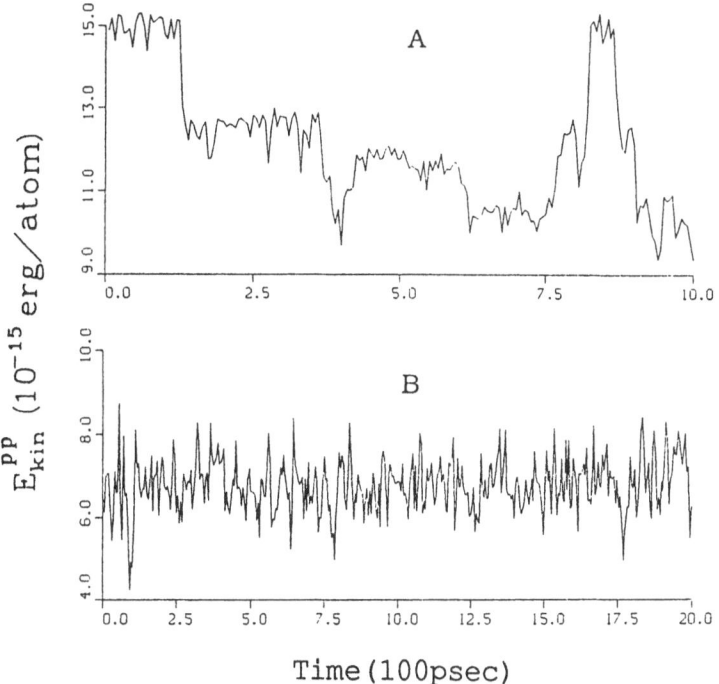

E_{kin}^{PP} (10^{-15} erg/atom)

Time (100psec)

Fig. 3. Different patterns of time-evolution of the short-time average of the kinetic energy at total energy 4 (see text).

between these different parts. The degree of the communication, it turns out, depends both on the total energy and on the initial conditions. The fact that those parts of the phase space which correspond to different values of the total angular momentum of the cluster do not communicate with each other follows from the basic conservation properties of Hamilton equations of motion. A relevant question which can be asked in this respect is: does the total angular momentum, at a given total energy, have any specific bearing on the "phase" properties of a small cluster. A preliminary indication is that the higher the total angular momentum the more rigid is the cluster. We, however, stress the preliminary nature of this inference. Work aiming at clarifying the role of the total angular momentum in defining the "phase" properties of a cluster system is in progress. Among other things, we hope to be able to answer the question about the ergodic properties of a cluster with specified total energy and total angular momentum. This question essentially reduces to the possibility (or impossibility) of further division into noncommunicating subareas of those parts of the phase and configuration spaces which correspond to a fixed total energy and a fixed total angular momentum.

It is intuitively appealing to associate the degree of nonrigidity of a cluster with the degree of its chaoticity as a dynamical system. To quantify the chaoticity of our 13-particle cluster we calculated for it the maximal Liapunov number. Note that the knowledge of this single number from the entire spectrum of the Liapunov numbers already provides quite a bit of information on the system. The zero value of the maximal Liapunov number, for example, is sufficient to conclude that the system is regular (nonchaotic). The only case when we obtained the value of this number close to zero (0.01 $psec^{-1}$) was for energy 1. Even for the rigid isomer at energy 2 (Fig. 2, panel A) the maximal Liapunov number is 0.52

psec^{-1} (see Table 1). As the degree of nonrigidity of the cluster increases so does the maximal Liapunov number. An important feature of this number is that it is not a monotonic function of the total energy. Its values for the cases of the nonrigid cluster considered above are 0.61÷0.83 psec^{-1} at energy 2, 0.55 psec^{-1} at energy 3, and 0.61 psec^{-1} at energy 4. The different values of this number at a fixed total energy seem to adequately label the cluster with respect to its "phase" or nonrigidity property. We put forward here a suggestion that the maximal Liapunov number can serve as a quantitative measure of the degree of nonrigidity of a cluster. The following observation based on the results presented in section II may lend support to this suggestion. It seems that the maximal Liapunov number may actually have a more absolute meaning independent of the value of the total energy (at least for a fixed cluster system). While its values between 0 and 0.5-0.6 psec^{-1} seem to correspond to a rigid 13-particle Lennard-Jones cluster, those above these values seem to indicate a nonrigid cluster with the limiting case of a liquid-like cluster for which this number assumes values close to 1 psec^{-1}. But we want to stress the speculative character of this last observation.

The "mixing experiment" clearly indicates that the degree of floppiness of a cluster depends not only on the frequency of visitation of its different isomeric forms but also on the very isomers involved in this visitation. As a result, the characterization of "phases" and the "phase change" phenomenon in small clusters should invoke not only the time scale argument used earlier[2,8] (the frequency of visitation vs the characteristic frequency of "internal" motion in the cluster) but also the particular sampling pattern favoring selected parts of the phase and configuration spaces. As we have seen, this pattern may strongly depend on the initial conditions. It seems that the maximal Liapunov number appropriately accounts for both factors.

The findings presented above have a direct bearing on the experimental studies of cluster systems. Recent measurements[13] indicate a relation between the chemical reactivity and the structure of small clusters. A full understanding of this relation cannot be achieved without taking into account the dynamical nature of cluster systems which is largely responsible for the specific structural changes in these systems. The relevance of our results is based also on the fact that they are obtained in the course of simulation runs long enough (up to 50 nsec) to be compared to the time scale of some real experiments. Although only the Lennard-Jones potential was used in this study we believe its principal findings are of general value.

IV. SUMMARY

The main goal of the study reported here is to elucidate the role of the detailed dynamics in what can be characterized as the "phase" and "phase change" behavior of small clusters. The study revealed that this behavior is not defined solely by the total energy of the cluster system but it may, and in most cases will, strongly depend on the initial conditions. The intricate interplay between the initial configuration of the cluster and the initial distribution of momenta, on the one hand, and the topology of the multidimensional potential energy surface, on the other, defines which of the different isomeric forms and with which frequency will be visited by the cluster in the course of its time-evolution. These forms and the frequency of their visitation eventually define the "phase" of the cluster as well as the structure(s) corresponding to this "phase". Thus the structure and structural properties of small clusters cannot be studied and understood (apart from the case of very low energies or temperatures) without inquiring into the detailed dynamical properties of these systems. The very notion of structure becomes meaningful only through a dynamical outlook. It turns out that the maximal Liapunov number (a dynamical characteristic) can serve as an adequate measure of such a "phase" property as the degree of nonrigidity of the cluster.

ACKNOWLEDGMENTS

We thank Dr. M. Davis for useful discussions and Dr. D. Li for carrying out some test runs.

REFERENCES

[1] See, for example, C. L. Briant, and J. J. Burton, J. Chem. Phys. 63, 2045 (1975) and references therein.
[2] J. Jellinek, T. L. Beck, and R. S. Berry, J. Chem. Phys. 84, 2783 (1986).
[3] F. Amar and R. S. Berry, J. Chem. Phys. 85, 5774 (1986).
[4] T. L. Beck, J. Jellinek, and R. S. Berry, J. Chem. Phys. 87, 545 (1987).
[5] See, for example, R. D. Etters and J. B. Kaelberer, J. Chem. Phys. 66, 5112 (1977) and references therein.
[6] N. Quirke and P. Sheng, Chem. Phys. Lett. 110, 63 (1984).
[7] H. L. Davis, J. Jellinek, and R. S. Berry, J. Chem. Phys. 86, 6456 (1987).
[8] R. S. Berry, J. Jellinek, and G. Natanson, Phys. Rev. A30, 919 (1984).
[9] R. S. Berry, T. L. Beck, H. L. Davis, and J. Jellinek, Adv. Chem. Phys. 70, Part II, 75 (1988).
[10] See, for example, L. Cesari, Asymptotic Behavior and Stability Problems in Ordinary Differential Equations, Springer-Verlag, Berlin, 1959.
[11] G. Benettin, L. Galgani, and J.-M. Strelcyn, Phys. Rev. A16, 2338 (1976).
[12] M. Casartelli, E. Diana, L. Galgani, and A. Scotti, Phys. Rev. A13, 1921 (1976).
[13] E. K. Parks, B. H. Weiller, P. D. Bechthold, W. F. Hoffman, G. C. Nieman, L. G. Pobo, and S. J. Riley, J. Chem. Phys. (in press).

SPECTROSCOPIC SIGNATURES OF FLOPPINESS IN MOLECULAR COMPLEXES

David J. Nesbitt[a,b]

Department of Chemistry and Biochemistry, University of
Colorado and Joint Institute for Laboratory Astrophysics,
National Bureau of Standards and University of Colorado
Boulder, CO 80309-0440

ABSTRACT

   The challenge of correctly inferring even the qualitative features
of the potential energy hypersurface from spectroscopic measurements is
heightened dramatically in studies of weakly bound molecular complexes
where large amplitude motion is present. This is especially true for
data obtained from low temperature, supersonic expansions where Boltzmann
distributions limit the range of internally excited states that can be
investigated. To stress this point, we present simulated spectra for two
model triatomic systems, a "pinwheel" and a "hinge," with nearly flat
potentials that support extremely large amplitude internal rotation and
bending, respectively. Even in these highly "floppy" molecular systems,
the exact quantum term values can be fitted remarkably well to a standard
semirigid, asymmetric top Hamiltonian, but one corresponding to a qualitatively different, vibrationally averaged molecular geometry. These
results indicate that simple eigenvalue analysis of jet cooled molecular
spectra in the absence of hyperfine resolution may not be sufficiently
sensitive to large amplitude angular motion, and that data from a variety
of techniques may prove necessary to assess the degree of molecular
rigidity.

INTRODUCTION

   The small size of atoms and molecules has made a direct determination of molecular structure difficult, and as a result our concept of
structure stems from a variety of elegant, albeit mostly indirect, spectroscopic techniques. For small molecules in the gas phase, the arsenal
of high resolution spectroscopic techniques is impressive, and often
allows the equilibrium structures of nearly rigid molecules in a given
electronic state to be deduced from analysis of rotation-vibration term
values measured to high precision.[1] For weakly bound molecular complexes,[2]
however, large amplitude motion of the nuclei can be significant, and one
may be forced to replace traditional concepts of structure by vibrational
wave functions for nuclear motion with finite spatial extent along the

---

[a] Staff Member, Quantum Physics Division, National Bureau of Standards
[b] Alfred P. Sloan Fellow

various coordinates of the intermolecular potential energy surface. The corresponding eigenvalues, therefore, reflect an <u>average</u> of the molecular potential over these displacements, complicated further by the fact that any experiment determines only a fraction of the total quantum states.

In early microwave studies of weakly bound complexes,[2] a measure of this large amplitude motion could sometimes be inferred from the extent of vibrational averaging of multipolar properties of the monomer subunits in the complex, as determined by analysis of the hyperfine structure. However, with a few exceptions,[3,4] these microwave studies probe typically only the lowest several rotational transitions in the ground vibrational state, and consequently sample only a relatively small portion of the potential energy surface. The recent surge of activity in the near infrared,[5] on the other hand, has permitted study of many vibrationally excited states in molecular complexes over a wide range of rotational levels. Although many more states can be accessed via these techniques, Doppler and/or predissociation broadening in the near infrared typically restricts one to rotational resolution, and therefore precludes a detailed hyperfine analysis of vibrational averaging effects even in systems with favorable nuclei. Most importantly, the study of weakly bound complexes by both microwave and infrared methods has been greatly facilitated by the use of supersonic expansions to obtain low temperature molecular beams. Although this permits efficient generation of weakly bound species and a welcome reduction in spectral congestion, it also limits the set of states which can be investigated due to selection rules and thermal population of initial levels.

STATEMENT OF THE PROBLEM

Stated most simply, the challenge for the near infrared spectroscopist studying weakly bound systems is 1) how to infer the most correct description of the PES from a necessarily limited set of spectroscopic observations, and 2) how to identify the presence of large amplitude motion, i.e., what are the "signatures" of floppiness in molecular complexes? The difficulty of this task is schematically indicated in Fig. 1, which shows a mapping between a specific potential energy surface and a partial spectrum of eigenvalues which it determines. The crucial point is that this map is not unique i.e., a partial determination of eigenvalues obtained from any real experiment does not (indeed can not even in principle[6]) completely specify the potential surface. The implication is that qualitatively different potential surface topologies, e.g. that of a relatively "floppy" or "rigid" molecular complex, could yield spectroscopic results which, based solely on an analysis of term values, would be essentially indistinguishable.

The thrust of this paper is therefore two fold: 1) to demonstrate that this effect can be dramatic for realistic model systems and 2) to suggest complementary investigations which may provide a more sensitive probe of large amplitude motion. The strategy of our analysis is as follows. We consider two extreme examples of "floppy" triatomic molecules with 1) nearly free internal rotation ("pinwheels") and 2) nearly free bending motion ("hinges"). The exact quantum eigenvalues of these potentials are obtained via numerical methods, and the states below a certain threshold energy are used to predict rovibrational spectra under low temperature jet conditions. The striking feature of these spectra is their quantitative similarity to rigid, near prolate asymmetric top spectra,[7] despite the extreme floppiness of the actual intermolecular potentials. Indeed, we then demonstrate that the rotational eigenvalues can be fit quantitatively as asymmetric tops according to a Watson Hamiltonian analysis.[7]

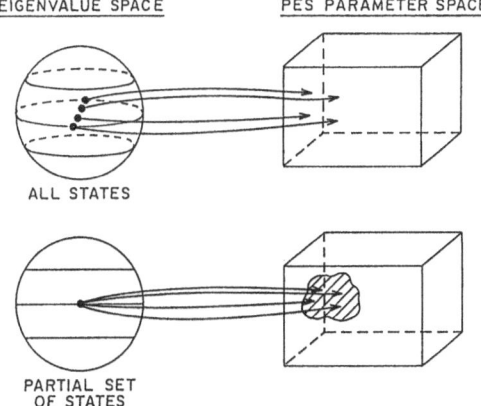

EIGENVALUE SPACE   PES PARAMETER SPACE

ALL STATES

PARTIAL SET
OF STATES

Fig. 1. Schematic representation of a map between the space of measured eigenvalues and the space of parameters determining a potential energy surface. With only a partial set of states observed, the map is seriously underdetermined, and allows for the possiblity that the same set of experimental data can be fit by many different potential energy surfaces. A question this paper addresses is, how different?

MODEL TRIATOMIC SYSTEMS

A fuller description of these calculations will be presented elsewhere; in the interest of brevity, therefore, only the key aspects of the numerical analysis will be summarized here.

Internal Rotors ("Pinwheels")

The molecular coordinates for a triatomic ABC molecule with nearly free internal rotation of fragment AB around its center of mass is given in Fig 2a. The Hamiltonian (excluding translation of the ABC center of mass) for a geometry of fixed internuclear separation (R) of centers of mass of AB and C and AB bond length (r) is

$$\hat{H} = B_1\hat{j}^2 + B_2\hat{\ell}^2 + V(\theta) \tag{1}$$

where $\hat{j}$ and $\hat{\ell}$ are the angular momentum operators for internal rotation of AB and end-over-end rotation of the complex respectively, B_1 and B_2 the associated rotational constants ($B_1 > B_2$), and the total angular momentum $\hat{J} = \hat{j} + \hat{\ell}$. The angular potential V is a function of the molecule-fixed angle θ, and can be expanded as a Legendre series. Matrix elements of this Hamiltonian in a space-fixed basis of $|Jj\ell\rangle$ can be obtained analytically for this Legendre expansion by angular momentum coupling algebra,[8] and the corresponding eigenvalues for each total J obtained explicitly by matrix diagonalization.

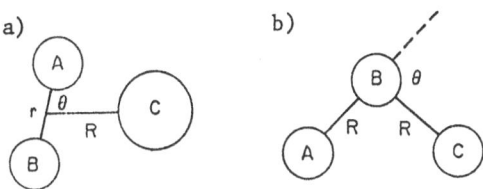

Fig. 2. Definition of molecular coordinates for triatomic ABC representing a) nearly free internal rotation ("pinwheel") and b) nearly free bending ("hinge").

To simplify the description further, we consider a homonuclear diatomic AB, and truncate the Legendre expansion after the first non-vanishing term in the anisotropy,

$$V(\theta) = V_o + V_2 P_2(\cos\theta) .\qquad(2)$$

As a function of V_2, the magnitude of this anisotropy, we can construct a correlation diagram for $j = 0$ and 1 connecting the free rotor ($V_2 \ll B_1$) and rigid T-shaped asymmetric top ($V_2 \gg B_1$) limits, as shown in Fig. 3a. For states correlating with $j = 0$ in the free rotor limit, the rotational spacings are quite regular for all values of V_2. With increasing $V_2 > 0$, however, the $j = 1$ level splits into Π and Σ states, ultimately correlating with $K = 1$ rotational levels of the vibrational ground state and $K = 0$ levels of the asymmetric stretch excited state of a T-shaped molecule. The doubling of the Π levels <u>increases</u> with <u>decreasing</u> V_2, reflecting the increased penetration of the AB wavefunction into near colinear geometries, and hence the more rapid interconversion between equivalent configurations on the potential surface. The key point is that for intermediate values of V_2 this additional doubling in the Π level structure mimics $K = 1$ doubling in an asymmetric top, but with even <u>greater</u> asymmetry than predicted for the rigid T-shape limit ($V_2 = \infty$).

This similarity motivates an attempt to fit the calculated term values to a conventional semirigid rotor Watson Hamiltonian, varying A, B, C, D_J and D_{JK}. For definiteness, the eigenvalues are calculated for $M_a = M_b = 1$ amu, $M_c = 20$ amu, $r = 0.74$ Å and $R = 3.1$ Å (this is a crude model of H_2 internal rotation in H_2HF complexes[9] corresponding to $B_1 = 61.6$ cm^{-1}, $B_2 = 0.965$ cm^{-1}). To be consistent with thermal populations in

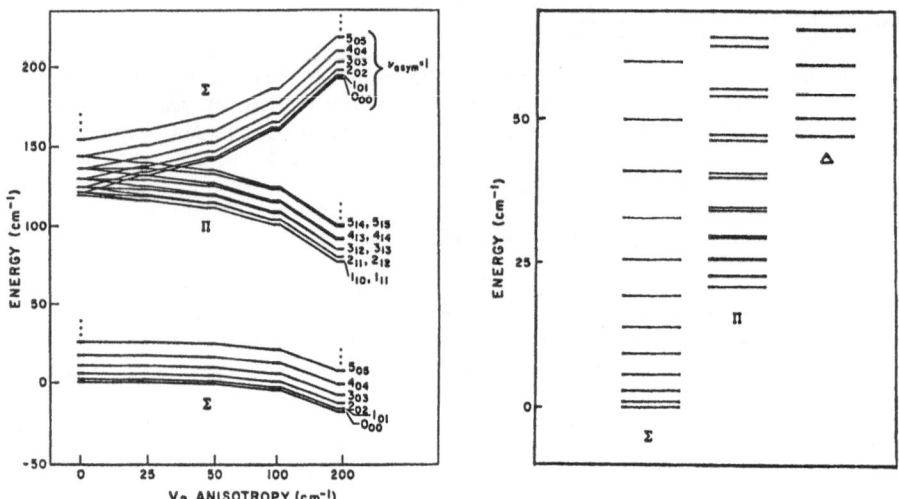

Fig. 3. a) Correlation diagram for a "pinwheel" ABC molecule with $M_a = M_b = 1$ amu, $M_c = 20$ amu, $R = 3.1$ Å and $r = 0.74$ Å. (This corresponds to $B_1 = 61.6$ cm^{-1} and $B_2 = 0.965$ cm^{-1}.) The diagram shows the exact eigenvalues as a function of the anisotropy of the potential i.e., the coefficient of $P_2(\cos\theta)$. The limits of $V_2 = 0$ and ∞, correspond to a free internal AB rotor and a rigid T-shaped molecule, respectively. b) Energy level diagram for a "hinge" ABC molecule with $M_a = M_b = M_c = 20$ amu, and $R = 1.0$ Å, as determined by rigid bender numerical analysis. The states are sorted via angular momentum around the A axis (Σ, Π, Δ,...) and strongly resemble the $K_a = 0,1,2...$ level patterns of a near prolate, asymmetric top.

Table 1. Semirigid Watson Hamiltonian fits[a] to nearly free rotor eigenvalues as a function of V_2 anisotropy

V_2	A_{rigid}[b]	A	B	C	D_J	D_{JK}	σ[c]
25	2.5	119	1.128	0.776	-4.1×10^{-4}	8.3×10^{-2}	0.089
50	4.2	115	1.070	0.853	-1.1×10^{-4}	5.3×10^{-2}	0.022
100	7.7	107	1.024	0.904	-1.9×10^{-5}	3.0×10^{-2}	0.004
200	14.3	96.3	0.996	0.931	-3.1×10^{-6}	1.5×10^{-2}	0.0006
400	24.3	83.5	0.981	0.943	-2.8×10^{-7}	7.2×10^{-3}	0.0001

[a] all units in cm^{-1}.
[b] for a _rigid_ planar molecule,[1] the rotational constants are related by $(A_{rigid} = [C^{-1}-B^{-1}]^{-1}$; the discrepancy between A_{rigid} and A is a measure of the "floppiness" of the molecule.
[c] states for both j = 0 and 1 up to J_{TOTAL} energies of 5 kT (T=10 K) have been included in the fit.

a supersonic jet, we consider only rotational eigenvalues up to 30 cm^{-1} (equal to 5 kT at T = 10 K) but include states correlating with both j = 0 and 1 levels since nuclear spin statistics[1] for H_2 prevent cooling of the ortho/para levels on the time scale of the expansion. Table 1 lists the standard deviation of the fits as a function of V_2, as well as the rotational constants obtained. The quality of the fits is in all cases good, and rapidly improves with increasing values of the V_2. It is important to note that the AB molecule is executing wide amplitude internal rotational motion for _all_ values of V_2 tested. This is readily seen by comparing the fitted values of A and the zero inertial defect[1] estimate of A from B and C, both of which should eventually converge to each other and B_1 in the limit of a rigid, T shaped structure ($V_2 = \infty$). The presence of large amplitude motion is also suggested by the significant K_a dependence (i.e. D_{JK}) of the rigid rotor constants.

As a qualitative presentation of a fit for $V_2 = 100$ cm^{-1}, a stick plot spectrum predicted at 10 K for $K_a = 0 \leftarrow 0$ and $K_a = 1 \leftarrow 1$ transitions in the best fit asymmetric top is shown in Fig. 4a for comparison with the exact ABC transition frequencies for $\Sigma \leftarrow \Sigma$ and $\Pi \leftarrow \Pi$ transitions. The agreement is well within the width of the lines in the plot. From such a parallel spectrum B and C are the only rigid top rotational constants that can be determined, and these two are insufficient to determine a unique structure[1] even for a perfectly rigid triatomic which requires specification of three orientations. However, the only T-shaped structure consistent with B and C is shown in the insert to Fig. 4a. As is anticipated from the above discussion, the effect of large amplitude motion in the AB molecule is to _overestimate_ the asymmetry of the equivalent rigid structure, hence the inferred AB bond length is more than _twice_ the true value. The key point is that analysis of a limited range of eigenvalues from low temperature spectra can be consistent to within experimental precision with two _qualitatively different_ potential energy surfaces.

Rigid Benders ("Hinges")

As second example we briefly turn to a triatomic molecule with a very soft bending coordinate (see Fig. 2b). For simplicity we consider a symmetric triatomic with 1Å bond lengths and three equivalent masses of

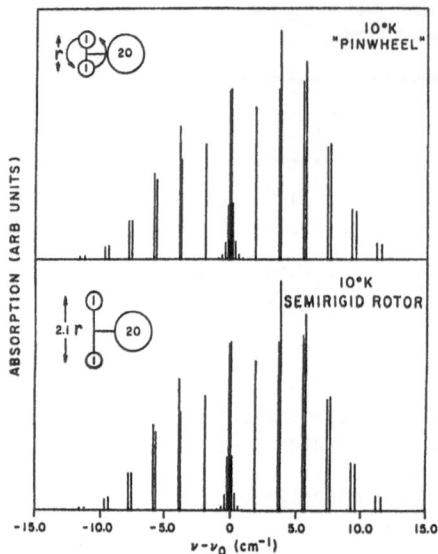
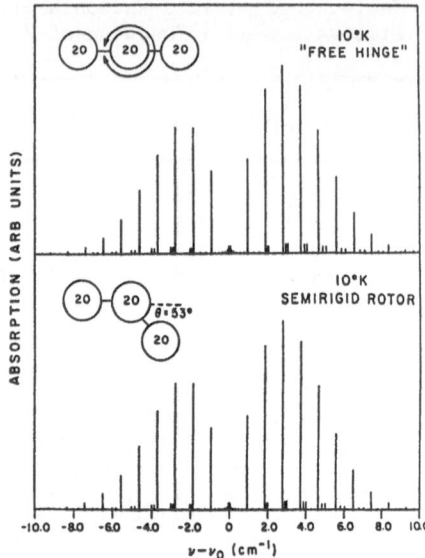

Fig. 4. Simulated spectral plot of the low energy $\Pi \leftarrow \Pi$ and $\Sigma \leftarrow \Sigma$ transitions in a 10 K "pinwheel" with V_2 = 100 cm^{-1}, and the corresponding K_a = 1 $\leftarrow$ 1 and 0 $\leftarrow$ 0 transitions in a 10 K semirigid rotor with constants least squares fit to the true eigenvalues. Intensities are determined from Boltzmann factors, rotational degeneracies, and near prolate top Hönl-London factors,[1] b) Simulated spectral plot of $\Pi \leftarrow \Pi$ and $\Sigma \leftarrow \Sigma$ transitions in a 10 K "hinge" characterized by the potential in eq. (3), and the corresponding least squares fit semirigid asymmetric top spectrum of K_a = 1 $\leftarrow$ 1 and 0 $\leftarrow$ 0 transitions. Intensities are calculated as for a), but would most likely show some discrepancies attributable to large amplitude motion.

20 amu. As a model for a weak bending potential, a high power dependence on the angle θ is used, i.e.

$$V\theta = N \cdot \theta^n \qquad (3)$$

where for the present calculation, n = 6 and N = 10 cm^{-1}. This potential is negligible for small values of θ, but rises rapidly as θ increases, forming an angular "box" inside of which the triatomic can execute wide hinge-like motion. The size of N determines the free angular travel of the hinge, approximately $\Delta\theta$ = $\pm100°$ of essentially unrestricted bending ("unrestricted" can be defined as the range of θ over which $V(\theta) \ll$ the lowest energy eigenvalue). In addition, an infinite potential is imposed on θ = $\pm180°$ in order to prevent masses A and C from passing through one another.

Numerical solutions for the quantum eigenvalues of this model triatomic for each value of total angular momentum J can be obtained by the rigid bender methods of Bender et al.[10] The first several energy eigenvalues for bending eigenfunctions of Σ and Π symmetry are shown in Fig. 3b. The Π bending vibrations in this linear but floppy molecule are strongly split by ℓ-doubling[11] for which a quadratic dependence on J(J+1) mimics the behavior of K doubling in a correspondingly bent triatomic molecule. The resulting similarity of these bending energy level patterns with the K_a = 0 and K_a = 1 levels of a near prolate asymmetric top is clearly evident, and suggests once again a trial fit of the exact

rotational eigenvalues to a Watson Hamiltonian.[7] For the purposes of
illustration, a predicted stick plot spectrum (K=1←1,0←0) for this hinge-
like triatomic is reconstructed from a best fit to the eigenvalues
appreciably populated in a 10 K supersonic jet ($E_{ROT} \lesssim 10$ kT) which
should be compared with the spectrum reconstructed from the exact quantum
eigenvalues (see Fig. 4b). As was the case with the nearly free internal
rotor, the agreement in each of the line positions is well within the
width of the lines in the plot. The only rigid rotational constants that
can be obtained from an asymmetric top least-squares fit to such a parallel spectrum are B and C. Although a knowledge of B and C does not
uniquely determine a structure even for perfectly rigid triatomics, the
values required to generate the predicted spectrum are consistent with a
symmetric triatomic with a 1.098Å bond length and 53.5° bond angle (see
Fig. 4b). Again, the key points are 1) analysis of the rotational and
vibrational eigenvalues from a low temperature spectrum of a molecule may
not necessarily reveal the presence of extremely wide amplitude bending
motion, and 2) the quality of fit to a semirigid rotor Hamiltonian is
little assurance that the molecule is indeed only semirigid!

DISCUSSION

The above two examples clearly demonstrate how large amplitude
motion along soft coordinates in a "floppy" molecule may prove difficult
to identify simply from the rotational analysis of states populated in a
supersonic jet. It is therefore worth considering briefly what experimental efforts might provide a more sensitive probe of the spectroscopic
"signatures" of wide amplitude motion.

One obvious but nonetheless important strategy is simply to determine a wider range of spectroscopic eigenvalues. For both of the examples presented in this paper, the fits obtained fail to extrapolate
accurately to higher J and particularly to higher K_a states. Indeed, a
large K_a dependence of the rotational constants has often been associated
with large amplitude motion in complexes such as HF[12] and HCl dimer.[13]
On the other hand, one of the strong advantages of supersonic jet
experiments is precisely this collapse of rotational state population,
and consequently a greater certainty of spectral assignment. This
suggests the need for experiments over a wide range of internal state
temepratures, and underscores the synergistic value in pinhole
($T_{ROT} \approx$ 1-5 K), slit ($T_{ROT} \approx$ 3-30 K) and equilibrium, cooled cell
($T_{ROT} \gtrsim$ 100 K) absorption methods.

The vibrational frequencies of the intermolecular modes provide
perhaps the most direct information on the coordinates of large amplitude
motion. These data are now becoming available by direct far infrared
observation,[14,15] as well as by detection of vibrational hot bands, difference and combination bands builts on high frequency intramolecular
modes in the near infrared. In our laboratories, these methods have been
used to observe bending modes in N_2HF,[16] CO_2HF,[17] H_2HF,[9] D_2HF,[18] as well
as all three van der Waals vibrational modes in ArHF,[19,20] ArHCl[21] and
NeHCl.[22] By way of example, Fig. 5 shows sample data for ArHCl, in which
transitions in both the perpendicular bend and van der Waals stretch combination modes are evident in the same spectral scan. Of particular
interest is the possibility of observing a progression of low frequency
vibrational modes to investigate rigorously the effects of anharmonicity.

As has been clearly shown in the microwave,[2] resolution of the
hyperfine structure in these infrared spectroscopies would provide useful
information on vibrational averaging of multipolar moments of the monomer
constituents. This capability has already been demonstrated in the far

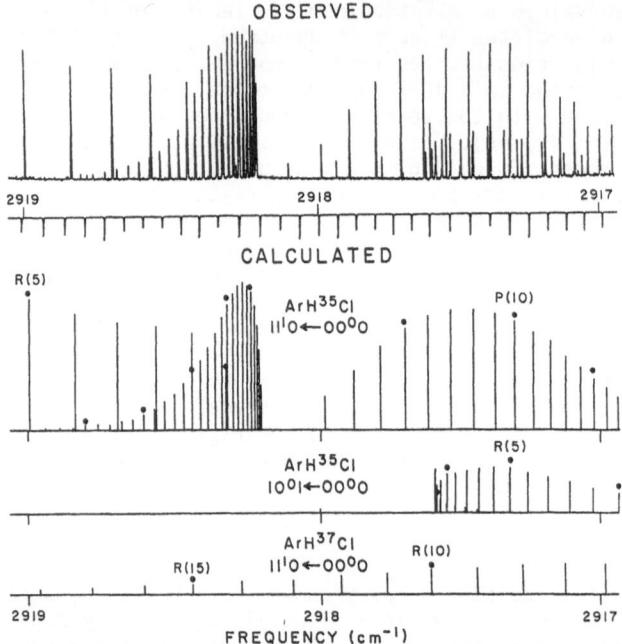

Fig. 5. a) Sample spectral scan through the origin of the perpendicular bend combination mode ($11^10 \leftarrow 00^00$) in ArHCl. Note the additional progression of lines near 2917.6 cm^{-1} attributable to direct excitation of the van der Waals stretch combination mode ($10^01 \leftarrow 00^00$) covered in the same single frequency scan. In an adjacent spectral region, transitions to the parallel bend combination mode are also observed and assigned. b) Calculated spectrum for the same region including contributions from ^{35}Cl and ^{37}Cl isotopes, perpendicular bend and van der Waals stretch modes.

infrared,[14],[15] as well as in the near infrared through the use of microwave-infrared double resonance techniques.[23] In light of the comparative ease of spectroscopic searching in the infrared versus the microwave, it may well prove more efficient first to identify the complexes and extract rovibrational information in the infrared, and then exploit microwave levels of resolution to investigate hyperfine phenomena.

The above methods all are based on a high resolution determination of the <u>eigenvalues</u> to provide information on "soft" coordinates in the potential energy surface. However, methods that are sensitive to the spatial distribution of the <u>eigenfunctions</u> may in fact sample these coordinates more directly. One traditional example of this is the use of H/D isotopic substitution in microwave studies of hydrogen halide complexes to look at zero point bending motion, or in complexes with molecular hydrogen for determination of the full potential energy surface.[2] Another example is the explicit vibrational dependence of dipole moments in complexes, studies of which are particularly feasible in the near and far infrared. The Coulomb explosion method, which has been an important part of the discussions at this conference, looks to be a very promising and direct probe of the square amplitude of the eigenfunctions, which would be particularly suitable for determining low frequency, large amplitude motion in floppy complexes.

In such a brief discussion there is room only to identify a few additional eigenfunction based probes of large amplitude motion that we have pursued in our infrared studies. First, the analysis of anomalies in the infrared _intensities_ can reveal the presence of significant changes in the vibrationally averaged molecular structure as a function of quantum state. For example, centrifugal straightening[17] (with increasing J) or rotational saturation[12] (with increasing K_a) of a low frequency bending coordinate in a complex will tend to increase or decrease, respectively, the projection of the transition moment along the inertial A axis and thereby influence the ratio of parallel to perpendicular intensities. Even more dramatically, the relative intensities of parallel/perpendicular transitions in hindered internal rotors can be very sensitive to the magnitude of anisotropy in the potential. An extreme example of this is NeHCl,[22] in which the fundamental transition is roughly an order of magnitude weaker than the bending combination transitions.

A second probe of large amplitude motion may be found in the magnitude of _Coriolis_ effects, which become progressivley important as the periods of low frequency vibrational and end-over-end rotational motion becomes comparable. Taking examples from our own work,[17] this can be clearly seen in the CO_2HF complex, in which the centrifugal distortion effects appear to be dominated by narrowing or straightening of the eigenfunction with increasing rotation. Coriolis interaction of stretch and bend vibrational modes in the inert gas hydrogen halide systems[19-22] is another example of this effect, which in the case of Ne and ArHCl complexes serves to mix all three van der Waals modes.

A final point to make is that the observation of _isomers_ in weakly bound complexes indicates that there indeed can be multiple minima in the potential energy surface. These potential minima can be energetically separated by a large enough barrier to produce two species kinetically stable on the time scale of a supersonic expansion, such as observed in N_2OHF and ON_2HF.[24] Alternatively, the barrier can be small enough that a more appropriate description is simply that of a vibrationally excited state, such as in the parallel bending vibrations in NeHCl[22] and ArHCl.[21] In yet another limit, the minima can be equivalent, and the barriers sufficiently small that the best description is of tunneling as in the HF[12] and HCl[13] dimers. In all cases, however, the presence of multiple minima suggests the possiblity of low energy paths for interconversion, the physical consequence of which would be large amplitude motion along those coordinates.

SUMMARY

The determination of the intermolecular potential energy surface in weakly bound complexes from spectral analysis of the eigenvalues is not unique. In particular, near infrared spectra of extremely "floppy" molecular species with large amplitude motion obtained under jet cooled conditions can be deceptively well fit to a conventional asymmetric top Hamiltonian. Two explicit examples of model triatomics have been treated, that a nearly free internal rotor ("pinwheel") and a rigid bender ("hinge"), in which the eigenvalues populated significantly at 10 K can be well predicted from a semirigid rotor analysis. The central conclusion of this study is that an excellent fit to a conventional, semirigid Hamiltonian does not exclude the possibility of large amplitude motion in cordinates not directly sampled by the transition. Additional data from a variety of experimental techniques, particularly those sensitive to the _eigenfunctions_ rather than _eigenvalues_, may prove essential to determine the correct qualitative features of the potential energy surface.

ACKNOWLEDGMENTS

This work has been supported by grants from the National Science Foundation (CHE86-05970 and PHY86-04504), Petroleum Research Fund, and Research Corporation. Further support from the Henry and Camille Dreyfus Foundation and the Sloan Foundation are gratefully acknowledged. P. R. Bunker is especially acknowledged for providing us with rigid bender analysis programs, as well as for many helpful discussions. Finally, W. C. Lineberger should be acknowledged for his encouragement of a timely submission of this manuscript.

REFERENCES

1) G. Herzberg, <u>Infrared and Raman Spectra</u> (Van Nostrand Reinhold, New York, 1945).
2) There are several relevant articles in <u>Structure and Dynamics of Weakly Bound Complexes</u>, A. Weber, Ed., Nato Advanced Scientific Institute, Series C <u>212</u> (1987).
3) A. C. Legon, D. J. Millen, and S. C. Rogers, Proc. R. Soc. London Ser. A <u>370</u>, 213 (1980).
4) W. J. Lafferty, R. D. Suenram and F. J. Lovas, J. Mol. Spectrosc. <u>123</u>, 434 (1987).
5) D. J. Nesbitt, Chem. Rev., in press.
6) As was pointed out to the author by Z. Vager, even a complete spectrum of eigenvalues, E_k, can not determine the Hamiltonian uniquely, which can be readily seen by expressing $\hat{H} = \sum_k E_k |k\rangle\langle k|$, where $|k\rangle$ is any arbitrary but complete, orthonormal set of states.
7) J. K. G. Watson, in <u>Vibrational Spectra and Structure</u>, J. R. Durig, Ed. (Elsevier, New York, 1977).
8) L. D. Landau and E. M. Lifshitz, <u>Quantum Mechanics</u> (Pergamon Press, Oxford, 1958).
9) C. M. Lovejoy, D. D. Nelson, Jr., and D. J. Nesbitt, J. Chem. Phys. <u>87</u>, 5621 (1987).
10) P. R. Bunker and J. M. R. Stone, J. Mol. Spectrosc. <u>41</u>, 310 (1972).
11) E. B. Wilson, Jr., J. C. Decius, and P. C. Cross, <u>Molecular Vibrations</u> (Dover, New York, 1955).
12) A. S. Pine, W. J. Lafferty, and B. J. Howard, J. Chem. Phys. <u>81</u>, 2939 (1984).
13) N. Ohashi and A. S. Pine, J. Chem. Phys. <u>81</u>, 73 (1984).
14) M. D. Marshall, A. Charo, H. O. Leung, and W. Klemperer, J. Chem. Phys. <u>83</u>, 4924 (1985).
15) D. Ray, R. L. Robinson, D.-H. Gwo, and R. J. Saykally, J. Chem. Phys. <u>84</u>, 1171 (1986).
16) C. M. Lovejoy and D. J. Nesbitt, J. Chem. Phys. <u>86</u>, 3151 (1987).
17) C. M. Lovejoy, M. D. Schuder, and D. J. Nesbitt, J. Chem. Phys. <u>86</u>, 5337 (1987).
18) C. M. Lovejoy, D. D. Nelson, Jr., and D. J. Nesbitt, J. Chem. Phys., submitted.
19) C. M. Lovejoy, M. D. Schuder, and D. J. Nesbitt, J. Chem. Phys. <u>85</u>, 4890 (1986).
20) C. M. Lovejoy, M. D. Schuder, and D. J. Nesbitt, Chem. Phys. Lett. <u>127</u>, 374 (1986).
21) C. M. Lovejoy and D. J. Nesbitt, Chem. Phys. Lett., submitted.
22) C. M. Lovejoy and D. J. Desbitt, Chem. Phys. Lett., submitted.
23) G. D. Hayman, J. Hodge, B. J. Howard, J. S. Muenter, and T. R. Dyke, J. Chem. Phys. <u>87</u>, 1670 (1987).
24) C. M. Lovejoy and D. J. Nesbitt, J. Chem. Phys. <u>87</u>, 1450 (1987).

NONSTATISTICAL BEHAVIOR IN THE VIBRATIONAL PREDISSOCIATION

DYNAMICS OF BINARY AND TERTIARY COMPLEXES

R.E. Miller

Department of Chemistry
University of North Carolina
Chapel Hill, N.C. 27514

ABSTRACT

Near infrared laser-molecular beam spectroscopy has been used to study a wide range of van der Waals and hydrogen bonded molecules. In this article we discuss (1) the geometrical structure of selected systems, (2) mode dependent vibrational predissociation of these species, (3) the determination of intermolecular vibrational frequencies from near infrared spectroscopy and (4) the effects of intermolecular vibrational excitation on the vibrational predissociation dynamics.

1. INTRODUCTION

Weakly bound complexes have recently been the subject of intensive study using a variety of infrared laser techniques. In the near infrared, where the majority of this effort has been focused [1-6], it is the intramolecular vibrations of the constituent monomer units that are excited by the laser. Experiments of this type are of interest since they provide accurate molecular constants for both the ground and excited vibrational states of these complexes, from which structural information can be readily obtained. In addition, dynamical variables, such as the vibrational predissociation lifetime [2] and the final state distributions of the fragments [7], are now becoming available. Far infrared spectroscopy has also been used to study vibrations associated with the weak intermolecular bond [8,9]. The motivation here is in the determination of accurate intermolecular vibrational frequencies which are sensitive to the details of the intermolecular potential surface. In this case, however, the molecule does not dissociate since the photon energy is insufficient to rupture the intermolecular bond. Clearly, data obtained from these two spectral regions are completely complementary.

In this article we discuss some of our recent results

which demonstrate the usefulness of near infrared spectroscopy in the study of weakly bound clusters. In addition to providing accurate molecular constants, structures and dynamical information on the vibrational predissociation of these complexes, these results can also, in some favorable cases, be used to obtain accurate values for the intermolecular vibrational frequencies through the observation of fundamental, combination and hot band transitions.

2. EXPERIMENTAL

The experimental apparatus used in this study is based on the opto-thermal detection method, which has been discussed in detail elsewhere [10]. This method makes use of a liquid helium cooled bolometer detector to monitor the energy of a highly collimated molecular beam, which is produced from a differentially pumped free jet expansion source. An F-center laser (10 mW output power) is crossed with the molecular beam and tuned into resonance with the molecular transitions of interest. This results in the vibrational excitation of the molecules in the beam, which gives rise to either an increase in molecular beam energy, for the case of stable species, or a decrease if the excited state of the molecule is dissociative. In the latter case the recoil of the fragments out of the beam results in a decrease in the molecular flux (and hence translational energy) reaching the bolometer. In practice, the laser is amplitude modulated and the bolometer signal is detected using a lock-in amplifier. Under these conditions the signals associated with stable and unstable species are 180° out of phase and therefore appear in the spectrum as signals with positive and negative amplitude, respectively. In all but one case (Ar-HF), the spectra presented here have been inverted to emphasize the transitions associated with the complexes.

Scanning of the F-center laser is accomplished using a PDP11/73 microcomputer to control the three tuning elements of the laser [10,11], while at the same time monitoring the laser frequency by recording the transmission spectrum of three confocal etalons and a water vapor cell, as well as the signals from the bolometer. The absolute calibration of the final spectrum is accomplished by using either a monomer transition observed in the molecular beam spectrum or a water vapor transition recorded in the gas cell. Interpolation between the fringes of a 150 MHz confocal etalon is used to obtain the relative frequency calibration of the spectrum. In this way the molecular beam transitions can be determined to within approximately 0.0002 cm^{-1}.

3. SPECTROSCOPY AND STRUCTURE OF HYDROGEN CYANIDE DIMER AND TRIMER

In view of the large dipole moment of hydrogen cyanide, the linear dimer and trimer are ideally suited for study using microwave spectroscopy. As a result, the structures of both these complexes have been accurately determined [12,13]. These systems are also well suited for study using the opto-thermal detection method since the C-H stretch of hydrogen

cyanide is strongly infrared active and lies within the
tuning range of the F-center laser. For the case of the
dimer, there are two distinct C-H stretches, one involving
the hydrogen bonded proton while the other corresponds to
excitation of the "free" C-H stretch. Figure 1 shows the
spectrum associated with the latter. As expected, the
spectrum is characteristic of a linear molecule, a fit to
which gives ground state rotational constants which are in
excellent agreement with those obtained from the microwave
studies. The high resolution is due to the highly collimated
nature of the molecular beam, the limit resulting from the
slight non-orthogonality of the laser-molecular beam crossing
[14]. Also evident in the spectrum are two monomer
transitions (P(1) at 3308.5204 and P(2) at 3305.5432 cm^{-1})
which, as expected, appear with opposite sign to those of the
dimer. This confirms the fact that the vibrationally excited
dimer dissociates in a time which is short with respect to
the flight time of the molecules from the laser crossing to
the bolometer.

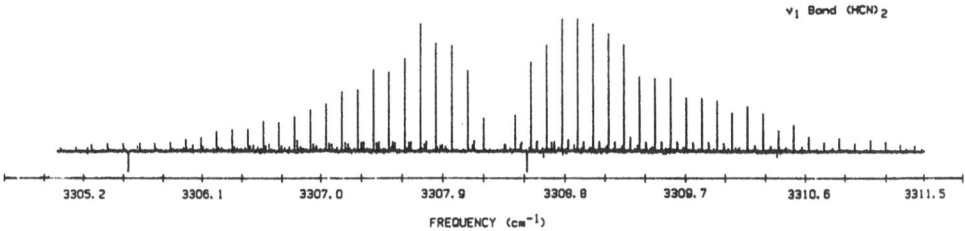

FIGURE 1

At somewhat higher source pressures, than those used to
record the spectrum of the dimer, transitions associated with
the linear trimer were also observed. For example, Figure 2
shows the "free" C-H stretch of the trimer which is located
at the low frequency end of the dimer spectrum. In fact,
several intense P-branch transitions of the dimer are evident
in the spectrum. A linear hydrogen cyanide timer clearly has
three C-H bonds, two of which are involved in hydrogen
bonding. The two modes associated with these are best
thought of as symmetric and asymmetric combinations of the
two C-H stretches [15]. A somewhat more detailed discussion
of these will be given in a later section.

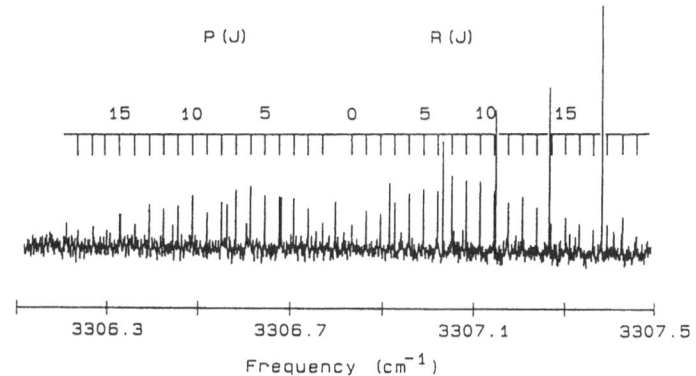

FIGURE 2

In the course of searching for the three C-H stretching
modes of the linear timer, the spectrum shown in Figure 3 was
also found. This spectrum can be understood in terms of a
planar, cyclic trimer in which there are now three non-linear
hydrogen bonds. This spectrum is characteristic of an oblate
symmetric top and can be used to determine the center of mass
separations of the three monomer units. Due to the symmetry
of this complex, only one infrared allowed vibrational band
is expected and observed. Figure 4 shows the experimentally
determined structures obtained for the two isomers of the
trimer onto which the van der Waals radii of the individual
atoms have been superimposed.

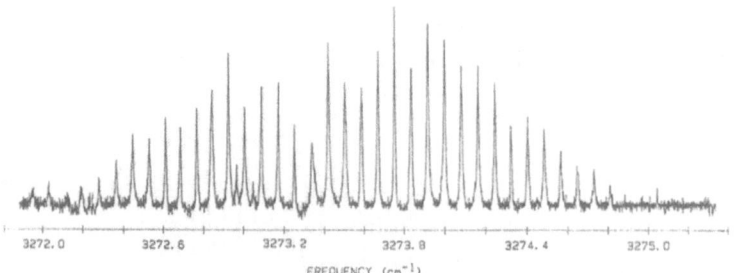

FIGURE 3

LINEAR TRIMER

CYCLIC TRIMER

FIGURE 4

Hydrogen cyanide trimer is one of the few systems for
which more than one isomer has been observed. It is
interesting to note that ab-initio calculations [15] for the
trimer also show two minima in the potential energy surface
corresponding to a linear and cyclic form. However, the
calculations indicate that the linear isomer is approximately
a factor of two more stable than the cyclic form. The fact

that both isomers are experimentally observed (with approximately equal abundance) at the low temperatures characteristic of the free jet expansion, suggests that the binding energies of these two isomers are, in fact, rather similar. For the tetramer, were the calculations indicate that the cyclic and linear isomers have comparable binding energies, we see no evidence for the linear form. This suggests that the cyclic isomer becomes more favorable as the size of the cluster is increased. This trend can be rationalized by noting that the bent hydrogen bonds become less strained as the number of molecules in the ring is increased. As a result, the additional hydrogen bond obtained in the cyclic isomer more than compensates for the fact that all of the bonds are somewhat strained. It is interesting to note that in the solid, hydrogen cyanide forms linear polymer chains. In this case, the effects of packing clearly outway the tendency for this system to form cyclic structures.

4.   MODE DEPENDENT VIBRATIONAL PREDISSOCIATION

Vibrational predissociation [1] of these weakly bound complexes occurs almost universally in conjunction with the near infrared excitation of their constituent monomer units. Much of the interest in this dynamical process stems from the fact that it occurs at energies which are low in comparison with most unimolecular reactions. As a result, quantitatively realistic theoretical calculations can be carried out for these systems [16,17]. In view of the low excitation energy involved and the weak coupling between the intramolecular and intermolecular degrees of freedom, it has long been expected that systems of this type might display strong mode dependent dissociation. Until recently, however, there was little experimental data to support this thesis. The hydrogen cyanide complexes are excellent systems in which to look for this dependence since the various C-H stretching modes clearly couple very differently to the weak intermolecular bond. In particular, the "free" C-H stretches of the dimer and trimer are effectively decoupled from the dissociation coordinate while the hydrogen bonded C-H modes are strongly coupled. This fact is reflected in the linewidths of the transitions for the two bands of the dimer. As previously indicated, transitions associated with the "free" C-H stretch are very narrow (instrument limited) indicating that the lifetime is long (>120 ns). On the other hand, for the hydrogen bonded stretch the transitions are broadened beyond the instrumental linewidth due to the much shorter lifetime (6.1 ns) associated with the excited state. This difference is clearly understood in terms of the geometric coupling arguments given above and is observed in a number of systems.

The situation is quite different, however, for the two hydrogen bonded modes of the hydrogen cyanide trimer. Although these modes both involve motions associated with hydrogen bonded protons (and have very similar vibrational frequencies) the lifetimes are found to differ by a factor of 100 [14]. It is interesting to note that a large difference in lifetimes (a factor of 22) has also be observed using real time measurement techniques for the two NO vibrational modes of nitric oxide dimer [7]. In view of the fact that the two

NO sub-units in the dimer are parallel, these two vibrational
modes are once again quite similar, corresponding to the
symmetric and asymmetric combinations of the NO vibrations.
These results suggest that vibrational symmetry plays an
important role in the vibrational predissociation dynamics of
these systems. In any case, it is clear that for systems of
the complexity considered here, vibrational predissociation
is a highly mode specific and non-statistical process.

5. INTERMOLECULAR VIBRATIONAL FREQUENCIES FROM NEAR INFRARED SPECTROSCOPY

In the above discussion we have concentrated on the
vibrational degrees of freedom of the complex which correlate
with the high frequency intramolecular modes of the
constituent monomer units, since it is these that are
accessible using the near infrared laser systems now
available. However, as indicated in the introduction, the
intermolecular vibrations [8,9] (associated with the weak
bond) are of considerable interest in that they directly
depend on details of the intermolecular potential. Due to
their low frequency, these vibrational modes lie in the far
infrared region of the spectrum where tunable lasers are much
less developed. For this reason we have explored the
possibility of using combination differences between the
intramolecular fundamentals and the associated intermolecular
hot and combination bands to determine these low frequencies.
As we will now see, this allows for accurate determination of
the intermolecular vibrational frequencies in not only the
ground, but also the excited, intramolecular vibrational
states.

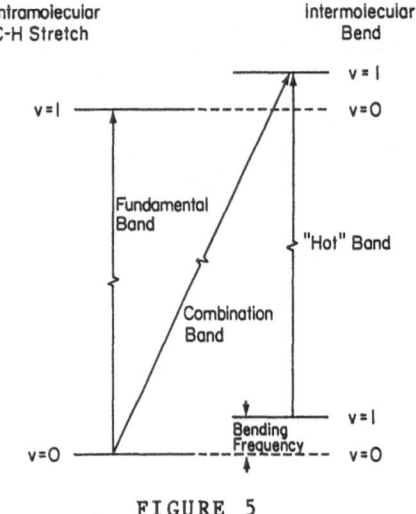

FIGURE 5

Upon closer examination of the hydrogen cyanide dimer
spectrum (Figure 1) several weak bands are also evident.
These bands are easily assigned to the two (symmetric and
asymmetric) low frequency bending hot bands of the dimer.
However, as indicated in Figure 5, these bands do not provide
sufficient information for the determination of frequencies
associated with the bends since no combination differences
can be formed. What is needed is the corresponding

combination band, which also shown in the figure. Using the l-type doubling splitting observed in the hot band transitions to estimate the frequency of the intermolecular bend, we are able to predict the approximate location of the combination band associated with the lowest frequency mode. The spectrum shown in Figure 6 was found within a wavenumber of this prediction. As expected, this band shows a strong Q branch since it involves the excitation of a bending vibration.

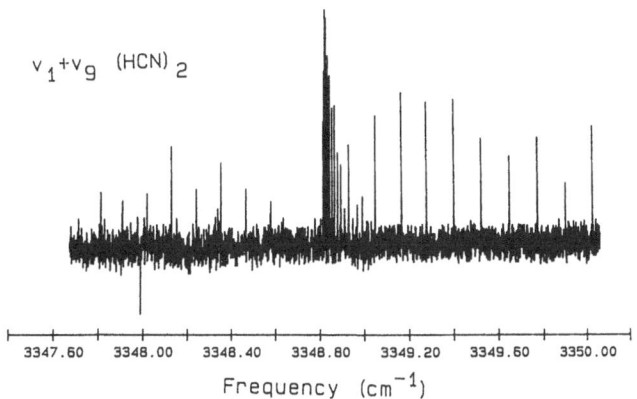

$v_1 + v_9$ (HCN)$_2$

Frequency (cm^{-1})

FIGURE 6

With all three of the vibrational bands shown in Figure 5, it is now possible to take combination differences to determine the bending frequency in both the ground and excited C-H stretching states, namely 40.7511(4) and 40.5520(4) cm^{-1}, respectively. The fact that the bending frequency is essentially unaffected by excitation of the C-H stretch is another indication that this intramolecular vibration is essentially decoupled from the intermolecular bond. To date we have been unsuccessful in our attempts to record the combination band in conjunction with the hydrogen bonded C-H stretch. The vibrational dependence in this case is expected to be somewhat larger and would clearly be of considerable interest in constructing an intermolecular potential surface which includes the intramolecular stretching dependence. Attempts are also being made to locate the combination bands associated with the higher frequency bending mode. In view of the relative ease with which near infrared lasers can be scanned, in comparison with those operating in the far infrared, this method of obtaining ground state intermolecular vibrational frequencies seems very attractive.

Another interesting aspect of this study is that the excitation of intermolecular bending vibrations does not influence the vibrational predissociation lifetime of the complex. Indeed, for the hot band transitions associated with the hydrogen bonded C-H stretch, the linewidths are the same as those of the fundamental. This is also seen in Ar-HF, where in a previous publication [2] we reported that the excited state Ar-HF does not vibrationally predissociate during its flight to the bolometer, which places a lower limit on the lifetime of $3*10^{-4}$ s. We have recently recorded the bending combination band (reported previously in direct absorption experments [18,19]) for this system (shown in

Figure 7) which again gives rise to a positive bolometer signal. These results suggest that small amount of intermolecular vibrational excitation does not appreciably alter the predissociation lifetime. This is most likely a result of the fact that the rate determining step is still the coupling of the intramolecular vibrational energy to the intermolecular coordinate. The fact that this coupling is unaffected by the excitation of the bending mode is just a restatement of the fact that the coupling between the two is very weak.

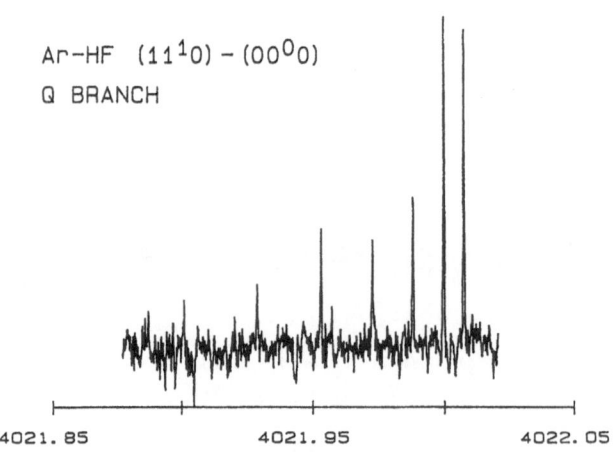

$Ar-HF \ (11^1 0) - (00^0 0)$

Q BRANCH

Frequency (cm^{-1})

FIGURE 7

6. ACKNOWLEDGMENTS

Support for this research is gratefully acknowledged from the National Science Foundation (CHE-86-03604) and from the Alfred P. Sloan Foundation.

References

[1]   R.E. Miller, J. Phys. Chem. 90 (1986) 3301; and references contained therein.
[2]   K.W. Jucks, Z.S. Huang and R.E. Miller, J. Chem. Phys. 86 (1987) 1098; Z.S. Huang, K.W. Jucks and R.E. Miller, J. Chem. Phys. 85 (1986) 6905.
[3]   C.M. Lovejoy, M.D. Schuder and D.J. Nesbitt, J. Chem. Phys. 86 (1987) 5337.
[4]   B.A. Wofford, J.W. Bevan, W.B. Olson and W.J. Lafferty, Chem. Phys. Lett. 124 (1986) 579.
[5]   A.S. Pine and W.J. Lafferty, J. Chem. Phys. 78 (1983) 2154.
[6]   G.D. Hayman, J. Hodge, B.J. Howard, J.S. Muenter and T.R. Dyke, Chem. Phys. Lett. 118 (1985) 12.
[7]   M.P. Casassa, J.C. Stephenson and D.S. King, "Structure and Dynamics of Weakly Bound Molecular Complexes", ed. A. Weber, NATO ASI Series 212 (1986) 513.
[8]   M.D. Marshall, A. Charo, H.O. Leung and W. Klemperer, J. Chem. Phys. 83 (1985) 4924.

[9]   D. Ray, R.L. Robinson, D.-H. Gwo and R.J. Saykally, J. Chem. Phys. $\underline{84}$ (1986) 1171; R.L. Robinson, D.-H. Gwo, D. Ray and R.J. Saykally, J. Chem. Phys. $\underline{86}$ (1987) 5211.

[10]   Z.S. Huang, K.W. Jucks and R.E. Miller, J. Chem. Phys. $\underline{85}$ (1986) 3338.

[11]   J.V.V. Kasper, C.R. Pollock, R.F. Curl, Jr. and F.K. Tittel, Appl. Opt. $\underline{21}$ (1982) 236.

[12]   A.C. Legon, D.J. Millen and P.J. Mjoberg, Chem. Phys. Lett. $\underline{47}$ (1977) 589.

[13]   R.S. Ruoff, T.D. Klots, C. Chuang, T. Emilsson and H.S. Gutowsky, 42nd sym. Mol. Spect., Ohio State Univ. (1987) 69.

[14]   K.W. Jucks and R.E. Miller, J. Chem. Phys., in press.

[15]   M. Kofranek, H. Lischka and A. Karpfen, Mol. Phys. $\underline{61}$ (1987) 1519.

[16]   J.M. Hutson, D.C. Clary and J.A. Beswick, J. Chem. Phys. $\underline{81}$ (1984) 4474.

[17]   N. Halberstadt, Ph. Brechignac, J.A. Beswick and M. Shapiro, J. Chem. Phys. $\underline{84}$ (1986) 170.

[18]   G.T. Fraser and A.S. Pine, J. Chem. Phys. $\underline{85}$ (1986) 2502.

[19]   C.M. Lovejoy, M.D. Schuder and D.J. Nesbitt, J. Chem. Phys. $\underline{85}$ (1986) 4890.

ROVIBRATIONAL SPECTROSCOPY OF ArCO VAN DER WAALS COMPLEX

Edward J. Campbell, Anne De Piante and
Steven J. Buelow

Chemistry and Laser Sciences Division
Los Alamos National Laboratory
Los Alamos, NM 87545 USA

INTRODUCTION

The information that can be obtained from the high resolution rovibrational spectra of small molecular clusters is very useful for determining intermolecular potentials and the effects of intermolecular interactions on intramolecular bonds. Recently, such spectra have been obtained via absorption measurements using F-center[1,2], frequency difference[3,4] and diode laser[5] sources. Because F-center and frequency difference lasers operate best in the 2-4 µm region, studies using these laser sources have involved high frequency vibrational modes (H-X, X=C,F,Cl,Br) or overtones. Although diode lasers cover a much broader spectral range (3-20 µm), they are less convenient to use for doing survey spectroscopy. We have developed several new techniques which improve the sensitivity and convenience of absorption measurements using diode lasers for the study of molecular clusters in supersonic expansions. Here we describe these techniques and their application for the measurement of the rovibrational spectra of ArCO near 5 µm.

EXPERIMENTAL

A schematic diagram of the apparatus used in our experiments is shown in Fig. 1. The output of a Laser Analytics diode laser is collected with a parabolic mirror and passed through a Spex 1 meter monochromator to select a single laser mode. A portion of the laser output (~5%) is split off, passed through a Fabry-Perot etalon (0.015 cm^{-1} FSR) and detected with an InSb detector. The signal from the detector is recorded by a transient digitizer, providing a relative frequency calibration for the spectra. The remaining radiation passes through a static gas cell and then into a vacuum chamber. Three spherical mirrors inside the vacuum chamber (r=450 mm), which are used in a White Cell[6] configuration, pass the laser radiation along the opening of a pulsed slit nozzle (2.5 cm x

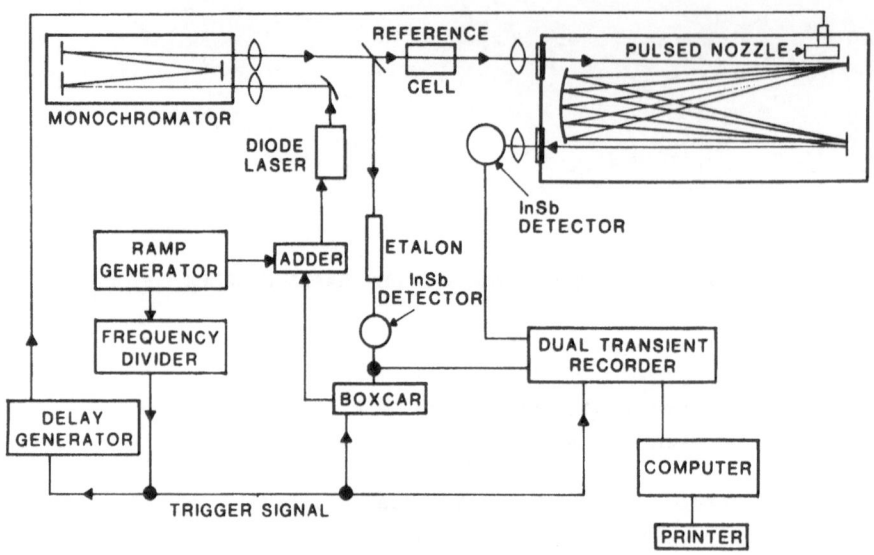

Figure 1. Schematic drawing of the experimental arrangement.

.005 cm). The radiation is mildly focused near the nozzle and passes 32 times within 1-2 cm of the nozzle slit. The path length in the high density region of the gas expansion is nearly a meter (Fig. 2). Under typical experimental conditions (75% Ar, 25% CO, 2 atm. backing pressure), the strongest ArCO features absorb 5-10 % of the laser radiation. The laser radiation passing through the expansion is detected by a second InSb detector and recorded by a transient digitizer.

The vacuum chamber is pumped by three mechanical pumps which have a total pumping speed of 1.5 m^3/minute. Cluster absorption can be observed as long as the background pressure in the chamber is less than 0.6 torr. Typically during our experiments the pressure is below 0.2 torr. Absorption lines of reference gases placed in the static gas cell provide absolute frequency calibration for the spectra. Usually this cell is not required since CO monomer absorption in the vacuum chamber is strong and produces numerous reference lines in the ArCO data. To reduce noise produced by amplitude fluctuations in the diode laser output, the diode laser is scanned very rapidly (200 cm^{-1}/sec). Signals from the ArCO absorption features which are about 0.003 cm^{-1} wide appear at electronic frequencies greater than 10 kHz - far above the low frequency fluctuations in the diode laser output. By averaging 2000 laser scans, absorbances as small as 3×10^{-5} can be detected.

During an experiment the diode laser is repeatedly scanned over a laser mode (.75 to 1.5 cm-1) every 5 milliseconds. The transient digitizers are triggered at the start of every hundredth laser scan. After this initial scan is completed, the nozzle is opened and a second laser scan is recorded on the second half of the transient recorder output (Fig. 3). The first scan is subtracted from the second, helping to correct for laser amplitude changes due to varying laser output over the mode and interference effects that can occur in the optics. Due to

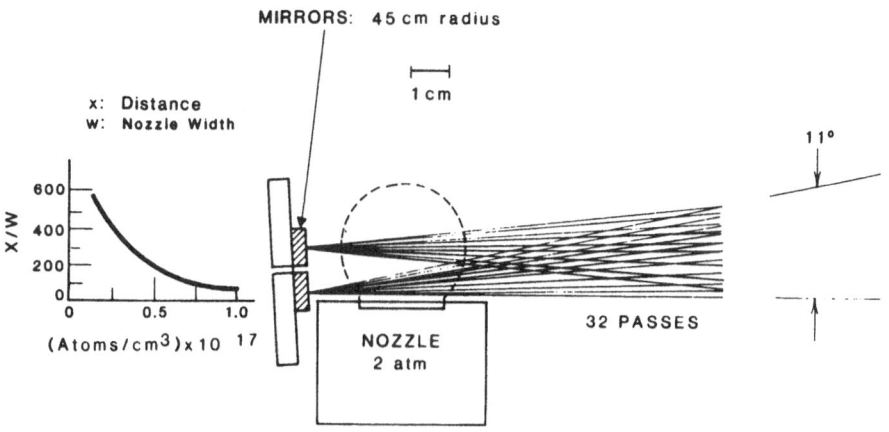

Figure 2. Scale drawing showing multipassing of diode laser radiation through nozzle expansion.

ohmic heating, the diode laser requires 10-15% of the scan time to return to the original lasing frequency at the end of each voltage ramp. It is during this time that the nozzle opens and reaches a steady state flow. The nozzle is kept open for a total of 7 milliseconds.

Although signal to noise can be quite high on a single scan (> 30), it is often necessary to average many scans to observe weak spectral features. To prevent drifting of the diode laser scan region during averaging, the scan region is stabilized using a boxcar integrator and voltage summing device. This is accomplished by sampling the etalon signal with a boxcar integrator at a specific time in the scan. If at this time the etalon signal is not at a zero crossing, the output voltage of the boxcar will be non-zero. This output voltage is summed with the ramp voltage scanning the diode laser, shifting the spectral region scanned by the laser so that a zero crossing of the etalon trace occurs at the sampling time of the boxcar. Thus the boxcar produces an error signal which prevents drifting of the diode laser. The linewidths[1] of the CO monomer features recorded in a single 5 millisecond scan of the diode laser are 0.0023 cm-1. When 2000 such scans are averaged, the linewidths increase to 0.0028 cm^{-1}. Thus, once locked to a zero crossing, the spectral region scanned by the laser drifts less than 0.0005 cm^{-1}.

The slit nozzle is capable of producing linewidths[4] less than 0.0015 cm^{-1}, therefore, the resolution in our experiments is currently limited by the diode laser. Although the diode laser beam does not cross the expansion perpendicularly, the multiple crossing angles contribute less than 0.0005 cm^{-1} to the total linewidth.

(1.)For this measurement, the vacuum chamber was pumped by two diffusion pumps (5000 liter/sec) to a pressure of 10^{-4} torr to minimize absorption by background CO.

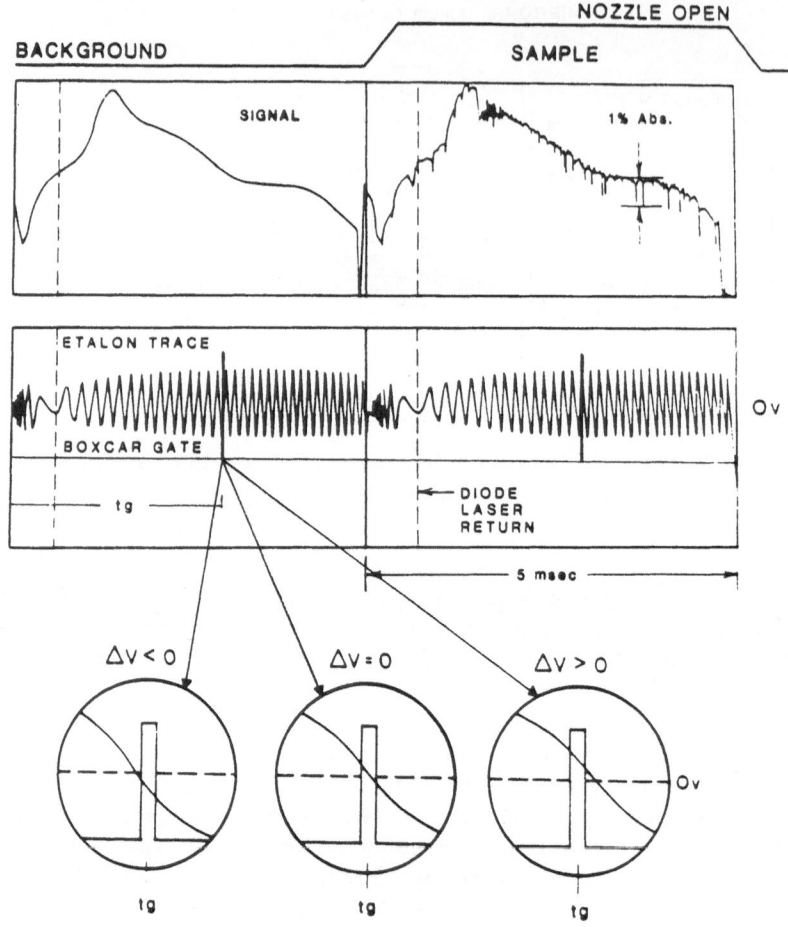

Figure 3. Traces of transient recorder output. Top panel shows
signal from radiation passing through vacuum chamber.
First half of trace corresponds to scan taken before
nozzle opens. Note that cluster absorbances are clearly
visible on second scan recorded after the nozzle opens.
Middle panel shows etalon signal (FSR=0.03 cm^{-1}). Bottom
figures show enlarged view of etalon trace near boxcar
sampling gate.

RESULTS AND DISCUSSION

We observe over 300 absorption lines near the CO monomer band center
at 2143.27116 cm^{-1} [7] which can be attributed to the ArCO complex. The
spectral region between 2130 to 2160 cm^{-1} also contains many lines for
pure CO clusters. Although the number of pure CO clusters in our Ar/CO
expansions can be reduced by reducing the proportion of CO in the Ar/CO
gas mixtures, they cannot be completely eliminated. Since the scan region
of the diode laser can be locked very accurately, it is possible to first
take a spectrum with an Ar/CO gas mixture, repeat the scan using only CO,
then subtract the two. Although the temperatures of the expansions may

not be exactly the same, the spectral features due to pure CO clusters reproduce quite well. The linewidths of features attributed to the ArCO complex are all less than 0.003 cm^{-1}. Within the limits of our resolution, we find no evidence of line broadening due to vibrational predissociation of the ArCO complex.

The ArCO spectrum is similar to one produced by a b-type transition in a nearly prolate, symmetric-top molecule. No lines are observed that can be attributed to an a-type transition. The most prominent features in the spectrum, two sub-bands corresponding to the K=0->1 and K=1->0 transition regions (K=K$_{-1}$), are found in the null gap of the CO monomer. The Q-branch of the K=0->1 region is shown in Fig. 4. A preliminary, least-squares analysis of these P, Q, and R sub-branches yields rotational constants for the K=0,1 sub-levels. They are presented in Table 1. Since the bending motion of the CO subunit and rotation of the CO about the Ar-CO bond axis may be strongly coupled, the energy levels can be represented by the expression [8]

$$E_K = \nu_K + \overline{B}_K J(J+1) - D_K[J(J+1)]^2 + \delta_{K,1}[1/4(B_1-C_1)J(J+1) - \delta J^2(J+1)^2]$$

where $\overline{B}_K = (B+C)/2$ and the term following the Kronecker delta accounts for the asymmetry splitting in the K=1 states. The spectrum contains a progression of such K-type sub-bands on each side of the CO monomer line center. The P, Q and R branches of the K=1->2 and K=2->1 transition

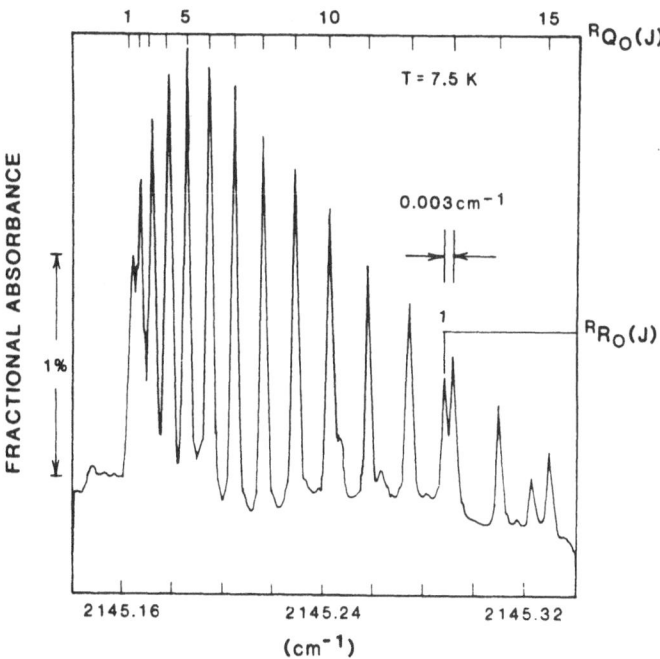

Figure 4. ArCO $^RQ_0$ band located in the CO monomer
null gap. Spectrum is an average of 300
5 millisecond scans.

Table 1. Effective Parameters for ArCO

| | $\nu=0$ | | $\nu=1$ | |
	K=0	K=1	K=0	K=1
$\overline{B}$ (MHz)	2072.2(7)[a]	2064.6(1.3)	2074.1(1.3)	2062.5(7)
B-C (MHz)		128.5 (6)		130.5(6)
$D/10^{-2}$ (MHz)	6.3(3)	7.3(3)	7.0(3)	6.4(3)
$\delta/10^{-3}$ (MHz)		-3.2(4)		-1.9(4)
$A-\overline{B}$ (MHz)	71643(12)		70973(12)	
$\Delta\nu$ (cm-1)	2142.8322(8)			
Δ_K (MHz)	1287(3)			

[a]Numbers in parentheses denote one standard deviation and apply to last digits of constants.

regions are all split by the asymmetry of the complex. Assuming that the origins, $\nu^0{}_K$, of the various K sub-levels can be expressed as

$$\nu^0{}_K = \nu + (A-\overline{B})K^2 + \Delta_K K^4,$$

we can solve for $\Delta\nu = \nu'-\nu''$, $(A-\overline{B})'$, $(A-\overline{B})''$ and $\Delta_K = \Delta_K' = \Delta_K''$ from the K=2->1, K=1->0, K=0->1 and K=1->2 sub-band origins, where the prime and double prime refer to the excited and ground states respectively. Values for these constants are also given in Table 1.

Calculations of potential surfaces [9] for the ArCO complex suggest that the CO subunit may not be strongly oriented in the complex. The calculations predict that only small-to-intermediate-sized barriers restrict the rotation of the CO in the plane of the complex. Although there have been no published studies of the spectra of ArCO, there have been several experimental and theoretical studies of the very similar ArN_2 complex [10,11]. This complex possesses both localized bending states as well as nearly free-rotor states of the N_2 sub-unit. For low K states, the energy level spacings resemble those of a slightly asymmetric top, while for very high K states, the energy level spacings are closer to those of a free rotor. Similar behavior is observed in our data. The energy level spacings for transitions involving K>1 levels bear a decreasing resemblance to those of a slightly asymmetric top. Preliminary analysis indicates that for the higher K levels, a

Hamiltonian for a semi-rigid molecular structure will not reasonably reproduce our ArCO spectra. In conclusion, we feel that to obtain meaningful structural information, it will be necessary to analyze the data using a hindered rotor Hamiltonian.

ACKNOWLEDGMENTS

We would like to thank Kathleen Mays for her assistance in recording and analyzing spectra. This work was supported by ISRD funds from Los Alamos National Laboratory.

REFERENCES

[1] T.E. Gough, R.E. Miller and G. Scoles, Appl. Phys. Lett., **30**, 338(1977).

[2] R.E. Miller, R.O. Watts and A.Ding, Chem. Phys. **83**, 155(1984).

[3] A.S. Pine, W.J. Lafferty and B.J.Howard, J. Chem. Phys.,**81**, 2939(1984).

[4] D.J. Nesbitt, *Structure and Dynamics of Weakly Bound Molecular Complexes*, edited by A. Weber, (D.Reidel Publishing Co., Dordrecht, 1987).

[5] G.D. Hayman, J. Hodge, B.J. Howard, J.S. Muenter and T.R. Dyke, Chem. Phys. Lett., **118**, 12(1985).

[6] J.U. White, J. Opt. Soc. Am., **32**, 285(1942),

[7] G. Guelachvili, J. Mol.Spectrosc., **75**, 251(1979).

[8] B.J. Howard, T.R. Dyke and W. Klemperer, J. Chem. Phys., **81**, 5417(1984).

[9] G.A. Parker and R.T. Pack, J. Chem. Phys., **69**, 3286(1978).

[10] G. Henderson and G.E. Ewing, Molec. Phys., **27**, 903(1974).

[11] G. Brocks and A. van der Avoird, *Structure and Dynamics of Weakly Bound Molecular Complexes,* edited by A. Weber, (D.Reidel Publishing Co., Dordrecht, 1987).

INFRARED SPECTROSCOPY OF MOLECULAR IONS*

Takeshi Oka

Department of Chemistry and Department of Astronomy
and Astrophysics, The University of Chicago
Chicago, IL. 60637 U.S.A.

1. INTRODUCTION

In various ways molecular ions are opposite to van der Waals molecules so far discussed in this workshop. In order to make the latter molecules are cooled so that they attach to each other. To make molecular ions you rip molecules apart and electrocute them. Van der Waals molecules are very weakly bound while molecular ions are strongly bound although some ions tunnel between a few well bound structures.

Let me start from a simple exercise. Consider the hydrogen molecule H_2. The simplest way to make an ion is to knock off an electron; we then have H_2^+. You can also add an electron to make H_2^-. Since H_2 is composed of two electrons and two protons, we can also do the same thing about protons. Thus by subtracting a proton we obtain H^- (deprotonated ion) and by adding a proton H_3^+ (protonated ion). In effect we obtain four ions starting from an ordinary molecule. Some of you may think that these

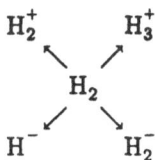

Figure 1. Four "stable" ions from the hydrogen molecule.

ions are exotic species and exist in a small amount only in special conditions of the laboratory. They are rare in the terrestrial atmosphere but some ions are very abundant in the Universe. For example, it has been well established by the theoretical studies of Chandrasekhar and his colleagues[1] that the opacity of the sun, that is, the deviation of the solar spectrum from blackbody radiation is due to bound-free and free-free transitions of H^- which exist in vast amounts in the chromosphere of the sun. The same situation applies to the tens of billions of main sequence

---

*This talk is dedicated to I. Plesser and his family.

stars in our galaxy which is also one of the tens of billions of galaxies in the Universe. The protonated hydrogen H_3^+ is also an important ion in dense molecular clouds as indicated by Herbst and Klemperer[27] when they introduced ion-molecule reactions as the important scheme for molecular formation in space. The role of H_3^+ as the protonator is so universal that Hiroko Suzuki called the chemistry of dense clouds the H_3^+ chemistry.[3] Thus, for example, the ubiquitous protonated carbon monoxide HCO^+ is produced by the proton hop reaction between H_3^+ and CO.

The new wave of molecular ion spectroscopy which is the subject of this talk was initiated by the discovery of the millimeter wave emission of HCO^+ by Buhl and Snider[4] in 1970. At the time of the discovery this was an unknown species and thus called Xogen, but its identity as HCO^+ was immediately conjectured by Klemperer[5], and experimentally confirmed by Woods and his students[6] five years later in 1975. The laboratory spectroscopy of H_3^+ was started the same year and the spectrum was discovered in 1980.[7]

2. Hydride Ions

Let us extend our exercise of Fig. 1 to molecules containing one heavy atom such as O, N, and C which are abundant elements in the Universe. We obtain the picture shown below. This picture is extremely rich and con-

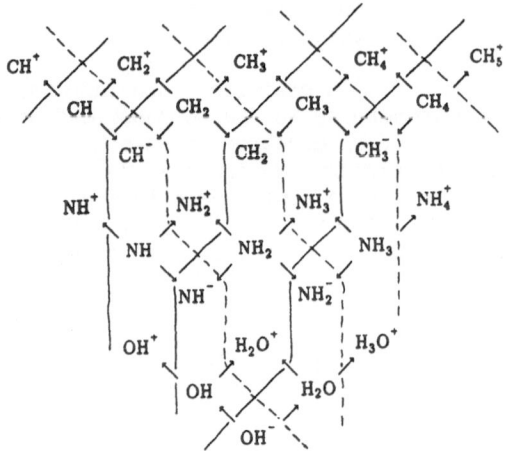

Figure 2. Hydride molecules, radicals and ions.

tains many fundamental species each of which deserves an extended explanation. Here we limit our discussion to molecular ions. The earlier high resolution spectroscopy of molecular ions was done through their electronic emission spectra. All diatomic cations in Fig. 2, i.e., CH^+, NH^+, and OH^+ and one tiatomic cation H_2O^+ have been studied by this method (see Herzberg's Faraday Lecture in 1971).[8] High resolution spectroscopy of other ions, i.e., OH^-, H_3O^+, NH^-, NH_2^-, NH_2^+, NH_3^+, NH_4^+ and CH_3^+ have all been done in the last five years. The method of laser infrared spectroscopy is generally applicable to study all these ions. The six carbo-ions CH_2^+,

CH_4^+, CH_5^+, CH^-, CH_2^- and CH_3^- are yet to be studied. This talk is not meant to be an exhaustive review and readers are referred to the review papers by Gudeman and Saykally[9] and of Sears[10] for a more complete summary. Also the proceedings of the recent Royal Society meeting on this subject[11] will be useful.

During our study of infrared spectra of eight molecular ions in Fig. 2, that is, OH^+, OH^-, H_2O^+, H_3O^+, NH_2^+, NH_3^+, NH_4^+ and CH_3^+, I was constantly amazed by the idiosyncrasy of each spectrum. No two of them are similar. However, we learned to group them in a variety of ways. The solid lines in the picture group them into isoelectronic species and the broken lines into isoprotonic species. Let us consider the isoelectronic species. The group of 9 species starting from CH_5^+ are 10 electronic system. Their united atom is Ne and like the neon atom their ground electronic states are all singlet. The group of 7 species starting from CH_4^+ are 9 electron species all isoelectronic to the fluorine atom and their ground states are all doublets. The 8 electron species starting from CH_3^+ are isoelectronic to the oxygen atom and their group states are triplets except for CH_3^+ for which it is singlet (the CH_2 ground state is almost singlet as many of you very well know). The 7 electron species is isoelectronic to the nitrogen atom; NH^+ has a doublet ground state but there is a very low lying quartet state also.

We can group these species in a different way shown with broken lines in Fig. 2. We then have isoprotonic species with one to five protons. For all these species all protons are equivalent within a species and therefore the total spin angular momentum quantum number I (corresponding to $I = \Sigma_i I_i$) is a very good though not rigorous quantum number. We therefore have ortho- and para- spin modifications for the two-proton (I=1,0) and the three proton (I = 3/2, 1/2) species and ortho-, para-, and meta-spin modifications for the four-proton (I=1,2,0) and five-proton (I=3/2, I=5/2, and 1=1/2) species. It has not been experimentally established yet that the five protons in CH_5^+ are equivalent. They are non-equivalent in the theoretically predicted C_s equilibrium structure but they are likely to be equivalent because the tunnelling motions between various structures will occur. Since different spin wavefunctioins have to combine with different vibration-rotation wave-funtions to satisfy the permutation-inversion symmetry of the total wave-function, the quantum number I plays an important role in the analysis of spectra.

Spectroscopic studies of each of the molecular ions are extensive work of many graduate student-years. All these ions are very light and their vibration-rotation spectra extends over several hundred wavenumbers. It takes quite some time to cover the whole region with high sensitivity. Today I will limit my discussion to two points. The first point is the qualitative features of spectra from which we obtain a qualitative under-standing on structure and dynamics of molecular ions. The second point is the large variations of molecular properties between the two most similar species, that is, isoelectronic-protonic species.

3. Key Features of Spectra

Spectroscopists rely a lot on accurately measured frequencies. However, in the process of analysis we also rely on the qualitative key features of spectra which give qualitative information on the symmetry of the quantum mechanical system which cause the spectra. This is par-ticularly so in molecular ion spectroscopy because we usually do not know the identity of the carrier of the spectra at the beginning. Let me give you a few examples.

Figure 3 shows the observed 15 lines of the ν_2 fundamental band of H_3^+. This spectrum extends over 600 cm^{-1} and does not have obvious

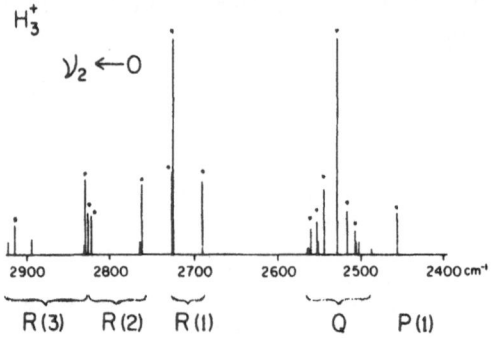

Figure 3. The ν_2 fundamental band of H_3^+. (Ref. 7)

symmetry or regularity. If I were Jim Watson,[12] I should see the hidden equilateral symmetry of H_3^+ in this spectrum. For me the key feature of the spectrum is the wide gap from 2690 cm^{-1} to 2562 cm^{-1} indicating the absence of the R(0) transition and this signifies that the J=0 K=0 ground rotational level is forbidden by the Pauli principle. This clearly indicates from symmetry argument[13] that the three protons in H_3^+ are equivalent, that is, the equilibrium structure of this ion is an equilateral triangle and that the ground electronic state is totally symmetric A_1'. If the three protons are not equivalent as the three oxygen atoms in ozone, we should see the R(0) transition.

Figure 4 shows two transitions R(5) and P(5) of the ν_3 band of the NH_4^+ ions. These traces are only a small portion (perhaps 1/500) of the total scans, but as soon as Mark Crofton and I[14] saw them we were convinced that we obtained the spectrum of NH_4^+ because the observed pattern

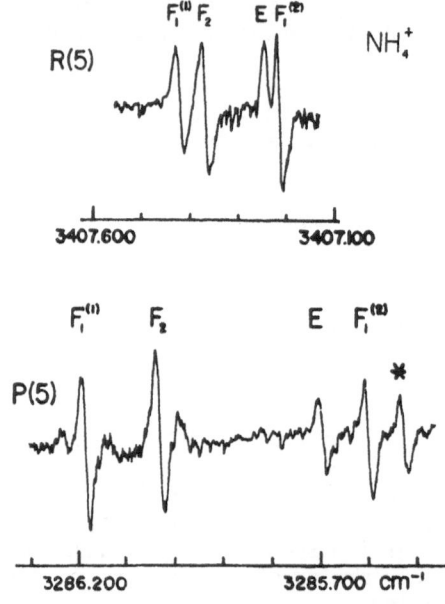

Figure 4. A portion of the ν_3 fundamental band of NH_4^+. (Ref. 14)

is unmistakably that of a species with tetrahedral symmetry. The lowest totally symmetric moment allowed for tetrahedral symmetry is in the form of the third order spherical harmonics $[Y_{3,2} - Y_{3,-2}]$ and whatever physical interaction is causing the splitting, the pattern should look like this.

The electron multiplicity also gives good fingerprints. For example, spectra NH_2^+, NH_3^+ and NH_4^+ sometimes appear intermixed but the triplet, doublet, and singlet structure of these lines helps us to identify them.

The spectral patterns of NH_3^+ and CH_3^+ give qualitative information on their structure and electronic states. We see in these patterns that K=0 levels are missing for every other J quantum number. This observation, together with the absence of other strong bands in the vicinity indicates clearly that both NH_3^+ and CH_3^+ are equilateral triangle and planar. The fact that even J levels are missing for CH_3^+ while odd J levels are missing for NH_3^+ shows that the symmetry of the ground electronic states for CH_3^+ and NH_3^+ are A_1' and A_2'', respectively.

4. Comparison of Isoelectronic-protonic Species

Among the species shown in Fig. 2, the species closest in their quantum chemical properties are species containing the same number of electrons and protons, isoelectronic-protonic species. In this talk I shall tentatively call them isoelectroprotonic species. The case of H_3O^+ and NH_3 is an example. I shall compare their characteristics determined from infrared spectroscopy. Both molecules have 10 electrons and three protons. Thus, like NH_3, H_3O^+ have the C_{3v} pyramidal structure and inversion (umbrella) motion. The potential for the H_3O^+ inversion motion is shown below. The vibrational energy states involved in the discussion have been accurately determined by Di-Jia Liu.[15] The only difference between H_3O^+ and NH_3 is the O and N nuclei and this difference appears in their structure and dynamics in the following way (I learned this way of looking at it recently from Mitchio Okumura). The nuclear properties do not affect quantum chem-

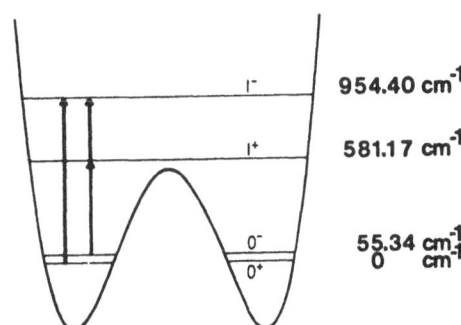

Figure 5. The inversion potential function of H_3O^+. (Refs. 15,16)

istry much except for the charge. Since the oxygen has an extra charge, the electrons are pulled in more to the center in H_3O^+. The bonds are tighter in H_3O^+ than in NH_3. This is clearly seen in the shorter bond-length (0.976 A in H_3O^+ versus 1.017 A in NH_3) and the higher hydrogen stretching frequencies (ν_3 = 3530 cm^{-1} versus 3337 cm^{-1}). This also increases the Sp_2 hybridization and H_3O^+ is more planar than NH_3. Thus the H-X-H angle is larger in H_3O^+ than in NH (111.3° versus 107.8°) and the barrier for the inversion motion lower ($\sim$800 cm^{-1} versus $\sim$2000 cm^{-1}, the height depends on the assumed shape of the potential). Thus latter differences appear in the large increase of the inversion frequency which has been very precisely measured (55.3481 cm^{-1} versus 0.79345 cm^{-1}). I might emphasize here that the appearance of inversion frequency of NH_3 in the microwave range is a great accident of nature and so far limited only to NH_3 and other amino and imino compounds. We now have analogous splittings in H_3O^+ and related compounds in the far infrared region.

In Fig. 2 we saw that the methyl anion CH_3^- is also isoelectroprotonic to H_3O^+ and $NH3$. The quantum chemical discussion here is perhaps not as simple because of the very low electron affinity of CH_3^- (0.08 eV). Observation of this species will be extremely interesting although the production and spectroscopy will be much more difficult because of the low electron affinity.

We can give similar discussion to other groups of isoelectroprotonic species. Thus, for example, NH_2^+ and CH_2 are very similar but because of the extra charge of NH_2^+, it is more tightly bound ($\nu_3 = 3360$ cm versus 3153 cm^{-1}) and more quasi-linear. Figure 6 shows the bending potential of NH_2^+ calculated by Peyerimhoff and Buenker.[17] You will note that the central barrier is just above the zero-point vibrational level. For such a molecule we need quantum mechanics for quasi-linear species.

Similar interesting comparisons can be made for members of other isoelectroprotonic groups in Fig. 2; $CH-NH^+$, $CH^--NH-OH^+$, NH^--OH, $CH_2^--NH_2-H_2O^+$, $CH_3-NH_3^+$ and $CH_4-NH_4^+$. I limited Fig. 2 to compounds containing O, N, C atoms because of their astrophysical interest but we can readily extend it to compounds containing the B and F atoms.

NH$_2^+$ $\tilde{X}\,^3B_1$ BENDING POTENTIAL

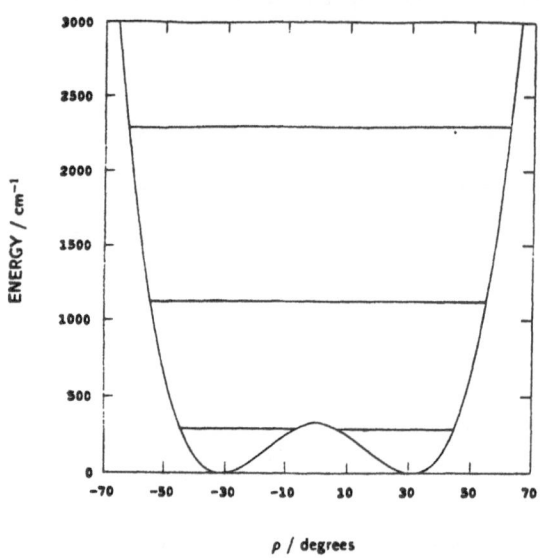

Figure 6. The bending potential for NH_2^+. (Ref. 17)

5. Production of Ions, Alchemy

For studying these ions, we use various tunable laser infrared sources and modulation techniques. We use non-linear optical techniques for the former and velocity modulation for the latter. This latter method developed by Gudeman and Saykally is extremely powerful.[18] And then the main thing, of course, is the production of ions. Ever since the initial molecular ion spectroscopy of N_2^+ by Wilhem Wien in 1922,[19] laboratory spectroscopists used discharge for ion spectroscopy. (The detection of CO^+ comet tail band by Deslandres in 1907[20] marks the first astronomical molecular ion spectroscopy). Chemistry in discharges, sometimes called plasma chemistry, is a blind chemistry, alchemy.

Here, I cannot resist my temptation to share with you the excitement I had yesterday in Jerusalem. This transparency (not shown) shows one of the great many pages of Isaac Newton's notebook which is in the National Library of the Hebrew University. I came to this Conference one day early to see the notebook. Newton started his chemistry in 1669 when he was 26 years old. Some historians of science say that the total number of hours Newton spent in alchemy experiments is comparable (or more) to the time spent in mathematics and physics. According to Humphrey Newton, Newton's assistant, in the 1680's "the fire in Newton's laboratory scarcely went out night or day while Newton himself would rarely retire before 2 or 3 o'clock, sometimes not until 5 or 6."

Our alchemy, plasma chemistry, is very similar to interstellar chemistry.[2] This is because the fraction of charged species ($\sim 10^{-6}$) is comparable in the laboratory plasmas and in molecular clouds and the chemical reactions are essentially binary reactions. For example, the cations in Fig. 2 are produced through the chain of hydrogen extraction reactions

$$X^+ + H_2 \longrightarrow HX^+ + H,$$

$$O^+ \longrightarrow OH^+ \longrightarrow H_2O^+ \longrightarrow H_2O^+$$

$$N^+ \longrightarrow NH^+ \longrightarrow NH_2^+ \longrightarrow NH_3^+ \dashrightarrow NH_4^+$$

and $\quad C^+ \diagdown CH^+ \longrightarrow CH_2^+ \longrightarrow CH_3^+ \diagdown CH_4^+ \longrightarrow CH_3^+.$

All these reactions have the large Langevin rate except the three reactions without arrows. Mark Crofton has introduced a clever use of He in the discharge which has enabled us to optimize our discharge to individual species in the chain. Helium is special among gases in that (1) it is chemically inactive (2) it increases electron temperature of the plasma (because of the high ionization potential of He) (3) its metastable states help ionization (Penning ionization) and (4) it does not trap protons (because of the low proton affinity of He).

Production of a large amount of specific carbo-ion is more difficult than that of other ions because of polymerization. Most ion molecule reactions between carbon containing species have Langevin rates and lead to a large amount of soot which hampers spectroscopy. Figure 7 shows a discharge tube we use for carbo-ion spectroscopy. A premixed gas (say He:H_2:HCCH=1000:7:2 with a total pressure in the cell of ~ 7 Torr) is introduced into the discharge through the multiple inlets. We note greenish blue emission at the inlets indicating fresh chemical reactions in the optical path. This is a water-cooled cell; we also use air-cooled and liquid nitrogen cooled plasma tubes.

Ignorance, simple-mindedness and recklessness are valuable requisites in this type of chemistry, and we surely have plenty.

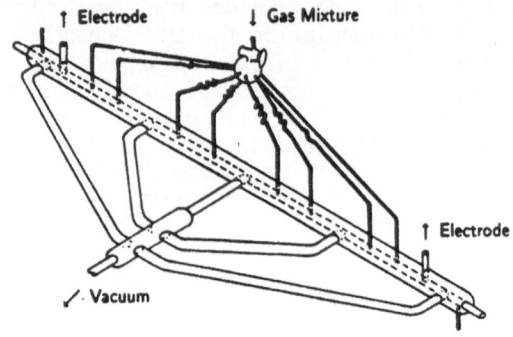

↑ Electrode   ↓ Gas Mixture

↑ Electrode

↗ Vacuum

Figure 7. A water cooled discharge cell (tarantula) used for carbo-ion spectroscopy.

6. Spectra of Carbo-Ions

Using a discharge tube similar to that shown in Fig. 7, Mark Crofton discovered in March 1985 extremely rich spectra of carbocations in the $3\,\mu$m region. These spectra kept us busy ever since and they will keep us occupied for the next several years. Since the C-H stretching vibrations of all carbo-ions fall in this region, spectra of many ions appear intermixed. So far we have identified three species, methyl cation CH_3^+, acetylene ion $C_2H_2^+$, and protonated acetylene (vinyl cation) $C_2H_3^+$. They are all very fundamental organic cations. Figure 8 shows a small portion of the spectra taken at liquid nitrogen temperature. Since acetylene

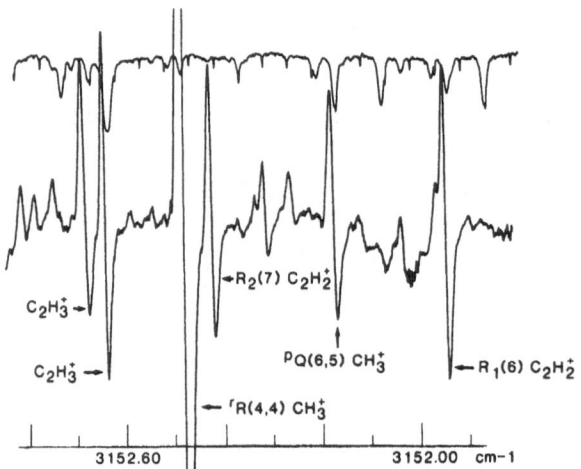

$C_2H_3^+$ →

$C_2H_3^+$ →

← $R_2(7)$ $C_2H_2^+$

$^PQ(6,5)$ CH_3^+

← $R_1(6)$ $C_2H_2^+$

← $^rR(4,4)$ CH_3^+

3152.60   3152.00 cm-1

Figure 8. A small portion of carbo-ion spectra.

freezes at the liquid nitrogen temperature, we use methane in the mixture ($He:H_2:CH_4$ ~1000;70;10 with the total pressure of ~7 torr). Out of these bushes of many spectral ines we extract groups of spectral lines. We use everything possible to help us in identifying a spectrum. The spectral pattern can be recognized if it is regular. Chemistry helps us to discriminate different ion lines. For example, we can empirically find a discharge condition which shows the $C_2H_2^+$ spectrum but barely shows the $C_2H_3^+$ spectrum. The width of a spectral line is proportional to $1/\sqrt{m}$ and if a line has a reasonable signal to noise ratio, we can discriminate lines of an ion containing one carbon atom from those containing more than one carbon atom; the lines of H_3^+ and H_eH^+ can also be identified from their linewidth.

We thus obtained a beautiful pattern of the ν_3-fundamental band of CH_3^+ in the bush as shown in Figure 9. The methyl cation CH_3^+ is the most

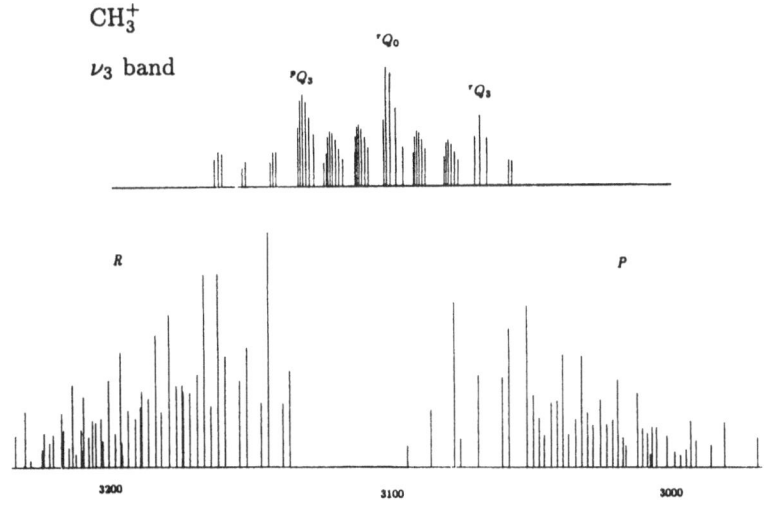

Figure 9. The ν_3 band of CH_3^+. (Ref. 21)

fundamental carbonium ion for which there had been no direct spectroscopic observations in any spectral region. More detail about this ion is found in our recent paper.[21]

We have subsequently detected the $\pi \leftarrow \pi$ ν_3-band of the acetylene ion $HCCH^+$ and the $A^2\pi_n \leftarrow X^2\Sigma_g^+$ electronic transition of C_2^-. Both of them are fascinating spectra which occupied us for quite sometime. Readers are referred to our recent papers. (Refs. 22, 23, 24).

7. <u>Protonated Acetylene $C_2H_3^+$</u>

The species on which we have spent much time and with which we are still struggling is the protonated acetylene $C_2H_3^+$. I believe this molecular ion is of particular interest to this workshop because two structures exist and the energy difference between them is theoretically predicted to be small. The two structures are shown below in Figure 10. Initially with SCF calculations the classical structure was predicted to be more stable[25] but as more CI calculations are introduced, the two structures became comparable in energy.[26] In the most recent results of Lee and Schaefer[27] the bridged structure is predicted to be more stable by 0.97 kcal/mol. In a recent paper on a Coulomb explosion experiment, Kanter, Vager, Both and Zajfman conclude that "the data clearly show the non-classical structure to dominate our sample of molecules".[28]

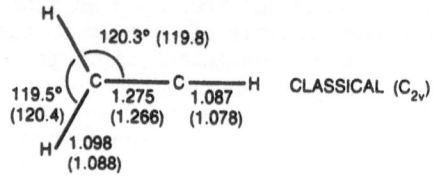

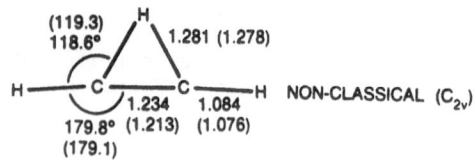

Figure 10. Theoretically predicted
structures of $C_2H_3^+$. (Ref. 27)

REFERENCES

(I have run out of given pages. My preliminary report on the spectrum of $C_2H_3^+$ is in Ref. 22. A detailed manuscript is in preparation.)

1. S. Chandrasekhar, Truth and Beauty, The University of Chicago Press, Chicago and London (1987).
2. E. Herbst and W. Klemperer, Ap. J. 185, 505 (1973).
3. H. Suzuki, Ap. J. 272, 579 (1983).
4. D. Buhl and L. E. Snyder, Nature 227, 862 (1970).
5. W. Klemperer, Nature 227, 1230 (1970).
6. R. C. Woods, T. A. Dixon, R. J. Saykally and P. G. Szanto, Phys. Rev. Lett. 35, 1269 (1975).
7. T. Oka, Phys. Rev. Lett. 45, 531 (1980).
8. G. Herzberg, Rev. Chem. Soc. 25, 201 (1971).
9. C. S. Gudeman and R. J. Saykally, Ann. Rev. Phys. Chem. 35, 387 (1984).
10. T. J. Sears, J. Chem. Soc., Faraday Trans. 83, 111 (1987).
11. Phil. Trans. R. Soc. Lond. A324 (1987).
12. J.K.G. Watson, S. C. Foster, A.R.W. McKellar, P. Bernath, T. Amano, F. S. Pan, M. W. Crofton, R. S. Altman and T. Oka, Can. J. Phys. 62, 1875 (1984).
13. L. D. Landau and E. M. Lifshitz, Quantum Mechanics, Pergamon Press (1977).
14. M. W. Crofton and T. Oka, J. Chem. Phys. 79, 3157 (1983).
15. D. J. Liu and T. Oka, Phys. Rev. Lett. 54, 1787 (1985).
16. D. J. Liu, T. Oka and T. J. Sears, J. Chem. Phys. 84, 1312 (1986).
17. S. D. Peyerimhoff and R. J. Buenker, Chem. Phys. 42, 167 (1979).
18. C. S. Gudeman, H. M. Begemann, I. Pfaff and R. J. Saykally, Phys. Rev. Lett. 50, 727 (1983).
19. W. Wien, Ann. Physik 69, 325 (1922), 81, 994 (1926).
20. H. Deslandres, C. R. (Paris) 145, 445 (1907).
21. M. W. Crofton, M.-F. Jagod, B. D. Rehfuss, W. A. Kreiner and T. Oka, J. Chem. Phys. 88, 666 (1988).
22. T. Oka, Phil. Trans. R. Soc. Lond. A324, 1298 (1987).
23. M. W. Crofton, M.-F. Jagod, B. D. Rehfuss and T. Oka, J. Chem. Phys. 86, 3755 (1987).
24. B. D. Rehfuss, D. J. Liu, B. M. Dinelli, M.-F. Jagod, W. C. Ho, M. W. Crofton and T. Oka, J. Chem. Phys., submitted.
25. J. A. Pople, Welch Foundation Conference XVI, Theoretical Chemistry, p. 11 (1973).
26. J. Weber, M. Yoshimine and A. D. McLean, J. Chem. Phys. 64, 4159 (1976).
27. T. J. Lee and H. F. Schaefer III, J. Chem. Phys. 76, 3437 (1986).
28. E. P. Kanter, Z. Vager, G. Both and D. Zajfman, J. Chem. Phys. 85, 7487 (1986).

ESTIMATION OF MICROWAVE AND INFRARED SPECTRAL

PARAMETERS OF SOME MOLECULAR IONS FROM

AB-INITIO ELECTRONIC STRUCTURE CALCULATIONS

R. Claude Woods

Department of Chemistry
University of Wisconsin
Madison, WI 53706

INTRODUCTION

The observation of the microwave spectra of new molecular ions has been a principal objective of the author's research program for a long time. Successful detection of the spectrum of a new ionic species may lead to a precise determination of its molecular structure, allow radioastronomical observation of it in the interstellar medium, and/or provide useful spectroscopic probes of the dynamics of plasmas in which it may be located. To search over a sufficiently wide spectral range at high enough sensitivity to detect a new ion, however, is a tedious task, which may require weeks of effort. To minimize this experimental burden accurate advance predictions of molecular geometric structures and thus rotational constants (reciprocals of moments of inertia) are highly desirable. With the computer facilities and public programming packages that are currently available we have found that *ab initio* calculations of sufficient quality to be very useful in this regard are quite practical. In this paper we will briefly describe results that we have obtained for several molecular cations and anions of interest using the Gaussian 82 programs of Pople and coworkers[1] with rather large basis sets and extensive treatment of electron correlation. The emphasis has been (1) to employ calculations that were of the highest quality that seemed practical, (2) to calculate all the spectroscopic constants, e.g., vibration-rotation interaction and anharmonicity, from the molecular potential function, and (3) to use identical calculations on known systems (typically neutrals) for calibration purposes to the maximum extent possible. These same calculations have also turned out to give quite accurate predictions of vibrational constants, and thus they can also be useful for locating and assigning infrared spectra of new molecular ions. Electric dipole moments and electric field quadients, necessary for estimating microwave intensities and nuclear quadrupole hyperfine splittings, respectively, have also been successfully and reliably computed in this work. The present paper will provide a brief summary of results to give the reader a feeling for which systems have been treated, at what level of theory and at how great an expenditure of computer time, and what accuracy has been achieved in the prediction of spectral constants. Full details of the calculations have been or will be published elsewhere.[2-5] The actual calculations have been carried out by K. A. Peterson, a member of the microwave research group at the University of Wisconsin.

The molecular systems that will be discussed can be grouped into three series. The first contains closed shell diatomics with 14 electrons, including the ions CF^+, NO^+, BO^-, and CN^- and the neutral reference molecules BF, CO, and N_2. The second set are the second row analogs of these with 22 electrons. The positive ions are SiF^+, PO^+, NS^+, and CCl^+, and the negative ones, AlO^-, SiN^-, CP^-, and BS^-, while the corresponding neutrals are AlF, SiO, PN, CS and BCl. The third type are linear triatomic protonated versions of some of these, including especially $HBCl^+$ and HNSi. Calculations have also been made on other ions of interest including open-shell systems, non-linear triatomics, and cluster ions, but these studies are still in progress and will not be described here.

COMPUTATIONAL METHODS

The potential energy functions have been calculated using Moller-Plesset perturbation theory with large basis sets of contracted Gaussian type orbitals (cGTO's). The latter always contain polarization functions (d-type atomic orbitals) and in some instances they also include f-type orbitals. The total number of cGTO's are in the range 66 to 100 for various cases. After an initial SCF step electron correlation is treated at a sequence of increasingly higher levels of theory: MP2, MP3, MP4DQ, and MP4SDQ. Only the last of these (fourth order Moller-Plesset with single, double and quadruple substitutions) has been found to be adequate for the purpose at hand, and it is these results that are described here. For each diatomic, electronic energies are computed at 8-10 internuclear distances, and then these points are fit to a sixth degree polynomial function in $\Delta r = r - r_e$. The most critical results are the equilibrium distance r_e, which determines the rotational constant B, and the quadratic potential constant which determines ω_e, the harmonic vibrational frequency, but the potential function is also employed to compute the centrifugal distortion (D), vibration-rotation interaction (α), and anharmonicity ($\omega_e x$) constants, which are very helpful in predicting and interpreting spectra. In the triatomics the stretching potential is sampled by calculations at up to 38 pairs of bond distances, ΔR and Δr, and fit to a two-variable polynomial function of sixth degree with six cross terms ($\Delta R^i \Delta r^i$). The bending degree of freedom was less fully explored, with only enough non-linear geometries considered to fit quadratic and quartic bending force constants. The calculated harmonic and anharmonic force fields were used to compute the various spectroscopic constants (B, ω's, x's, D, and α's) using standard perturbation theory results. In order to calculate the electric dipole moments and the field gradients at the nuclei correlated electronic wave functions were obtained using the CI-SD method (configuration interaction with singles and doubles) with the same large basis sets. Since a CI-SD calculation takes about three times as long as the corresponding MP4SDQ one, this was normally practical only at the equilibrium internuclear distance. Calculated dipole moments from this approach agree with experiment in the known cases to 0.2 D or better, while the quadrupole coupling constants eqQ appear to be reliable to within 1 MHz for first row nuclei and to within 2 MHz for second row nuclei.

CF^+, NO^+, BO^-, CN^-, BF, CO, N_2

The primary set of calculations on the fourteen electron diatomics used a basis set with (11s, 7p, 2d) primitives contracted to [8s, 5p, 2d] cGTO's on each atom for a total of 66 basis functions. All 14 electrons were correlated in the post SCF steps. Results for all but BO^- have already appeared in a recent publication.[2] These calculations were done on an IBM 3081 computer at the IBM Palo Alto Science Center and required about 9 min. of CPU time per energy point. The r_es for both the positive ions and the neutrals of this series are already known from high resolution spectroscopy, as are the ω_es and other spectroscopic constants. The raw MP4SDQ values of r_e are within 0.0025 A° of the experimental values for CO, N_2, CF^+, and NO^+, while for BF the discrepancy is 0.007 A°. If the errors in r_e are plotted in the natural sequence BF, CF^+, CO, NO^+, N_2, they form a fairly smooth curve, and if the *ab initio* values for each ion are corrected by the errors seen in the cases of the two adjacent neutrals in this sequence, the r_e(corr.) values are within 0.002 A° of experiment. Since the relative uncertainty in the rotational constant, and hence in a microwave transition frequency, is twice that in r_e, this would have corresponded to errors of about 0.4% in frequency if the r_e(corr.) had been used to predict microwave spectra of the ions. This corresponds to a wide, but practical search range. Predictions of r_e that are 5 times less accurate than this become very easy to make from *ab initio* calculations, but the search range then becomes almost unworkably large for the high sensitivity conditions usually required. The most important results of this work are the corresponding predictions of the two negative ions' r_e (and B) values. For the very important CN⁻ ion our MP4SDQ value of r_e agrees almost exactly with another recent large *ab initio* calculation, a CEPA-1 study by Botschwina.[6] The ω_es, or harmonic vibrational frequencies also agree very well with experiment in the known members of this series (within about 0.4%), except for NO⁺ (~2% in error). The latter problem is due to a UHF instability that occurs in NO⁺. Tests for this kind of instability are available, so such problems can be foreseen if they exist in unknown cases. The excellent predictions of vibrational spectra, including the anharmonic corrections, that are possible with these calculations are illustrated in Table I, which gives MP4SDQ and experimental term energies of the lower vibrational levels of several of these molecules. Calculations of the vibrational spectra using the CEPA-1 method have been successful for many important small molecules and ions,[7-10] and results such as those in Table I indicate that these MP4SDQ calculations give accuracies in vibration constants that are very competitive with those of the CEPA-1 method. The accuracy of the CI-SD predictions of quadrupole coupling constants with this basis set are suggested by the following comparisons of theory (before slash) and experiment (after) for eqQ (in MHz): ¹¹BF (-4.71 / -4.5 [11]), ¹⁷CO (4.07 / 4.59 [12]), ¹⁴N₂ (-5.681 / -5.39 [13]), and NO⁺ (-7.20 / -6.76) [14].

A second set of calculations on this series has been carried out to study the basis set dependence of r_e and ω_e calculations at the MP4SDQ level. Five basis sets of increasing flexibility containing 30, 36, 44, 54, and 68 cGTO's, respectively, were employed, and only the valence electrons were correlated. The largest basis set contained f orbitals, and was in that sense better than the basis set of the preceding paragraph. When errors (MP4SDQ minus experiment) in r_e or ω_e are plotted in the same sequence (BF, CF⁺, CO, NO⁺, N₂), the curves as expected become smoother and closer to zero as basis set size increases. A smooth curve means that the correction method will work very well. The results for r_e from the basis containing f orbitals are actually superior to those in the previously discussed calculations. In particular BF pulls into line with the other four cases, and the errors all fall in the range +0.2% to +0.3%, so the r_e(corr.)s for the ions are good to ± 0.1% with this basis set. Details of all these calculations will be published in a new paper that contains our work on the 22 electron cations and neutrals.[3]

AlF, SiF⁺, SiO, PO⁺, PN, NS⁺, CS, CCl⁺, BCl

Similar calculations have also been carried out on the series of 22 electron second row analogs. For the neutrals and positive ions two basis sets (A and B) were used. Basis set A had [7s, 4p, 2d] cGTO's on the first row atoms and [9s, 6p, 2d] on the second for a total of 66 functions, while set B could be described as [8s, 5p, 2d, 1f] (first row atoms) and [10s, 7p, 3d, 1f] and contained a total of 93 functions. Only the valence electrons were correlated. Calculations were done on a VAX 8600 and required 0.5 h per point for basis A and 2 h per point for basis B. The goal of the work was to predict the spectroscopic constants of the four positive ions, but after the basis A work had been done on BCl, CCl⁺, and CS, precise spectroscopic values for CCl⁺ were published[15] and became available for comparison to theory. Details of our work on these three molecules at the basis set A level were contained in Ref. 2 and all the other results of this section will be described more fully, with complete references to experimental and other theoretical work in Ref. 3. Results for BCl, CCl⁺, and CS for set A were excellent, comparable to those for the first row systems already discussed. The errors in r_e vary smoothly across this series of three, so the r_e(corr.) differs from experiment by only -.0019 A°. (Even for the raw MP4SDQ r_e the discrepancy (relative to experiment) is only + 0.0032 A°.) Table I also makes clear the excellent predictions of vibrational term energies that were achieved with basis set A.

Table I. Term energies at MP4SDQ level for 14 electron molecules (in cm^{-1}).

		G(0)	G(1)	G(2)	G(3)	G(4)
BF						
	MP4SDQ[a]	699	2080	3438	4774	6087
	Experiment[c]	698	2077	3431	4762	6070
CO						
	MP4SDQ[a]	1077	3207	5308	7380	9421
	Experiment[c]	1082	3225	5342	7432	9496
CF$^+$						
	MP4SDQ[a]	896	2667	4412	6130	7822
	Experiment[c]	893	2659	4400	6114	7803
CS						
	MP4SDQ[b]	638	1904	3155	4391	5613
	Experiment[c]	641	1913	3172	4418	5652
CCl$^+$						
	MP4SDQ[b]	588	1754	2907	4049	5178
	Experiment[c]	587	1752	2903	4041	5166

[a]With 66 function basis set and all electrons correlated.
[b]With basis set A (66 functions, see text) and valence electrons only correlated.
[c]For full references to experimental values, see Ref. 2.

When basis set A calculations were extended to the other three neutrals it became clear that the errors in r_e increase as one moves from CS back to AlF, reaching about 1% at that end of the series. On the other hand the errors for the six experimentally known molecules (5 neutrals plus CCl$^+$) do vary quite smoothly when plotted vs. their position in the sequence given in the subtitle of this section, suggesting that the correction method would be successful and very helpful when applied to the remaining ions. In view of the previous remarks about the accuracy required for practical microwave searching, the correction method become critical when there are errors of the order of 1% in the *ab initio* values of r_e. The errors in ω_e are also greater for AlF, SiO, and PN (reaching -5% for SiO) than they had been for CS, CCl$^+$, and BCl. In an attempt to improve matters the calculations with basis set B were carried out. These did indeed result in reduced errors in the r_es for all six known species (typically reductions of 40-50%), and the ω_es for AlF, SiO, and PN were also improved. On the other hand the ω_es for the other three known species become considerably less well predicted. Most troublesome was the fact that the CCl$^+$ errors did not fall on the smooth curve of the neutral molecule errors nearly so well as they had for basis set A. This means that the correction method seems to work less well for set B than for set A, at least for CCl$^+$. The differences between the set B and set A values of r_e and ω_e were then plotted across the whole sequence of nine molecules and a fairly systematic alternation between the ions and neutrals was observed. This indicates that the expansion of the basis set affects ions and neutrals in a systematically different way. At this point the previously mentioned study of the first row analogs with a series of five different basis sets was carried out, and it indicated that generally the better the basis set the smoother the curve, the less the alternation between ions and neutrals, and the more successful the correction method. This would suggest that basis set B would provide more reliable predictions for SiF$^+$, PO$^+$, and NS$^+$ than basis set A. Conversely, the comparison of CCl$^+$ to experiment had supported basis set A as the more reliable of the two. Only when experimental data become available for some of these three ions can this question be fully resolved. In any case the differences between the two sets of corrected values are only about 0.1% for r_e and about 1% for ω_e, so the corrected results from either basis are probably a great improvement over the raw *ab initio* values. All the spectroscopic constants have been calculated for all nine molecules, as have the CI-SD electronic properties (dipole moments and quadrupole coupling constants). They all agree very well with available experimental values. Full details will be contained in Ref. 3.

AlO^-, SiN^-, CP^-, and BS^-

A similar study has been carried out with the four 22 electron anions interspersed into a sequence with the same five neutrals: AlF, AlO^-, SiO, SiN^-, PN, CP^-, CS, BS^-, and BCl. In this case basis A, already described, was used along with basis set C, which contained a total of 89 cGTO's. Set C was an augmented version of set A, but differed from set B in that it contained diffuse functions on each atom, rather than f orbitals. In this case both basis sets gave reasonably similar results for r_e and ω_e for the five neutrals, and no experimental information is available for any of the 4 ions. When the r_e (or ω_e) predictions for set C minus those for set A are plotted vs. position in the above series, there is again a pattern of alternation between the ions and the neutrals. In this case one could argue that it is a consequence of the fact that diffuse functions are more critical for anions than for neutral molecules. If one accepts this argument, then it is to be expected that corrected values of r_e (and ω_e) derived from basis set C will be more reliable than those from set A. The differences between the two sets of corrected values are of the same magnitude as those between sets A and B for the positive ions (except r_e for BS^- where they differ by ~0.2%). Again either corrected result (from basis set A or C) should be much more reliable than the raw ab initio values. All the work on the negative ions series, as well as the previously mentioned work on BO^-, will be published in a separate paper.[4] This will include full predictions of all the spectroscopic constants and CI-SD calculations of dipole moments and quadrupole coupling constants.

$HBCl^+$ and HNSi

Calculations on $HBCl^+$ were done with a basis set of 76 cGTO's, and only the valence electrons were correlated. The MP4SDQ calculation required 1 hr for each of the 38 energy points on a VAX 8600. In this case equivalent calculations on the diatomic molecules BH and BCl are used to generate corrections to the ab initio r_e values. The band origins of the three fundamental bands of $HBCl^+$ are predicted to occur at 2788, 716, and 1123 cm^{-1}, when harmonic and anharmonic contributions are considered. The stretching vibration-rotation constants α_1 and α_3 are also calculated from the anharmonic potential function, and α_2 is estimated by referring to the value for HCP. A final corrected estimate for B_0, the effective rotational constant in the vibrational ground state, is $18,890 \pm 80$ MHz for the most common isotopic form. A CI-SD calculation with the same basis set at the calculated equilibrium geometry required 3 h on the VAX 8600 and yielded a value for the dipole moment (3.27 D) and values for all the chlorine and boron quadrupole coupling constants. Calculations at the MP2 level showed that the other possible isomer $HClB^+$ is actually only a saddle point in the potential. There is no barrier along the bending coordinate to rearrangement back to the stable $HBCl^+$ isomer. A manuscript giving full details of the work on $HBCl^+$ has already been accepted for publication.[5] Recently we have also carried out a similar MP4SDQ study of the isoelectronic neutral molecule HNSi. This transient molecule has still not been observed spectroscopically. It may be related to the molecules of the previous section by considering it to be protonated SiN^-.

REFERENCES

1. J. S. Binkley, M. J. Frisch, D. J. DeFrees, K. Raghavachari, R. A. Whiteside, H. B. Schegel, E. M. Fluder, and J. A. Pople, "GAUSSIAN 82," Carnegie-Mellon University, Pittsburgh (1983).
2. K. A. Peterson and R. C. Woods, J. Chem. Phys. 87:4409 (1987).
3. K. A. Peterson and R. C. Woods, to be published.
4. K. A. Peterson and R. C. Woods, to be published.
5. K. A. Peterson and R. C. Woods, J. Chem. Phys., in press (1988).

6. P. Botschwina, Chem. Phys. Lett. 114:58 (1985).
7. P. Botschwina, J. Mol. Spectrosc. 120:23 (1986).
8. P. Botschwina, J. Mol. Spectrosc. 118:76 (1986).
9. P. Botschwina, Chem. Phys. Lett. 107:535 (1984).
10. P. Botschwina and P. Sebald, J. Mol. Spectrosc. 110:1 (1985).
11. F. J. Lovas and D. R. Johnson, J. Chem. Phys. 55:41 (1971).
12. B. Rosenblum and A. H. Netherot, Jr., J. Chem. Phys. 27:828 (1957).
13. T. A. Scott, Phys. Rep. C 27:89 (1976).
14. W. C. Bowman, E. Herbst, and F. C. DeLucia, J. Chem. Phys. 77:4261 (1982).
15. M. Gruebele, M. Polak, G. A. Blake, and R. J. Saykally, J. Chem. Phys. 85:6276 (1986).

PHOTODISSOCIATION DYNAMICS OF THE H_3^+ MOLECULE

Jose M. Gomez Llorente and Eli Pollak

Chemical Physics Dept.
Weizmann Institute of Science
Rehovot, 76100 Israel

Abstract

A classical trajectory study of the photodissociation dynamics of H_3^+ is presented. Total angular momentum barriers confine a large number of classical states in the interaction region for infinite time at energies above the dissociation limit. The microcanonical density of states was evaluated using Monte Carlo methods. The quantal decay mechanism of the bound states embedded in the continuum is tunneling through the total angular momentum barriers. A sudden approximation is provided to evaluate the decay rates and product energy distributions. Fourier transforms of the classical mechanical dipole moment correlation function are presented and compared with the experimental coarse grained spectrum.

1. Introduction

In a highly intriguing paper, Carrington and Kennedy[1] reported a measured photodissociation spectrum of highly excited H_3^+. In their experiment, H_3^+ is formed in a discharge, accelerated to a reaction vessel where it is irradiated by a CO_2 laser. The fragment H^+ is then measured as a function of laser frequency which is made tunable by Doppler shifting. 20000 lines were observed in a spectral region of 850-1100 cm-1. H^+ product translational energies were measured in the range of 0-4000 cm-1. The linewidths measured are very narrow, 1-100 MHz. When the spectrum is coarse grained Carrington and Kennedy find surprisingly that the spectrum may be well characterized by 'clumps' with a spacing of approximately 50 cm-1 between the clumps.

These experimental results pose a number of theoretical questions. From the translational energy distribution it is clear that one is dissociating states whose energy is far above the dissociation limit. Moreover, if the absorption spectrum consisted only of bound to continuum transitions then it would be structureless. Since this is not the case one concludes that all the lines are associated with transitions between extremely long lived resonance states. Thus one must find the dynamical mechanism which is responsible for these long lived resonances. This is not a trivial problem. All the well known mechanisms for trapping are either not applicable or won't give lifetimes in the experimental range of 10^{-7} to 10^{-9} seconds. Specifically, a van der Waals predissociation mechanism is not applicable because H_3^+ is a strongly bound molecule and the time scale separation found in van der Waals molecules does not exist in H_3^+. The usual unimolecular theories of decay do give lifetimes that are long relative to a molecular vibration that is lifetimes of the order of a psec. If stretched far one can find for triatomics lifetimes of the order of 10^{-10} sec but not longer. It is therefore a challenge to understand the mechanism underlying these resonances. Any mechanism must be robust enough to account for the enormous number of experimental lines.

Since it is obvious that the lines are formed from highly excited H_3^+ it is not easy to see how one can solve an exact quantum mechanical theory for this system. One must suggest alternatives which will account for the lifetime and product translational energy distributions. Finally, one would like to understand the mechanism responsible for the clumps in the coarse grained spectrum.

In this paper we report and review some of our recent advances in solving these problems. We note that Child and coworkers[2,3] have also attempted to answer some of the questions we are posing. They used approximate quantum mechanical treatments, they model the dissociation via an atom rigid rotor model. They totally neglect the possibility of exchange. The reason for such a severe approximation is obvious, it is virtually impossible to provide anything more exact if one limits oneself to quantum mechanics. Our approach has been that because the photodissociation is in the high energy region where the density of states is high one should expect a classical mechanical theory to be very good. Classical mechanically one can, at least in principle, integrate the equations of motion exactly.

In section II we provide a classical mechanical picture of the binding mechanism[4] and provide an estimate for the number of resonance states.[5] In section III we show how one can combine classical mechanics with semiclassical tunneling probabilitites to obtain estimates for the lifetimes and product energy distributions of the photodissociation process.[6] In Section IV we discuss the clumps found in the spectrum.

2. Total angular momentum barriers

The concept of an angular momentum barrier in diatomic scattering is well known. If the diatomic interaction potential V at zero total angular momentum is attractive then for a given total angular momentum J of the diatom the effective potential is $U_J = V + J^2/2I$ where I is the moment of inertia of the diatom. As the diatomic distance becomes smaller, the angular momentum term increases so that the structure of U_J is that of a well separated from the asymptotic region by a barrier, whose height is J dependent. If one initiates a classical trajectory below the barrier height in the internal region it will not dissociate. Quantum mechanically, one finds a shape resonance whose lifetime is determined by quantal tunneling through the barrier.

We have recently pointed out that a similar effect exists in triatomic collisions (and in more complicated systems).[4] The potential energy surface of a triatomic molecule is a function of three internal variables, for example the distances R_{BC} between atoms B and C, R_{A-BC} the distance between atom A and the center of mass of BC and the angle γ between the two distance vectors. It has been shown[7,8] that the minimum of the Hamiltonian of the three body system at a given total angular momentum J and a fixed configuration of the three body system is

$$U_J = V + J^2/2I \qquad\qquad\qquad (1)$$

where V is the interatomic Born Oppenheimer potential energy surface and I is the moment of inertia of the triatomic system. One can now further minimize Eq. (1) with respect to the coordinates. In analogy with the diatomic case, we minimize Eq. (1) with respect to R_{BC} and γ such that $U_J^{min}(R_{A-BC})$ is the minimum of the Hamiltonian subject to only two constraints: J and R_{A-BC} are kept fixed.

Plots of U_J^{min} for the H_3^+ system on the DIM potential surface[9] are shown in Fig. 1. Note the similarity to the centrifugal barriers found in diatomic scattering. What is more important though is that because the effective potential is a minimum it will trap for infinite time, classical trajectories whose energy is higher than the asymptotic dissociation energy. Specifically, if U_J^{min} has a barrier at R*, then any classical trajectory initiated with total angular momentum at $R_{A-BC} < R^*$ will never be able to reach the region $R_{A-BC} > R^*$; it will be a bound classical trajectory. Quantally these bound orbits will appear as resonances whose lifetimes are determined by the tunneling probability through the total J barrier.

In the next section we will show how such lifetimes can be computed and that they are in good agreement with the experimental observations of Carrington and Kennedy. Here we review recent Monte Carlo numerical simulations of Berblinger et al.[5] that show that the classical number of states embedded in the continuum is large enough to be compatible with the large number of experimentally measured lines.

The classical density of states at a given energy and angular momentum vector is defined as

$$\rho(E,J) = \frac{1}{h^6} \int d\Gamma \delta(E-H)\delta(\underline{J}-\underline{J}') \qquad\qquad (2)$$

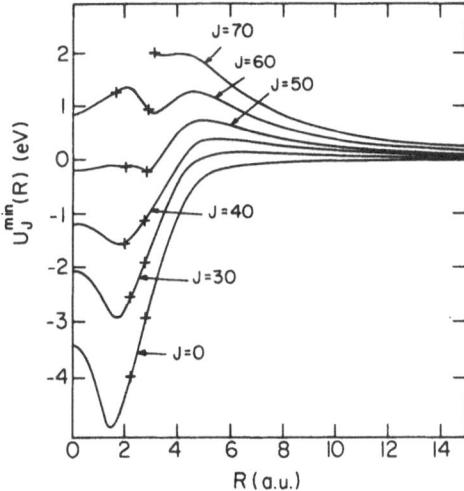

Figure 1. Total angular momentum barriers for the H_3^+ system. R is the H^+ to the center of mass of H_2 distance. The crosses on each curve show the crossover from a minimum of U_J in a collinear configuration of H_3^+ (large R) to a T-shaped configuration (small R).

where the volume element is 12 dimensional, 6 coordinates and conjugate momenta. One can then show that the density of states at a total angular momentum J and energy E can be reduced to a seven dimensional integral. This integral is evaluated numerically by Monte Carlo techniques.[5] The number of states with total angular momentum J and energy E denoted $W_J(E)$ is then evaluated by integrating the density up to energy E. The number of classical states whose energy is greater than the triatomic dissociation energy is evaluated by integrating the density of states from the dissociation energy up to the total angular momentum barrier height. This is the classical number of resonance states. The details of this computation are provided in Ref. 5. In Fig. 2 we show a plot of the total number of resonance states as a function of J. There are close to 80,000 classical states, much more than needed to account for 20,000 experimental lines.

Finally we note that the total angular momentum can drastically change the structure of the H_3^+ molecule. Inspection of Fig 1 shows that the minimum of the effective potential changes from an equilateral structure at low J to a collinear structure at high J, where the enormous spinning of the molecule flattens it out. We have run classical trajectories at J=60 close to the energy of the minimum of the potential and found that in fact the molecule behaves as a linear triatomic molecule.

3. Tunneling lifetimes

The method we used to evaluate tunneling lifetimes is an extension of the trajectory surface hopping technique.[10,11] The idea was originally proposed by Waite and Miller,[12] who also tested it out for some model cases where it worked well. We have further developed this method[6] providing an algorithm which makes it especially easy to apply to unimolecular dissociations. Conceptually, instead of arrival at a crossing of surfaces one has an arrival at a classical turning point. Instead of a Landau-Zener transition probability one uses a suitably defined semiclassical tunneling probability.

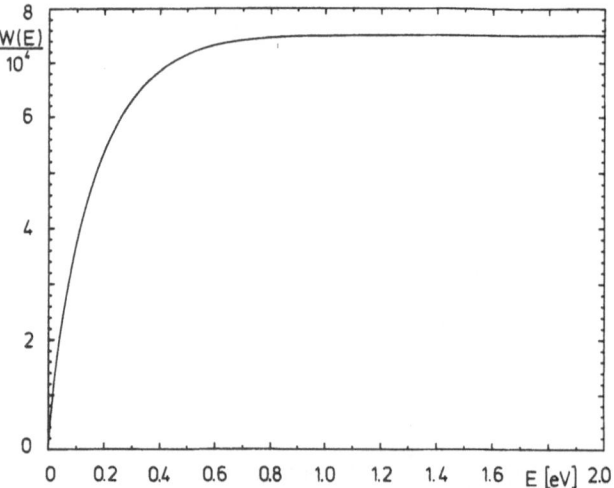

Figure 2. Total number of resonant states (between E=0 and the J-dependent barriers) as a function of energy, summed over all J.

Waite and Miller use this idea to compute an average survival probability. For a three degree of freedom system such as H_3^+ it is though very expensive to compute a converged microcanonical averaged survival probability. Instead we propose a method which directly computes the microcanonical "high pressure" unimolecular rate constant. This gives the same results as obtained from a survival probability if the usual assumption of exponential lifetime distribution is satisfied.

Tunneling is always initiated on the caustics of the classically allowed motion. Inspection of Fig. 1 suggests that since the total angular momentum barriers are located at relatively large distances one can assume that the caustics are well approximated by points such that $\dot{R}_{A-BC}=0$ along a classical trajectory. For a given classical trajectory we denote all such turning points by R_{tp}. We have shown in Ref. 6 that if motion in the interior region is ergodic, then the unimolecular tunneling rate is

$$k_{uni}(E) = \lim_{T \to \infty} \frac{1}{T} \sum_i P_i(R_{tp}) \qquad (3)$$

where T is the integration time of the trajectory and $P_i(R_{tp})$ the tunneling probability at the i-th turning point. As one integrates the trajectory for longer and longer times it goes through additional turning points and samples more regions of phase space. Ergodicity assures us that if we integrate for long enough times we will get the converged unimolecular rate.

This method is easily extended to energies where dissociation is classically allowed. The classical rate may be obtained by reflecting the trajectory back into the "internal" region as it crosses the barrier. If by imposing a reflection at the barrier top, motion remains ergodic then Eq. (3) still holds but of course the "tunneling probability" is just the classical unit value.

We evaluated the tunneling probability at the turning points via a sudden approximation. The sudden approximation assumes that the phase space variables perpendicular to the tunneling degree of freedom remain frozen during the tunneling process. Denote the instantaneous value of the perpendicular degrees of freedom at the i-th turning point as r_i, p_{r_i}, γ_i, p_{γ_i}. The imaginary action is then evaluated as

$$A(R_i) = 2 \int_{R_i}^{R_i^O} dR\{2m[H_s(r_i,p_{r_i},\gamma_i,p_{\gamma_i},R)-H_s(r_i,p_{r_i},\gamma_i,p_{\gamma_i},R_i)]\}^{1/2} \qquad (4)$$

where R_i^O is the outer turning point and the full Hamiltonian at $\dot{R}=0$ (H_s) is taken as the effective potential for the tunneling process. The tunneling probability is then evaluated using the usual semiclassical formula

$$P(R_i) = \{1+\exp(A(R_i)/\hbar)\}^{-1} \qquad (5)$$

Given the turning points and the tunneling probabilities one can also compute the products translational energy distribution. The final translational (or more generally internal) energy associated with the i-th turning point may be obtained by integrating the classical trajectory from the outer turning point out to the asymptotic region. Since motion is practically separable from the outer turning point outwards, one can simplify and obtain the final translational energy E_T^i from the potential energy at the outer turning point. The probability for observing E_T^i

is simply the tunneling probability. The average products translational energy is

$$\langle E_T \rangle = \sum_i E_T^i P(R_i) / \sum_i P(R_i) \qquad (6)$$

In Fig. 3 we plot the value of the unimolecular rate as a function of time T using the sudden approximation for four different energies at a total angular momentum J=40. Convergence is found although it is slow. The convergence is ca. 10% although the integration time is extremely long ~7·10^{10} sec. It is obvious that the single trajectory used to obtain these rates cannot be back integrated although energy is conserved to within five significant figures. However we find from various indicators that the motion in the interior region is ergodic. For example, the ratio of turning points in each of the three possible arrangement channels is 1:1:1 within the statistical noise limit. Power spectra are grassy.[5,13] Since we are here interested in a microcanonical average it really makes no difference whether we run one long trajectory or many segments of shorter trajectories, as long as our integration is accurate for times of the order of traversal from one turning point to the next. Numerically, it is much more convenient to run one very long trajectory and not worry about back integration.

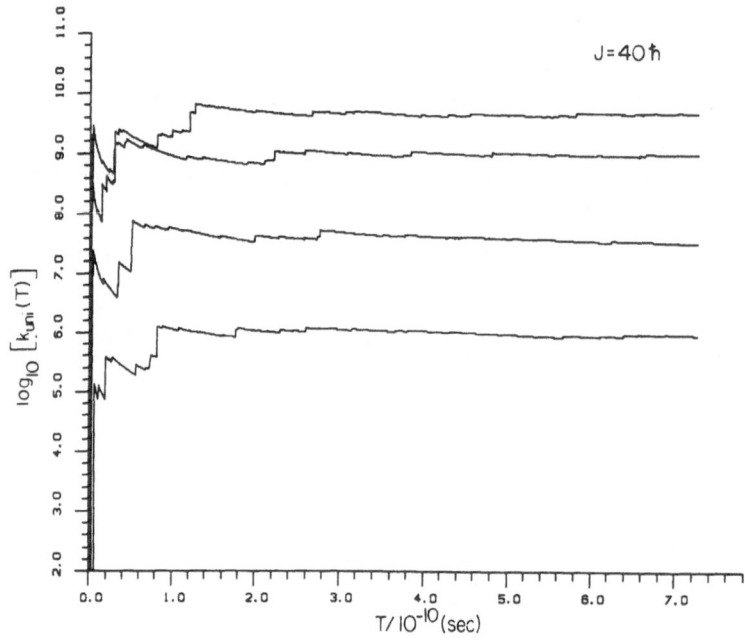

Figure 3. Tunneling rates for unimolecular decomposition of H_3^+ at four energies: 0.2993, 0.3252, 0.3532 and 0.3673 eV.

For each of the energies shown in Fig. 3 we also computed translational energy distributions. These are always strongly peaked towards the maximally allowed translational energy so as to maximise the tunneling probability. In Fig. 4 we show a profile of the total angular momentum barrier, the four energies sampled and the average products translational energy as a function of energy.

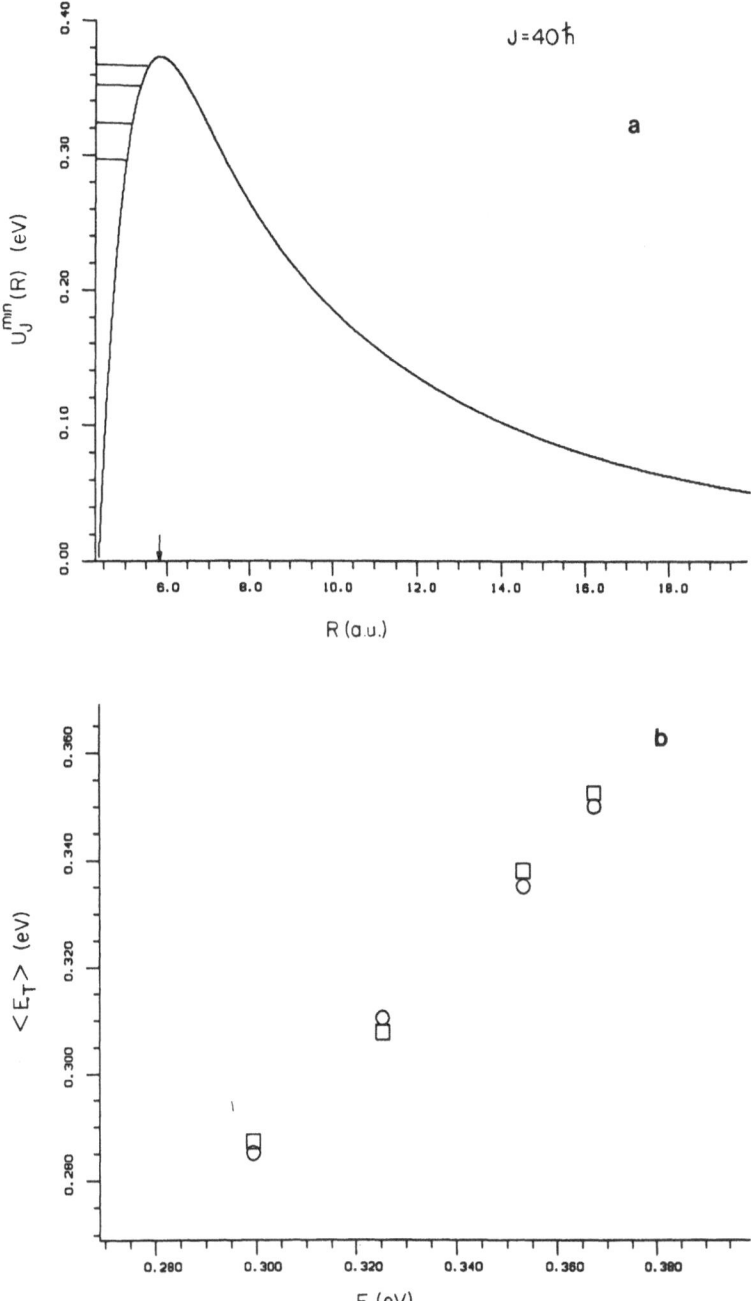

Figure 4. (a) Blowup of the J=40 total J barrier. The solid lines show the four energies we studied. (b) Average product translational energy as a function of E at J=40. Circles are based on the sudden approximation, squares are the results of a separable approximation, cf. Ref. 6.

Note that the average is almost identical to the energy. Vibrational and rotational excitation are extremely small.

Since the decay rate is determined by tunneling it changes exponentially with the energy at a fixed total angular momentum. At J=40 the experimentally accessible lifetime range spans a short range of energies of the order of .03 eV. As J decreases, the barrier thickens and this range decreases even more. The translational energies found at J=40 are rather high. For example, the coarse grained spectra of Carrington and Kennedy include only translational energies which are less than 1000 cm-1. We have computed lifetimes at lower J, for J=20 we find that the energy range within the experimental lifetime window is 0.01 eV while the average product's translational energy is 250 cm^{-1}. Here it is interesting to note that Badenhoop and Schatz[14] find that the average total angular momentum of H3+ formed from the reaction of $H_2+H_2^+$ where H_2^+ is highly vibrationally excited is J=25.

The high sensitivity of lifetime and product translational energy to E and J implies that the averaging over E and J inherent to the experiment is actually much smaller than one might have originally expected. In the next section we will consider the implications of this averaging on an attempt to simulate the "clump" spectra via classical trajectories.

4. The dipole moment correlation function

The classical trajectory computations presented in the previous sections have answered many of the questions posed by the experiment. The long lifetimes are associated with total angular momentum barriers. These barriers form enough resonance states to be compatible with the large number of experimental lines. The range of products translational energy accessed in the experiment is easily accounted for by the range of total angular momentum barrier heights. The one remaining intriguing question is a theoretical explanation for the clumps found in the coarse grained spectra. In this section we will discuss the methodology which we believe should be used to provide a classical mechanical framework for these spectra.

The experimental setup is such that for a given laser frequency ω one is observing in principle the transition from all possible initial states to all possible final states whose lifetime is within the experimental window and such that the energy difference $E_f-E_i=\hbar\omega$. In other words the experimental spectrum is ideally given by the expression

$$S_E(\omega) = \sum_{i,f} |T_{if}|^2 \delta(E-E_i)\delta(E+\hbar\omega-E_f) \tag{7}$$

where T_{if} is the transition amplitude from state i to state f and the delta function is the dirac delta function. It is easy to show that this expression may be written as the fourier transform of the microcanonical time correlation function of the T operator.[15] The classical limit of Eq. (7) is just the Fourier transform of the classical microcanonical correlation function associated with the classical limit of the T operator.[15,16] Since the experiment is an infrared absorption experiment it would seem that the T operator is just the dipole moment operator. For the dipole moment operator one has the selection rule that only $\Delta J=\pm 1$ transitions are allowed. Thus the experimental spectrum will be a sum of classical spectra corresponding to discrete values of the total angular momentum.

With this scheme the experimental clump spectrum would be reproduced by a double summation. For each J one must integrate over all classical spectra at the energy window corresponding to the experimental lifetime. Then one must sum over the total angular momenta. We used a dipole moment function which is very similar to the one used by Berblinger and Schlier.[17] Exact details will be provided elsewhere.[18] Since our numerical integrator was stable only up to times of order we could resolve our spectra accurately only to within a frequency window of ca. 100 cm^{-1}. We find that with this scale the energy averaging is trivial, the spectra are numerically identical for the allowed energy range at a fixed J. Thus all that needs to be done is compute spectra for a number of J's. This is shown in Fig. 5. We find that the peaks of the spectra shift with J however the location of the main peak is at ca. 600 cm^{-1} well below the CO_2 laser frequency range, while the shift of the peaks as a function of J is ca. 10-15. cm^{-1}, much smaller than the 50 cm^{-1} experimental clump spectra. In addition, the width of the peaks is also much larger than the widths of the clump spectra.

All trajectory computations thus far were coplanar. H_3^+ is a symmetric top molecule, thus coplanar trajectories correspond to K=J where is the projection of the total angular momentum on the axis perpendicular to the instantaneous plane of H_3^+.[19] The experimental accesses all-possible K values. We are at present studying the dipole correlation function for an initial condition with K=0 case. Here we expect the peak at a given J to move to higher frequency and the spacing between peaks at different J values to increase. Whether this will suffice for an explanation of the clump spectrum remains still to be seen.

5. Discussion

In this paper we reviewed and presented some new results on the classical mechanical analysis of the photodissociation dynamics of the H_3^+ molecule. The main problem which has still remained unresolved is a theory for the clump spectra. One source of concern is the fact that we

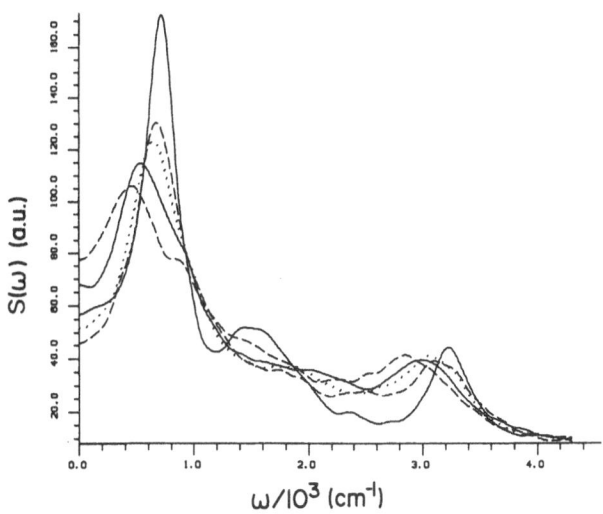

Figure 5. Dipole moment spectra for H_3^+ as a function of total angular momentum. (J=5,10,15,20,30). The frequency and intensity of the main peak increase with J.

could not find a numerical integrator which would stay stable for time scales of the order of the spacing found in the clump spectra. The reason for this is obvious, we are dealing with a chaotic system which is governed by exponential divergences. It is imperative to find a numerical integration method which would stay stable even in such unstable systems. This might seem impossible, but recent advances in the study of chaotic systems indicate[20,21] that one might be able to propagate 'averaged' trajectories which follow the true dynamics for relatively long times.

A second source of worry is how good is the classical approximation? We have recently studied[22] a much simpler two degree of freedom classically chaotic system where we find excellent correspondence between classical and microcanonical spectra and correlation functions, provided that the quantal density of states is large enough. Given the high density of states in the H_3^+ system, we believe that this is not a very big problem.

Thirdly one always worries over the potential energy surface. We have used a DIM surface[10] which has many good features to it but it is not an accurate surface. Recent computations by a number of groups[23] seem to indicate though, that the lifetimes and total J barriers are quite insensitive to detailed features of the potential energy surface. While one can never be positive, it doesn't seem likely that the chaos sampled by the experiment will be too sensitive to details. The clumps come from some long range correlation.[13] It would be very surprising to us if such a long range correlation were strongly dependent on details of the surface.

The reader who has followed us till now will note that we have not yet fully understood the clump spectra. We leave this as a challenge for the future!

Acknowledgment

We gratefully acknowledge numerous stimulating discussions and correspondence with Professors Brumer, Naaman, Prior, Schlier, Shapiro and Tennyson. This work has been supported by generous grants from the US Israel Binational Science Foundation and the Minerva foundation.

References

1. A. Carrington and R.A. Kennedy, J.Chem.Phys. 81, 91 (1984).
2. M.S. Child, J.Phys.Chem. 90, 3595 (1986).
3. R. Pfeiffer and M.S. Child, Mol.Phys. 60, 1367 (1987).
4. E. Pollak, J.Chem.Phys. 86, 1645 (1987).
5. M. Berblinger, E. Pollak and Ch. Schlier, J.Chem.Phys., in press.
6. J.M. Gomez Llorente and E. Pollak, Chem.Phys., in press.
7. W.J. Chesnavich, J.Chem.Phys. 77, 2988 (1982).
8. K. Rynefors and S. Nordholm, Chem.Phys. 95, 345 (1985).
9. R.K. Preston and J.C. Tully, J.Chem.Phys. 54, 4297 (1971).
10. J.C. Tully and R.K. Preston, J.Chem.Phys. 55, 562 (1971).
11. R. Düren, J.Phys.B 6, 1802 (1973).
12. B.A. Waite and W.H. Miller, J.Chem.Phys. 73, 3713 (1980); 74, 3910 (1981).
13. J.M. Gomez Llorente and E. Pollak, Chem.Phys.Lett. 138, 125 (1987).
14. J.K. Badenhoop and G.C. Schatz, J.Chem.Phys. 87, 5317 (1987).
15. R.D. Levine, Adv.Chem.Phys., in press.
16. M. Wilkinson, J.Phys.A 20, 2415 (1987).
17. M. Berblinger and Ch. Schlier, Mol.Phys., in press.
18. J.M. Gomez Llorente and E. Pollak, to be published.

19. G. Herzberg, Infrared and Raman Spectra (van Nostrand, New York, 1945) p. 24.
20. D. Auerbach, P. Cvitanovic, J.-P. Eckmann, G. Gunaratne and I. Procaccia, Phys.Rev.Lett. 58, 2387 (1987).
21. E.J. Kostelich and J.A. Yorke, University of Maryland, preprint.
22. B. Eckhardt, G. Hose, J.M. Gomez Llorente and E. Pollak, to be published.
23. Ch. Schlier, J. Tennyson, private communication.

9. ...

10. ...

11. ...

12. ...

STRUCTURE OF SMALL MOLECULES VIA
THE COULOMB EXPLOSION METHOD

Zeev Vager

The Weizmann Institute of Science

76100 Rehovot, Israel

ABSTRACT

The so-called "Coulomb explosion" (CE) technique offers a potentially powerful method of studying gross geometric features of a great variety of molecules. During the past few years, new ideas and developments in these techniques have come to fruition. The use of the Coulomb-explosion technique combined with a radically new multi-particle detector, extremely thin film targets, and low-excitation ion source has enabled, for the first time, direct measurements of the complete stereostructure of several small polyatomic molecular ions. These measurements produce 3-dimensional images of the nuclear positions within each individual molecule in a beam. Combining the results of an ensemble of such images yields the nuclear density within the molecule. These developments have presented us with a unique opportunity to directly explore complex nuclear motions within molecules.

A collaborative work of Argonne National Laboratory and the Weizmann Institute has resulted in a new method that ultimately should enable the measurement of the square of wave functions of the nuclear-coordinates of selected states in small molecules. The direct observation of the "nuclear-density" of gas-phase molecules and clusters is of importance in several fields, like astrophysics, plasma-physics, physical-chemistry and theoretical chemistry.

The principle of the method is as follows. The first stage is the preparation of the studied molecules (ideally in a single quantum state) in a beam with a speed of above 2% of the speed of light ($\sim$200 keV/amu). The second stage is the "minimal multiple scattering" stripping. The molecules impinge upon an extremely thin target of Formvar[1] ($\sim$30Å or less) and all the binding electrons and other loose molecular electrons scatter away within a few times 10^{-17} of a second. To a very

To be published in "The Structure of Small Molecules and Ions", Proc. Int. Workshop in Memory of Prof. Itzhak Plesser, Neve-Ilan, Israel, Dec. 1987, eds. R. Naaman and Z. Vager, Plenum Press (1988).

good approximation, the nuclear part of the wave function does not change at all while the Hamiltonian switches from the original molecular Hamiltonian to an almost pure Coulomb repulsion regime. The next stage is the development of the nuclear wave function through the well-known Coulomb Hamiltonian into the asymptotic region of the final state. This stage (the CE) takes about 10^{-15} of a second. The final state is fully measured in a multiparticle position and time-sensitive (MUPPATS) detector[2] (situated about 6 meters downstream) which yields the final velocities of all of the fragment ions. The final velocities for each individual molecule can be transformed (optimally via the Schrödinger equation but so far only via "Coulomb-trajectory" calculations) into the initial "R-space" configuration. Each such configuration of an individual molecule is a sample of the probability density (PD) of the nuclear-coordinates (the wave-function squared integrated over the electronic coordinates) of the studied molecular state.

Up to now we have developed an experimental set-up including these extremely thin Formvar targets which provide efficient electron strippers with low multiple scattering characteristics and the MUPPATS detector which provides accurate velocity information on multiparticle events. We have also made an effort to prepare beams of selected molecular states. This is discussed by Ron Naaman in these proceedings.

Currently we have information on several molecular ions including: CH_5^+, CH_4^+, CH_3^+, NH_4^+, NH_3^+ $C_2H_3^+$, $C_3H_3^+$, $C_3H_4^+$, and C_3^+. Discussion of the C_3^+, C_3^{++}, $C_3H_3^+$ and $C_3H_4^+$ is presented by Elliot Kanter in these proceedings. The measurements[3] on CH_4^+ and its interpretation will serve here as an illustration of the quality and characteristics of the Coulomb-explosion results.

The CH_4^+ ions were prepared inside the high-voltage terminal of Argonne's 4.5-MV Dynamitron by low-energy electron impact with methane gas at a pressure of about 50 mTorr. The ions were steered toward a foil target composed of 0.3 $\mu g/cm^2$ Formvar supported on a fine nickel mesh. Measurements on multiple scattering, thickness calibrations[4], and methods of target preparation[1] are published elsewhere.

Downstream from the target, the dissociation fragments were deflected electrostatically in the horizontal (or "X") direction. Figure 1 shows a two-dimensional projection on the plane of the detector of the images of fragments from the Coulomb explosions of $\sim 10^4$ CH_4^+ ions. One can observe, in order of increasing deflection, the positions of the C^+, C^{2+}, C^{3+}, and C^{4+} ions and a large disk corresponding to the accompanying protons. For each event analyzed, a five-fold coincidence is recorded with one carbon ion and four protons. The position and time of arrival of each ion are converted on-line into the individual ions velocity vectors. This 15-dimensional velocity, v_c, v_1, v_2, v_3, v_4 (a point in V-space) is reduced further by subtraction of the center-of-mass velocity (which serves as an overall resolution test of the system) and a choice of orientation. Thus, 15-6=9 "body" degrees of freedom are left.

The aim of analyzing CE data is to get a parameterized function describing the probability density of the correlated nuclear coordinates (R-space). One is faced with the following problems:

1. There is no way to analytically deduce the initial R-space configuration from the final measured V-space coordinates. We must therefore numerically integrate trajectories from assumed R-space configurations.

2. A single Coulomb trajectory calculation, though simple, takes about 2 seconds on an IBM 3081.

3. Existing and future CE experiments are expected to have $10^6 - 10^7$ events. It is inconceivable to transform each measured event to its R-space configuration.

4. In the above example of CH_4^+ there are 9 structural degrees of freedom. It is absurd to think in terms of computing trajectories for 10 boxes per degree of freedom (10^9 boxes!).

5. When the molecules contain some identical atoms then one talks about "equivalent" and "non-equivalent" sites of atoms. How does one distinguish such atoms in V-space?

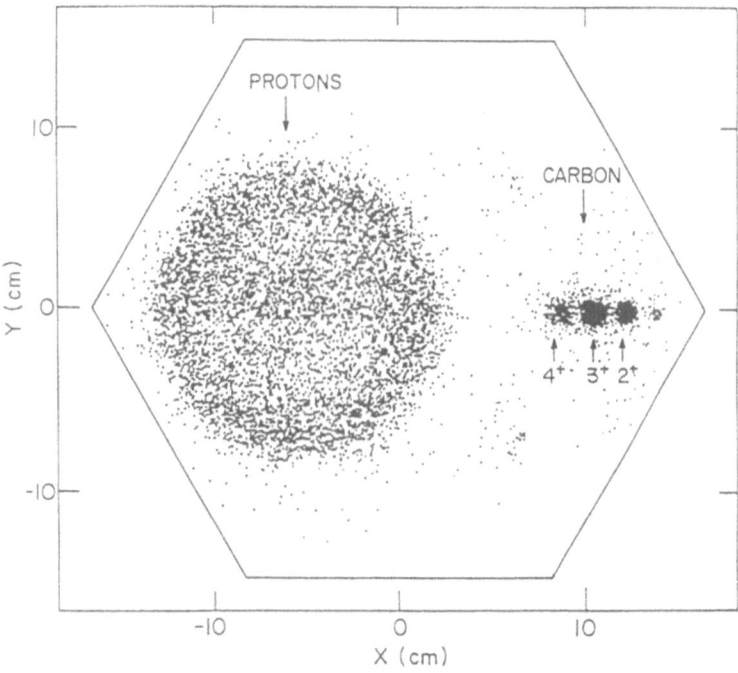

Fig. 1

A density plot of the projection on the plane of the MUPPATS detector of the images of fragment ions resulting from the Coulomb explosion of CH_4^+ ions. The hexagon denotes the active area of the detector. The protons, and various charge states of carbon ions, have been separated by electrostatic deflection in the x direction.

It is important to notice that the Coulomb interaction of any charge state of the carbon with four (equivalent) protons does not change the symmetry properties of the ensemble of molecules measured at the final V-space in respect with the initial distribution in R-space. Therefore it is expected that the use of symmetry coordinates in V-space (as well as R-space) will result in a simplified description of the density of measured events. The following coordinates are chosen:

$$A_1 = \frac{1}{4}(r_1 + r_2 + r_3 + r_4)$$

$$T_x = \frac{1}{4}(r_1 + r_2 - r_3 - r_4)$$

$$T_y = \frac{1}{4}(r_1 + r_3 - r_2 - r_4)$$

$$T_z = \frac{1}{4}(r_1 + r_4 - r_2 - r_3)$$

$$B_x = \frac{1}{2}(\alpha_{12} - \alpha_{34})$$

$$B_y = \frac{1}{2}(\alpha_{13} - \alpha_{24})$$

$$B_z = \frac{1}{2}(\alpha_{14} - \alpha_{23})$$

$$E_b = \frac{1}{\sqrt{12}}\{2(\alpha_{12} + \alpha_{34}) - (\alpha_{13} + \alpha_{14} + \alpha_{23} + \alpha_{24})\}$$

$$E_t = \frac{1}{2}(\alpha_{13} - \alpha_{14} + \alpha_{24} - \alpha_{23})$$

where $r_k \equiv CH_k$ distance in velocity space and $\alpha_{k\ell} \equiv H_k CH_\ell$ angle in velocity space.

The above coordinates describe the vibrational modes of methane (T_d symmetry). If the maximum of the density distribution is found away from the zero of almost any of those coordinates (except for A_1) than a "symmetry breaking" occurred. If we also knew that the measured molecules (CH_4^+) were prepared in the ground-state then the position of the maximum probability is a direct measure of the geometry of the expected Jahn-Teller distortion.

Before further discussion we should look at some Coulomb explosion data of CH_4^+ and NH_4^+ along some of the above coordinates. For simplicity, we choose to use only the five angle degrees of freedom (B_x, B_y, B_z, E_b and E_t). Given a point in the 5-dimensional space we define a density by the number of events measured in which their coordinates fall within a hypersphere with the selected point as a center and with a radius of 0.25 radians (This crude resolution may smear some sharp features). All permutations of protons of every measured event are taken into account (and thus the statistical error is not simply given by the Poisson statistics for the number of events within the cell but by the proper weighting of the multiplicity of every event). Once the "measure" of density is defined it is possible to look at the distribution near the maxima. It was found that all the significant maxima are very near $E_b = E_t = 0$ therefore we will restrict the description to the variability of the density distribution in the B_x, B_y, B_z coordinates. Probably the most informative cut through the distribution is the density on a plan which includes the B_x axis and bisects the angle between the B_y, B_z unit vectors.

As seen in Figure 2, the "peak" on the positive x axis is of a C_{2v} symmetry. It is not really a peak but an extremum which shows as a peak in this special cut. An equivalent point is situated at the same distance from the origin but on the negative B_x axis. It is lower than the real two equivalent peaks below the B_y, B_z plan which are of a C_s symmetry. The T_d point is at the origin and is of much lower probability. An instructive illustration is shown in Figure 3. This is exactly the same cut as Figure 2 but with events from an NH_4^+ Coulomb explosion measurement. The NH_4^+ ion is isoelectronic with the CH_4 molecule. It is expected to manifest a T_d symmetry. Indeed Figure 3 shows essentially a "Gaussian" peak around the origin. Coming back to CH_4^+, there are altogether 12 equivalent peaks of the C_s nature. Figure 4 shows three of them along a plan perpendicular to the (1,1,1) axis. The center of the distribution belongs to the C_{3v} symmetry. Though this point is higher in probability than the T_d probability it is lower than the C_{2v} extrema density. Figure 4 shows that there is a very "probable" transition between triplets of the C_s peaks around the C_{3v} point. The four "triplets" C_s groups are connected between themselves through the six C_{2v} extrema.

How is all this connected to information we know from theory on the CH_4^+ system? It is best to relate this to the results of a latest discussion on the potential energy surfaces of CH_4^+ [5]. A good list of other references to such calculations is given in this reference. Our special interest is in the C_{2v}, C_{3v} and C_s structures. In the calculations there are extrema of the potential surface corresponding to these structures. Only the C_{2v} is a true minimum of the potential surface. In comparison the CE densities shown in Figs. 3, 5 has also extrema at the above structures, but the maximum structures are of a C_s symmetry.

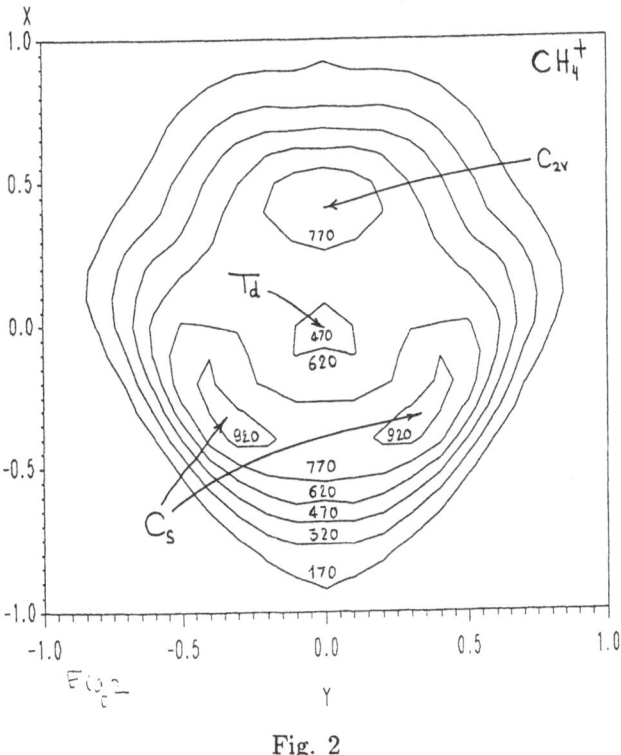

Fig. 2

Density of events in V-space (see text) after CE of CH_4^+. x is along the B_x axis and y is along $(B_y + B_z)/\sqrt{2}$. The scales are in radians.

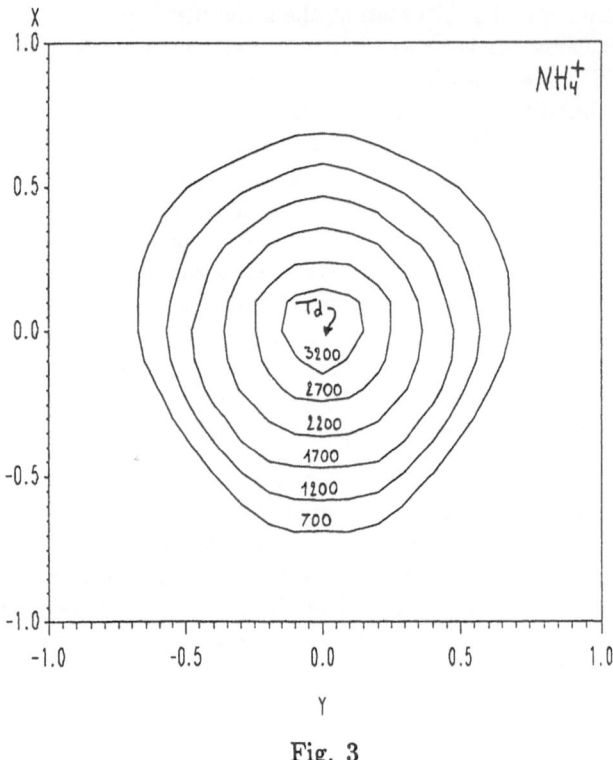

Fig. 3

The same as Fig. 2 but for events coming from the CE of NH_4^+.

The clue to the understanding of this is the method of preparation of the CH_4^+ ions. The low energy bombardment of electrons on methane gas does not guarantee either a vertical transition or a statistical distribution. Let us examine these two extreme cases. For a statistical distribution in such many degrees of freedom system we expect that the minimum of the energy surface would be populated with the highest probability. Therefore for such a case the highest density should have been in the C_{2v} structure. For the case of a vertical transition from the G.S. of CH_4 we expect that the different levels would be populated with their corresponding Frank-Condon coefficients (FCC). It is concievable that the 12 C_s locations would have a larger FCC than the C_{2v} states. The fluctual behaviour of such a group of excited states might be consistent with the densities shown in Figs. 2 and 4. In particular, the contours below the B_y, B_z plan in Fig. 2 can be visualized as a wave function square peaked at the two C_s location going back and forth above the true minimum of the C_{2v} where the probability is smaller because of higher momentum.

This example demonstrates the density in the nuclear coordinates of a truly dynamical situation. It stresses the quality of information that one can get from the Coulomb-explosion experiments if one could control with more skill the production mechanism of the measured species. This is the trend that we are trying to pursue in future experiments.

The above qualitative discussion was intended to illustrate the potential of the method rather then a rigorous data analysis. We do believe that the reduction of the data in terms of nuclear probability distribution accounting for the finite

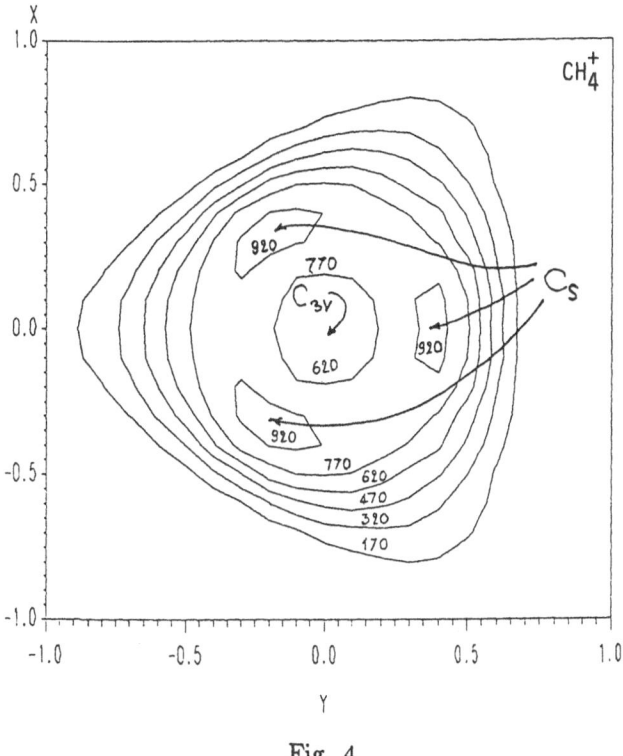

Fig. 4

The same as Fig. 2. The x-y plan is perpendicular to the $(1, 1, 1)$ axis in the B_x, B_y, B_z space. It passes through the C_s maxima.

statistics in terms of experimental errors is now completely understood and will be carried out in the near future[6].

REFERENCES

1. Ultrathin Foils for Coulomb-Explosion Experiments, G. Both, E.P. Kanter, Z. Vager, B.J. Zabransky and D. Zajfman, Rev. Sci. Instrum. <u>58</u>, 3 (1987) p. 424.

2. MUPPATS - A multiparticle 3D Imaging Detector Systems, A. Faibis, W. Koenig, E.P. Kanter and Z. Vager, Nucl. Instrum,. anf Meth. <u>B13</u> (1986) p. 673-677.

3. Direct Determination of the Stereochemical Structure of CH_4^+, Z. Vager, E.P. Kanter, G. Both, P.J. Cooney, A. Faibis, W. Koenig, B.J. Zabransky and D. Zajfman, Phys. Rev. Lett. <u>57</u>, 22 (1986) p. 2793.

4. D. Zajfman et al., to be published.

5. R.F. Frey and E.R. Davidson, J.C.P. Accepted.

6. Z. Vager, to be published.

PREPARATION OF MOLECULES IN A SINGLE STATE FOR COULOMB EXPLOSION

MEASUREMENTS

H. Kovner, A. Faibis and Z. Vager

Department of Nuclear Physics
Ron Naaman
Department of Isotope Research, Weizmann Institute of
Science, Rehovot, Israel

ABSTRACT

In recent years extensive work was done both at Argonne National
Laboratories (ANL) and at the Weizmann Institute (WI) studying the Coulomb explosion (CE) technique as a method for investigating the structure
of molecular species. It was demonstrated that this novel method for
measuring nuclear density distribution within small polyatomic molecules,
is unique in its ability to provide structural information on many species, and therefore has practical importance.

We intend to extend and develop the method so that the structure of
species with well defined internal energy will be studied as a function
of the internal energy. These measurements are of major importance in
understanding the structure and dynamics of floppy van der Waals complexes, molecules and ions. A description of the methods under development
is presented. The methods are based on the preparation of cold negative
ions, neutralization by laser photodetachment, and photodissociation
which allows for state selection.

INTRODUCTION

Applying the Coulomb explosion (CE) method one has in principle the
ability to obtain the structure of polyatomic molecules by measuring the
density function of the nuclei. This means that the method provides us
with the unique opportunity of observing X^2 when X is the nuclei wavefunction. The ability to measure structure by obtaining the full density
function of the nuclei is of importance mainly for species whose structures are difficult to deduce from spectroscopy. Among those are floppy
molecules and ions, for which no a-priori valid rotational-vibrational
model Hamiltonian exists. For such systems, as was discussed before (1),
spectroscopy may not provide enough information for conclusive structure
determination. In systems having large amplitude motions the structure
may be strongly state dependent. The average configuration of such molecules may vary strongly with the total angular momentum, or with vibrational excitation. Therefore it is important, in order to use the full

113

strength of the CE method, to measure molecules in a well defined quantum state.

Traditionally the sources used in accelerators are built to produce high currents of charged atoms. The Pelletron used at the Weizmann Institute utilizes either a Dua-Plasmatron or a sputtering source to produce negative ions. The ions are mass selected and then accelerated to energies of about 12 MeV. After the acceleration stage one can either flip charges or neutralize the ions in a field free region (the "Terminal"). Commonly a gas or foil stripper is used for the process, which when applied for molecules, produces products in many vibrational and rotational (sometimes even electronic) states. Starting from this "soup" of species, we want to be able to measure, in the CE experiments, molecules in a well defined quantum state. In what follows methods will be described and discussed that are currently being developed at the Weizmann Institute for achieving this goal.

DISCUSSION

Preparing an ion in a single quantum state is complicated since the ionization procedure usually involves very energetic processes. The same is true for neutralization of ions. In order to produce species with well defined internal energies one has to insure that the complete energetics are under control. This means starting with a molecule with relatively well defined internal energy, applying a photon for the ionization/neutralization process, and measuring the kinetic energy of the electrons emitted.

In ANL the CE of positive ions is investigated, hence one can ensure the preparation of ions in a single vibrational state by applying photoionization, and measuring the structure of the ions in coincidence with electron energy analysis. This technique provides resolution of few meV (2).

At the WI the CE of neutrals is studied. They are produced by electron detachment from accelerated negative ions. Hence a single state preparation is more complicated and it is not possible to apply a single stage method. In what follows two types of state preparation will be described. In the first a combination of source and neutralization processes is required to obtain neutrals with essential a single vibrational and very few rotational states populated. In the other scheme some of the special features of the CE technique itself are used to measure quantities of species in a well defined vibrational state.

Photodetachment

In the first approach towards single state preparation we applied electron photodetachment (EPD) for the neutralization process, instead of the gas or solid stripper. The photodetachment is selective due to the fact that a known amount of energy is deposited in the negative ion. In principle by combining it with electron energy analysis, and doing coincidence experiments one can perform a CE measurement of neutral molecules prepared with a well defined internal energy (3). The accuracy of the internal state definition depends on the resolution of the electron energy analyser. When this method is used in combination with accelerated species, the large velocity in the center of mass ensures high resolution combined with high collection efficiency. However because of the construction of the Pelletron accelerator at the WI, collection of the photodetached electrons will entail putting an energy analyzer in the terminal, where it is not feasible to perform this type of measurement. Due

to this technical problem the most straightforward use of the EPD method
is to prepare cold negative ions that have vertical transitions to a very
few vibrational states in the prepared neutral molecule. Many such systems have been reported (4), and the current setup can be used for their
investigation by the CE method without further modifications.

In figure 1 a neutral OH signal produced by EPD is presented as
function of the light frequency, where the OH⁻ is produced either with
the duoplasmatron source (NEC) or by the sputter ion source. For comparison the curve produced by a conventional EPD machine is given, taken
from reference 4.

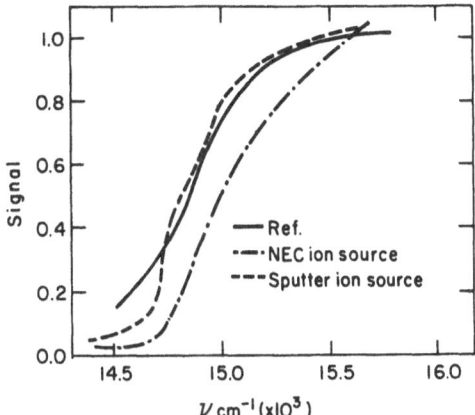

Fig. 1. Threshold photodetachment curve of OH⁻ produced by the
NEC (duoplasmatron) ion source (- -) or by the sputtering ion source (---) in the Pelletron accelerator. For
comparison the curve from a "conventional" EPD machine
is presented taken from reference 4 (solid line).

In figure 2 the OH signal is presented as function of the arrival
time and the energy. Since in our setup the laser is colinear with the
OH⁻ beam some of the neutrals can be produced before the terminal while
others can be formed after the free field region of the terminal. In
both cases the particles arrive with lower energies than those formed in
the terminal. The neutrals that were produced first, were not accelerated to the full extent that the the acceleration length allows. Neutrals
produced after the terminal were decelerated by the negative field existing in the second part of the accelerator. One can, however, distinguish
between the two types of neutrals by their arrival time at the detector.
The species produced before the terminal will have a longer flight time
and will arrive after those produced later.

These results demonstrate clearly the ability to produce neutral
species in the accelerator by laser photodetachment; however, without
combining it with an electron energy analyzer, it is not possible to use
it as a general method for preparing molecules in a single state. The
method is restricted to species having similar structure in the negative
ion and neutral states.

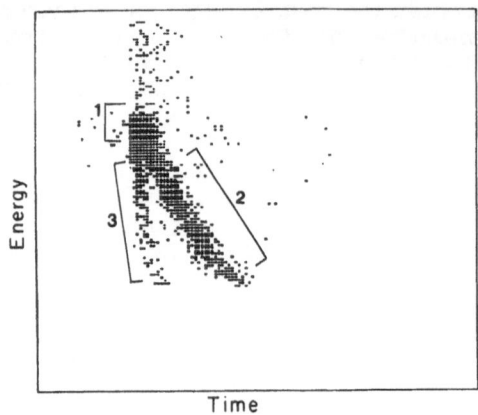

Fig. 2. Time-energy spectrum of OH molecules produced by EPD, by
a frequency doubled Nd:Yag laser (532nm). The region
marked 1 includes neutrals produced in the pelletron
terminal. They have highest energy possible, and shor-
test time of flight to the detector. In region 2 neutrals produced before the terminal arrive. In the area
marked 3 those OH neutrals produced after the terminal
arrive- these have the same energy as in 2 but shorter
flight time.

Photodissociation

Since detection of detached charged species (electrons) is difficult
due to the accelerator configuration, it would be interesting to detach a
neutral fragment, and measure its kinetic energy in coincidence with the
polyatomic species that are studied by the CE method. In other words if
we could detach a light atom from the initially prepared cold negative
ion, and if it would be possible to measure the atom's kinetic energy, it
would then be possible to measure the structure of polyatomic species
with a well defined energy. This is precisely the scheme for our photo-
detachment-photodissociation arrangement.

In the case of the dissociation experiments, where both atoms and
polyatomic fragments are produced, a coincidence measurement is performed
on the two fragments. The relative kinetic energy of the fragments can
be observed by the distance between them on the detector. By choosing
events in which the two fragments appear at a given separation, the internal energy of the polyatomic species is defined. This process is
equivalent to measuring the electron energy in electron detachment experiments.

In this experiment a negatively charged complex is prepared in a
pulsed nozzle combined with an electron source. In the expansion with
helium, vibrationally and rotationaly cold ions are produced, as well as
helium-molecule complexes. Depending on the experiment either a cold
negative ion can be selected for acceleration, or an ion attached to a
single helium atom. In the terminal two lasers can be used for the neutralization plus dissociation of the molecule, or a single laser wavel-

ength can be applied to detach an electron from the molecule-helium complex. In the former case the photofragment kinetic energy will be given by its location on the position sensitive detector. In the second case the same idea will be applied for measuring the He kinetic energy.

Having a spatial resolution of about 1mm on the detector, and a flight tube of about 30 meters from the terminal to the detector, we can achieve resolution of several wavenumbers in energy in the definition of the internal energy of the polyatomic molecule. Hence a method is in our hands to study the structure of polyatomic species having well defined internal energy.

REFERENCES

1. For example: C.M. Lovejoy and D.J. Nesbitt, J. Chem. Phys. 86, 3151 (1987); and D.J. Nesbitt in this proceedings.

2. D.W. Turner, C. Baker, A.D. Baker, and C.R. Brundle, Molecular Photoelectron Spectroscopy (Wiley, New York 1970).

3. (a) T.M. Miller, D.G. Leopold, K.K. Murray, and W.C. Lineberger, Bull. Am. Phys. Soc. 30,880 (1985). (b) D.G. Leopold, K.K. Murray, A.E.S. Miller, and W.C. Lineberger, J. Chem. Phys. 83,4844 (1985).

4. P.S. Drazaic, J. Marks, J.I. Brauman in "Gas Phase Ion Chemistry" Vol. 3 p:167 (Academic Press NY).

WHAT CAN COULOMB EXPLOSIONS TEACH US ABOUT CLUSTERS?

E. P. Kanter, A. Faibis, and L. Tack

Physics Division
Argonne National Laboratory
Argonne, IL 60439

INTRODUCTION

In recent years, there has been a considerable interest in the study of gas-phase atomic clusters. Among the motivations for this work is the fact that such clusters provide a bridge between molecular and condensed matter physics, and thus provide an opportunity to study the structural and electronic rearrangements involved in this transition. Small carbon cluster ions are particularly interesting because of their importance in many chemical processes such as catalysis and combustion and there has been a long history of work with these ions. While there now exists a substantial body of work on the subject[1], all of the experimental structural information to date is indirect. Typically, such experiments probe gross features by studying changes in stability with increasing size through measurements of relative abundances, ionization potentials, chemical activity, or fragmentation energies. Recently, more sophisticated spectroscopic techniques have provided some information on electronic states and further offered hopes of probing the nuclear vibrational motions within such systems.[2] Nevertheless, some of the most important findings (such as shell structure) still remain controversial because of the incomplete nature of the body of data. As has been reported at this workshop, Coulomb explosion experiments have recently provided detailed images of the structures of several small molecules. It has thus been tempting to ask: Can such methods be applied to clusters? We have explored this issue by studying the molecule C_3^+.

Although C_3^+ is considered to be the fundamental building block of the larger carbon clusters[3], the geometry of this molecule is unknown. Several authors have reported results of ab initio calculations in which the optimized geometry of the ion is deduced, assuming a linear configuration[4], in agreement with the structure of the neutral C_3

molecule.[5] We report here the results of a series of measurements exploiting the Coulomb Explosion Method (CEM) to study the geometric structure of the C_3^+ ion. Our results indicate a cyclic structure for this ion. For comparison, measurements were also made of the carbon geometries of small hydrocarbon cations of the form $C_3H_n^+$ (n=1-4). Ring structures, of varying rigidities, are observed for all of these except $C_3H_4^+$ which exhibits a linear geometry. We also present results for doubly-charged species.

EXPERIMENT

The CEM is a technique which provides direct information on the nuclear densities within a molecule by imaging individual molecules. The technique, which has been described in detail elsewhere[6], involves the foil-induced dissociation of a fast (MeV) beam of molecules. Through their mutual Coulomb repulsion, the resulting highly-charged atomic ions convert their Coulomb energy (which can be several hundred electron volts) into kinetic energy of relative motion. For a molecule containing N atoms, measurements of the 3N velocity components after the explosion provides information on the 3N spatial components within the original molecule.[7-8]

In the experiments reported here, the molecular ions were formed by bombardment of allene (C_3H_4) gas, at a pressure of ~50 mTorr, by ~9-35 eV electrons. For each molecular species that we studied, the electron bombarding energy was adjusted to lie just above (~0.5 eV) the measured threshold for production of the ion of interest. The ions were then accelerated by the Argonne 4.5-MV Dynamitron accelerator to an energy of 4.5 MeV, magnetically mass-analyzed, and then, after an ~8-μsec flight time, dissociated in a thin (~30-Angstroms) Formvar film. The resulting carbon fragment ions were charge-state analyzed by deflection in a uniform electric field. After a flight path of ~6 meters, the ions were detected, in triple coincidence, by the MUPPATS detector system.[9] Depending on the original spatial orientation of the molecule before the explosion, the ions can be separated by up to several centimeters on the detector surface and several tens of nanoseconds temporally. For each molecule, this system records the x- and y-coordinates where each fragment ion impinges on the surface of the detector (with 160-μm position resolution) as well as the relative time-of-arrival of each ion (600-psec time resolution). From this information, we deduce the 3 components of the velocity (relative to the molecular center-of-mass) of each fragment ion and there-fore record 9 parameters per molecule. Because of their differing deflections in the electrostatic field after the target, the various charge states of carbon ions are dispersed across the detector surface. After correction for this deflection, we deduce the three body-fixed coordinates v_a, v_b, and v_c, representing the lengths of the three relative velocity vectors for the carbon ions after dissociation. For clusters, the problem of relating these velocities to the original bond lengths, is

a difficult computational task because of the many-body nature of the interaction potential. However, by comparing the distributions of these velocity coordinates for several molecules, we can explore comparative
geometries quite easily. We discuss here only the symmetry features
observed in the coordinate system defined by these velocities, herein
described as "V-space". To further simplify the analysis, we will restrict our attention here to molecules in which all three carbon dissociation fragments emerge as C^{2+} ions and thus, aside from the differing final
velocities acquired due to structural features, these fragments are all
equivalent. For the protonated ions ($C_3H_n^+$) we neglect the effect of the
protons on the V-space geometries. For molecules such as these, with symmetric proton geometries, the net momentum carried by the protons is
small, as evidenced by our experimental data, and can be neglected for
<u>qualitative</u> comparisons of the carbon geometries. A more comprehensive
analysis of the complete spatial geometry of C_3^+ is currently in progress
and will be reported elsewhere.[10]

DISCUSSION

To compare the carbon geometries of these molecules in V-space, we
first construct the ordered lengths $v_a \leq v_b \leq v_c$. Consider the quantity

$$R_s = \frac{v_b + v_a}{v_c} \qquad (1).$$

This quantity is limited to the range $1 \leq R_s \leq 2$, corresponding to the
extremes of linear ($R_s=1$) and equilateral ($R_s=2$) geometries. Figure 1
shows the distribution in this quantity derived from the V-space data for
C_3^+. While it is not possible to deduce the spatial geometries of these
molecules directly from the V-space data without further complex analysis,
comparisons of such distributions for different structures provide a very
rapid and simple means of recognizing similarities in the
stereochemistries of such molecules.

There are several molecules in the series $C_3H_n^+$ (n=1-4) for which there
exist previous experimental and theoretical information concerning the
probable geometries. The structure of $C_3H_4^+$ produced via electron impact
ionization on allene has been studied by laser photodissociation[11] and
photoion-photoelectron coincidence investigations.[12-14] These results
suggest that although the protons can rearrange to form different
structural isomers, the carbon chain remains linear. The experimental
evidence for the linear structure is further supported by ab initio
calculations by Frenking and Schwarz.[15] Similarly, there have also been
studies of $C_3H_3^+$ produced by electron impact on allene (and also propyne).
ICR experiments[16-17] have proposed that both cyclic and linear isomers are
produced. The branching ratio for producing the two isomers is about 2:1
(cyclic:linear) for impact on propyne and assumed to be similar for the

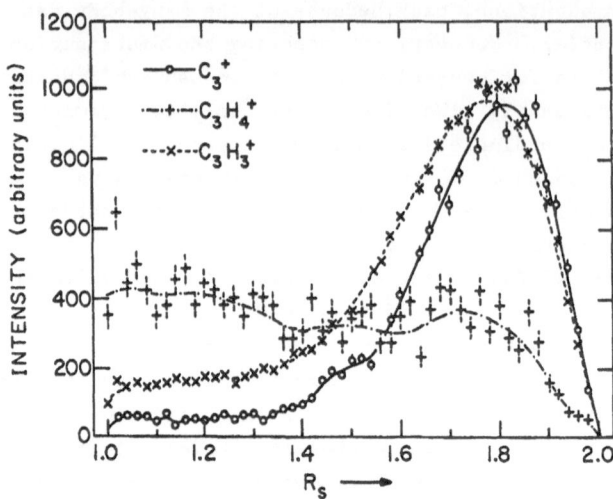

Fig. 1. Measured distributions for the ratio R_s, as described in text, for
the carbon fragment ions (all C^{2+}) resulting from the dissociation
of 4.5-MeV C_3^+, $C_3H_3^+$, and $C_3H_4^+$. The ordinate represents the actual
number of molecules for the case of C_3^+. The data for $C_3H_4^+$ are
normalized to yield the same integral as the C_3^+ data while the
distribution for $C_3H_3^+$ has been arbitrarily scaled. Error bars
represent the effect of sampling statistics only. The lines are
drawn to guide the eye.

case of allene. The ab initio calculations of Radom, et al.[18] predict the
cyclic species to be the more stable by at least 60 kcal/mole.

For comparison, Fig. 1 also shows the R_s-distribution obtained for
$C_3H_4^+$ by considering only the carbon geometry. Here we find the R_s-
distribution shifted toward unity, implying a more linear geometry. This
observation is consistent with the results described above for this ion
and the known structure of the neutral allene species. Additionally, we
have studied the same distribution for $C_3H_3^+$. As can be seen from Fig. 1,
this distribution is more ring-like, again consistent with previous
results, and quite similar to the distribution we obtain for C_3^+. If we
assume that the distributions found for C_3^+ and $C_3H_4^+$ represent those of
cyclic and linear structures respectively, then the $C_3H_3^+$ data can be best-
fit by combining these in the proportions 0.71±0.03 cyclic and 0.29±0.03
linear. This is in good agreement with the 2:1 ratio extracted from the
ICR experiments.

Additional data were also collected for the doubly-charged species
$C_3H_3^{++}$. Figure 2 shows a comparison of the ratio, R_s, for this molecule
and the singly-charged species. Here, the data seem to indicate that the
double-charged molecule is extremely floppy with a more uniform spread
between linear and cyclic geometries. It should be noted that the
observed floppiness of the doubly-charged species in comparison to the
singly-charged cannot be an experimental artifact. The blurring effects

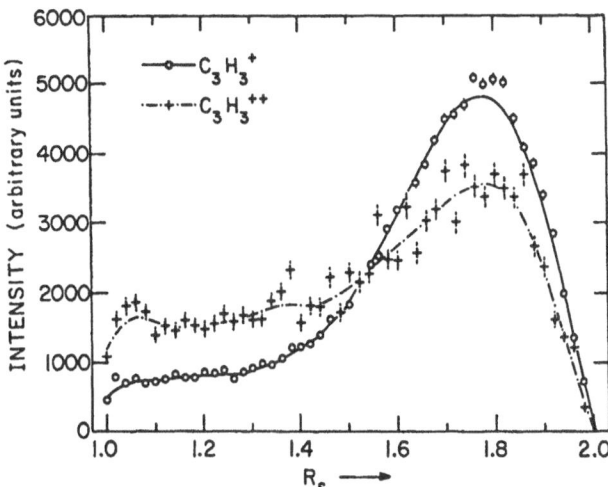

Fig. 2. Measured distributions for the ratio R_S, as described in text, for the carbon fragment ions (all C^{2+}) resulting from the dissociation of 4.5-MeV $C_3H_3^+$, and 9.0-MeV $C_3H_3^{++}$. The data have been arbitrarily scaled. Error bars represent the effect of sampling statistics only. The lines are drawn to guide the eye.

of experimental resolution (e.g. detector resolution, multiple scattering and energy loss in the stripper foil, etc.) all decrease with the increased beam energy (9.0 MeV) used in our study of the doubly-charged ion.

The C_3^+ distribution, which is peaked near $R_S=2$, implies a geometry in which the two <u>smallest</u> sides are equal to the largest side, characteristic of a ring structure. We must stress that quantitative conclusions about the spatial geometries cannot be drawn from these simple distributions in V-space and in particular the shifts of these distributions away from $R_S=2$ does not necessarily imply non-equilaterality in the original structures. Nevertheless, these comparisons do provide strong evidence that the geometry of the C_3^+ ion is more like the carbon geometry within the cyclic $C_3H_3^+$ than that within the linear $C_3H_4^+$ molecule. While it is conceivable that these results could be due to the observation of highly-excited vibrational motions resulting in distorted geometries, the clear lack of any linear geometries ($R_S=1$) in the C_3^+ data ensemble makes this a very remote possibility. Furthermore, the good agreement we find between our electron ionization thresholds, and those reported by the other authors[19], strongly suggests that we are indeed observing the ground state structures of these molecules. Additionally, experiments with C_3^+ ions prepared in a duoplasmatron source fed with CH_4 yielded identical results. More extensive exploration of the potential surface seems to indicate that the C_3^+ molecule is indeed bent[20], in agreement with these observations. Further analysis of our data, incorporating a mapping to spatial configuration space, is underway and should provide more detailed information on these structures.

Because of the strong coupling of their vibrational modes, and the intrinsically complicated many-body nature of polyatomic vibrations in general, molecular clusters represent a particularly difficult challenge to conventional experimental techniques.[21] We believe that the data presented here demonstrate an important new approach to the study of such molecules.

ACKNOWLEDGEMENTS

We wish to thank Dr. W. L. Brown for suggesting this problem and Dr. K. Raghavachari for sharing his preliminary results with us. This work was supported by the U. S. Department of Energy, Office of Basic Energy Sciences, under Contract W-31-109-ENG-38.

REFERENCES

1. see e.g. PDMS and Clusters, E. R. Hilf, F. Kammer, K. Wien, eds., Springer-Verlag, Berlin, (1987).
2. O. Chesnovsky, S. Yang, C. L. Pettiette, M. J. Craycraft, Y. Liu, and R. E. Smalley, Chem. Phys. Lett. 138:119 (1987).
3. W. L. Brown, R. R. Freeman, K. Raghavachari, and M. Schluter, Science 235:860 (1987).
4. W. Kuhnel, E. Gey, and H.-J. Spangenberg, Z. phys. Chemie, Leipzig 263:641 (1982).
5. I. Plesser, Z. Vager, and R. Na'aman, Phys. Rev. Lett. 56: 1559 (1986).
6. D. S. Gemmell, Chem. Rev. 80:301 (1980) and references therein.
7. E. P. Kanter, Z. Vager, G. Both, and D. Zajfman, J. Chem. Phys. 85:7487 (1986).
8. Z. Vager, E. P. Kanter, G. Both, P. J. Cooney, A. Faibis, W. Koenig, B. J. Zabranksy, and D. Zajfman, Phys. Rev. Lett. 57:2793 (1986).
9. A. Faibis, W. Koenig, E. P. Kanter, and Z. Vager, Nucl. Instrum. and Meth. B13:673 (1986).
10. A. Faibis, E. P. Kanter, G. Natanson, L. Tack, and Z. Vager, to be published.
11. P. N. T. Van Velzen and W. J. Van der Hart, Org. Mass Spect. 16:237(1981).
12. H. M. Rosenstock and K. E. McCulloh, Int. J. Mass Spect. Ion Phys. 25:327 (1977).
13. R. Stockbauer and H. M. Rosenstock, Int. J. Mass Spect. Ion Phys. 27:185 (1978).
14. A. C. Parr, A. J. Jason, R. Stockbauer, and K. E. McCulloh, Int. J. Mass Spect. Ion Phys. 30:319 (1979).
15. G. Frenking and H. Schwarz, Int. J. Mass Spect. Ion Phys. 52:131 (1983).
16. P. J. Ausloos and S. G. Lias, J. Am. Chem. Soc. 103:6505 (1981).
17. D. Smith and N. G. Adams, Int. J. Mass Spect. Ion Proc. 76:307 (1987) and references therein.
18. L. Radom, P. C. Hariharan, J. A. Pople, and P. v. R. Schleyer, J. Am. Chem. Soc. 98:10 (1976).
19. H. M. Rosenstock, K. Draxl, B. W. Steiner, and J. T. Herron, J. Phys. Chem. Ref. Data 6:Suppl. No. 1 (1977).
20. K. Raghavachari, private communication and to be published.
21. R. J. Saykally, Science, 239:157 (1988).

PROBING OF NUCLEAR WAVES IN TRIPLET H_2 (n=3)

W. Koot, W.J. van der Zande, P. Post and J. Los

FOM-Institute for Atomic and Molecular Physics
Kruislaan 407, 1098 SJ Amsterdam

INTRODUCTION

The wave function of the nuclei that are forming the
molecule is one of the basic ingredients to describe their
vibrational and rotational motion in the electronic
environment. Not only do they show the quantization of
rotational and bound vibrational motion, but they are also
important in the description of transitions between molecular
states through the Franck-Condon and Honl-London factors. In
all these cases, integrals containing the nuclear wavefunctions
are involved, so the wavefunctions are only probed indirectly.
We will describe an experiment in which the vibrational
wavefunctions are probed in a direct way[1], which opens new
possibilities to test the theoretical description of excited
molecules.

EXPERIMENTAL SETUP

Figure 1 contains a schematic outline of the experimental
setup and the relevant potentials of the hydrogen molecule we
have investigated are given in figure 2.

In our experiment we first prepare a beam of neutral H_2 in the metastable $c^3\Pi_u^-$ state. This is done by charge exchanging a fast beam of H_2^+ ions, extracted from an electron impact source, in cesium vapour[2]. Since the charge exchange to the $c^3\Pi_u$ state is almost resonant, this can be done with a large efficiency. After some distance, in which the unstable states that are formed have decayed, the resulting neutral beam is crossed with a CW intra-cavity dye laser which can be tuned between 15500 cm^{-1} and 17500 cm^{-1}. By choosing the right wavelength, we can selectively excite one rovibrational level of the n=3 manifold formed by the $g^3\Sigma_g^+(3d\sigma)$, $h^3\Sigma_g^+(3s\sigma)$, $i^3\Pi_g(3d\pi)$ and $j^3\Delta_g(3d\delta)$ states. When the excited state decays to the repulsive $b^3\Sigma_u^+$ state via emission of a photon, the molecule will dissociate releasing its residual energy as kinetic energy. With our experimental setup, we can measure this released kinetic energy ε, using a time and position sensitive detector[5].

COUPLED POTENTIALS

Because the excited 3d Rydberg-electron is only loosely bound to the internuclear axis, the projection Λ of the electronic angular momentum is not a very good quantum number and the different $3d\lambda$ states are heavily mixed[3,4]. The $3s\sigma$ state is coupled to this complex through its interaction with the $3d\sigma$ state. Therefore the proper eigenstate Ψ that is excited has the form

$$\Psi^+ = a_{s\sigma}^+ \chi_{s\sigma} \,|\, 3s\sigma^+\rangle + a_{d\sigma}^+ \chi_{d\sigma} \,|\, 3d\sigma^+\rangle + a_{d\pi}^+ \chi_{d\pi} \,|\, 3d\pi^+\rangle + a_{d\delta}^+ \chi_{d\delta} \,|\, 3d\delta^+\rangle \qquad (1a)$$

for + parity levels and

$$\Psi^- = \qquad\qquad\qquad\qquad\qquad a_{d\pi}^- \chi_{d\pi} \,|\, 3d\pi^-\rangle + a_{d\delta}^- \chi_{d\delta} \,|\, 3d\delta^-\rangle \qquad (1b)$$

for - parity levels. For each rovibrational level, the nuclear wavefunctions have been approximated by the vibrational wavefunction $\chi_\lambda(R)$, calculated in the potential well of the pure $|\lambda\rangle$ state, multiplied by the amount of λ-character $a_\lambda^\pm$ present in the excited state. The $a_\lambda^\pm$ are known from the spectroscopy for each level[4].

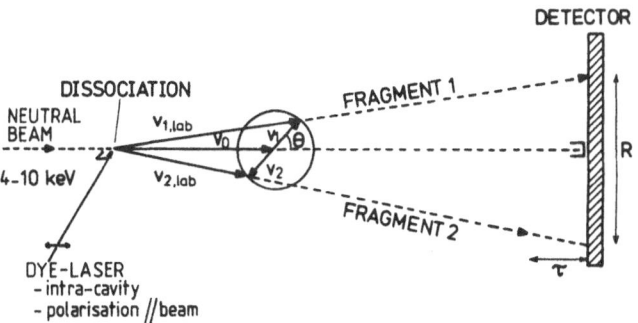

Figure 1. Experimental setup. The position (R) and time (τ) difference of the fragments arriving on the detector is measured. From these quantities the released kinetic energy ε and recoil angle θ can be calculated.

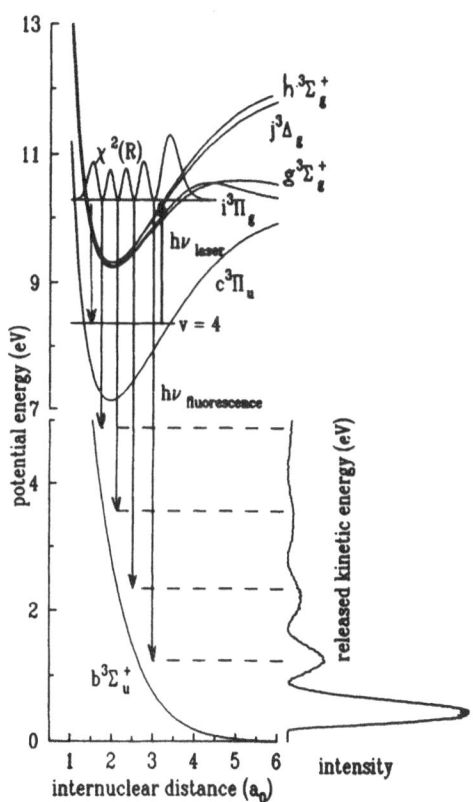

Figure 2. Potential diagram of H_2 showing the relevant potentials. Also drawn are the excitation and decay transitions.

Because the molecule can emit photons of different energies at various internuclear distances a broad kinetic energy distribution will be obtained. As indicated in figure 2, this kinetic energy distribution reflects the shape of the vibrational wavefunction since the transition probability is weighted by the probability to find the molecule at a specific internuclear distance. This last factor is determined by the vibrational wavefunctions.

In an exact treatment, the observed intensity is given by

$$N(\varepsilon) \propto v(\varepsilon)^3 \, |\langle \chi_\varepsilon | \mu | \chi_v \rangle|^2 \qquad (2)$$

where χ_ε and χ_v are the continuum and bound vibrational wavefunctions and μ is the (R-) dependent transition dipole moment. $v(\varepsilon)$ is the frequency of the emitted photon which is connected to the released kinetic energy. Using the completeness of the functions χ_ε, Eq. (2) can be cast into[1,6]

$$\mu(R) \, \chi_v(R) = \int \pm \left\{ N(\varepsilon) \, v(\varepsilon)^{-3} \right\}^{1/2} \chi_\varepsilon(R) \, d\varepsilon \qquad (3)$$

The transformation given by Eq. (3) can be applied to the experimental spectrum because the $b^3\Sigma_u^+$ potential curve is very well known and the $\chi_\varepsilon(R)$ can be accurately calculated. In this way the product of vibrational wavefunction and transition dipole moment can be obtained from the experimental spectrum.

Figure 3 shows an example of an experimental spectrum (3a) and the result of the transformation (3b). The agreement with a calculated wavefunction is good. The difference in amplitude between the experimental transformation and the theoretical wavefunction is caused by the transition moment and we can conclude that μ has to be R dependent. We have measured such kinetic energy release spectra for a number of excited levels in the different n=3 potentials. We hope to find from the resulting vibrational wavefunctions the behaviour of the transition dipole moment and to see the effect of the breakdown of the Born-Oppenheimer approximation for the highly excited states involved.

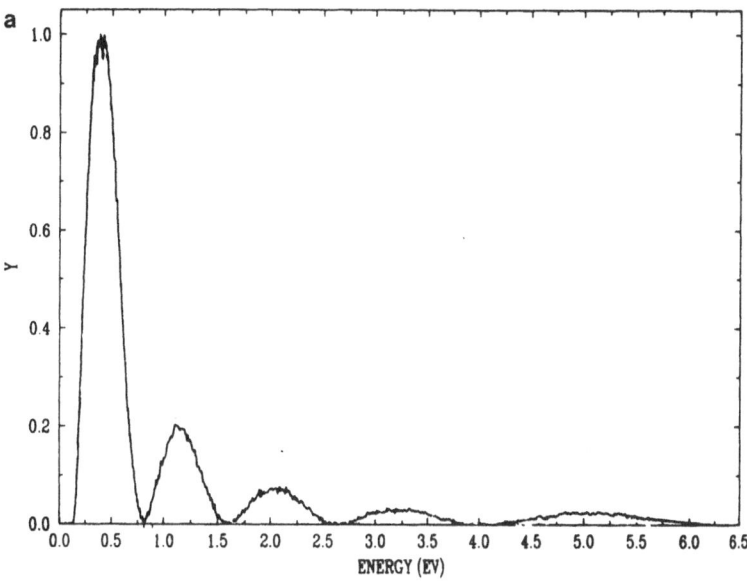

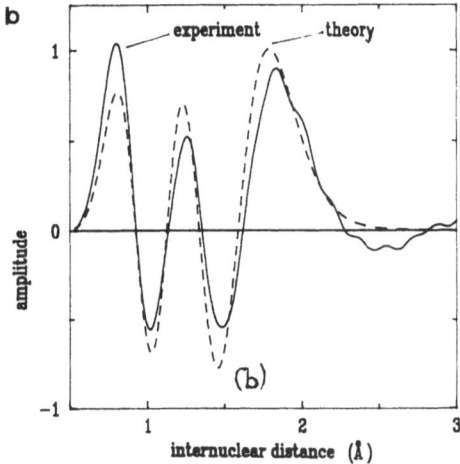

Figure 3. (a) released kinetic energy spectrum with the laser tuned to the $c^3\prod_u^-$ (v=4,N=1) -> $g^3\Sigma_g^+$ (v=4,N=1) transition. (b) solid line: transfromation of experimental result, using Eq. (3); dashed line: wavefunction calculated in potential.

In principle the excited state can also emit a photon back to the $c^3\Pi_u^+$ state, which is predissociated by the $b^3\Sigma_u^+$ state through rotational coupling[7]. This process will produce a sharp peak in the ε spectrum at a very high energy because all the internal energy of the $c^3\Pi_u^+$ state is converted into kinetic energy. Figure 4 shows a spectrum where both decay processes are observed together. In this way we have a complete picture of the various decay channels of the excited state, apart from radiative decay back to the metastable $c^3\Pi_u^-$ state which we can not observe because it does not dissociate.

The high energy peak in Fig. 4 has a tail to the low energy side due to the lifetime for photon emission to the $c^3\Pi_u^+$ state: the flight length to the detector is reduced when the molecule lives some time before it dissociates, which will then appear as a dissociation with a lower kinetic energy. For the $j^3\Delta_g^-$ level shown in Fig. 4, the $3d\delta$ character is responsible for the transition to the $c^3\Pi_u^+$ state. From the measured lifetime and the known δ-character, we can calculate the absolute dipole moment for this transition. It was found to be 2.4 ± 0.2 e a_0.

The ratio between transitions to the $c^3\Pi_u'$ and the $b^3\Sigma_u^+$ state is governed by the product of two factors:

1) the bound-bound resp. bound-free Franck-Condon factors multiplied by v^3. This factor can easily be calculated from the known potentials.

2) the square of the transition dipole moments from the different $|\lambda\rangle$ basis states weighted by their amplitude $a_\lambda^\pm$.

The variation of the ratio between the two decay channels for the different rotational and vibrational quantum levels seems to be in gros agreement with the variation of the $a_\lambda^\pm$ as found by Keiding et al[4].

CONCLUSION

We have found and applied a method to probe vibrational wavefunctions directly, using radiative dissociation via a well known repulsive potential. We expect it to be a sensitive test to the theory.

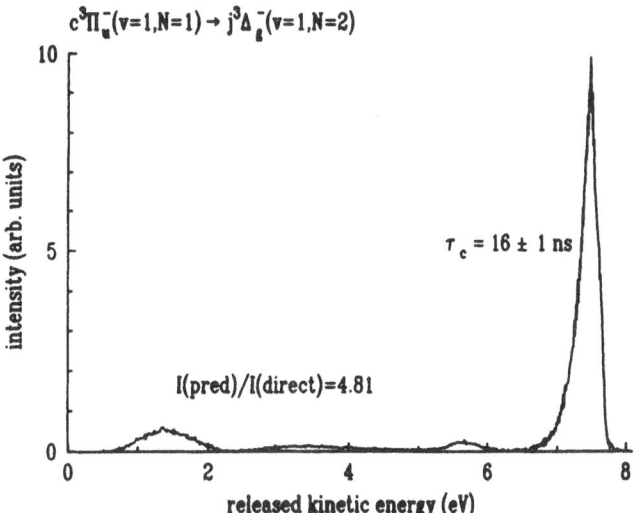

Figure 4. Released kinetic energy spectrum showing both the decay to the $b^3\Sigma_u^+$ state (first broad humps) and to the $c^3\Pi_u^+$ state (large peak at high energies). The peak from the latter process is lifetime broadened.

We were also able to monitor all dissociation channels of the n=3 complex in H_2 simultaneously. From these measurements we could find radiative lifetimes and therefrom absolute transition dipole moments.

For a complete account of the different experiments we refer to forthcoming publications[8].

REFERENCES

1) W. Koot, W.J. van der Zande and J. Los, Phys. Rev. Lett. 58, 2746 (1987).

3) R. Jost and M. Lombardi, Mol. Phys. 37, 1605 (1979).

4) S. Keiding and N. Bjerre, private communication.

5) D.P. de Bruijn, and J. Los, Rev. Sci. Instrum. 53, 1020 (1982).

6) A.L. Smith, J. Chem. Phys. 49, 4813 (1968).

7) G. Comtet and D.P. de Bruijn, Chem. Phys. 94, 365 (1985).

8) W. Koot, W.J. van der Zande, P. Post and J. Los, in preparation.

MULTIPHOTON IONIZATION OF CLUSTERS: REACTIONS AND SPECTROSCOPY

A. W. Castleman, Jr. and R. G. Keesee

Department of Chemistry
The Pennsylvania State University
University Park, PA   16802   USA

ABSTRACT

   Clusters display properties intermediate between the gas and the
condensed phase and studies of their varying properties as a function of
degree of aggregation provide a unique way of bridging atomic and molecular
science on the one hand with that dealing with condensed matter and
surfaces on the other. The application of supersonic molecular beam
techniques, coupled with both resonant and non-resonant multiphoton
ionization, time-of-flight mass spectrometer methods in conjunction with
reflecting electric fields will be discussed in terms of investigating
variations of spectral features, ionization potentials, reactions, and
dissociation processes.

INTRODUCTION

   Cluster research offers the exciting prospect of bridging the gap
between the gaseous and condensed phase by probing the details of
condensation and nucleation phenomena at the molecular level. Studies of
spectroscopic shifts upon successive clustering are especially interesting
with regard to the onset of liquid or solid-like features in the spectra
(1-6). Resonance enhanced ionization provides a unique way of assigning
the species to which a given molecule of interest is bound. Investigations
of the processes of cluster ionization and dissociation are of particular
interest since they contribute to a further understanding of the evolution
of changes in ionization potentials as a system approaches the bulk work
function. The details of intramolecular energy flow and energy disposal
following ionization can also be revealed.

   Studies of the molecular properties, reactions, and behavior of
clusters generally require ionization in one of the steps as either a
probe and/or a method of detecting clusters through mass spectrometry.
Although ionization can be accomplished through electron impact as well
as single photon techniques, resonance enhanced multiphoton ionization
often enables selective ionization of clusters in particular states. More
detailed and specific information can be obtained through resonance-
enhanced ionization spectroscopy, and is the preferred method when such
processes can be readily accomplished. Herein, examples are drawn from
three studies. In the first example, the spectroscopic shifts of two
probe molecules, phenylacetylene and p-xylene, clustered by a series of

rare gases, CO_2, H_2O, N_2, O_2, NH_3, and CCl_4, are investigated. In a
second example, involving p-xylene clustered with argon, the change in
ionizaton potential as a function of the degree of aggregation is considered. The third example draws upon studies of neutral species via mass
spectrometry involving an ionization step. The dynamics of cluster
dissociation following multiphoton ionization as obtained with time-of-
flight mass spectrometry for ammonia, alcohol, and xenon clusters are
explored, as well as the role of rotational tunneling in the dissociation
of small atomic ions, namely Ar_3^+.

SPECTRAL SHIFTS

Resonance enhanced multiphoton ionization through the specific
excitation of an electronic state of a chromophore contained within a
cluster is a powerful method of ascertaining the properties of clusters
in relating these to their counterparts in the condensed and isolated gas
phases. Generally, the clustering of atoms or molecules onto a chromophore result in a perturbation of the electronic states of that chromophore. The spectral shift of a given electronic transition from that of
the isolated chromophore is a measure of the relative differences between
the lower and upper states of the energetic perturbation induced by
clustering. This is analogous to the spectral shift of electronic
transitions observed for molecules in solutions or matrices from their
gas-phase transitions. A red shift implies that complex formation has
reduced the energy difference between the two states, whereas a blue shift
indicates an increase in the difference. The magnitude and direction of
the shift are due to a combination of effects, including dispersive and
repulsive interactions, hydrogen bonding, and electrostatic forces
involving such processes as dipole-induced-dipole or dipole-dipole
interactions.

We have employed both one- and two-color resonance enhanced
multiphoton ionization to investigate the $S_1 \leftarrow S_0$ π-electron transition
in phenylacetylene and p-xylene as clusters with various solvents. Single
color multiphoton ionization studies of the perturbed $L_b(^1B_2)$ states of
phenylacetylene (PA) bound with Ne, Ar, Kr, and Xe were all found to
induce a lowering of the S_1 resonance with respect to the ground state.
These observed red shifts have been attributed to dispersive interactions
with solvent molecules (7,8). In general, a spectral shift is governed by
three factors; (i) short-range electronic repulsive interactions which
result in a blue shift, (ii) electronic dispersive interactions which
result in a red shift, and (iii) differences of zero point energies
between the excited and ground states. Our results support the finding
(7) that in aromatic molecule-rare gas atom systems the spectral shifts
are dictated by atom polarizability, i.e., the important role of
dispersive forces in the perturbation of the S_1 excited state. Figure 1
displays the results of our study which shows a direct linear dependence
of the spectral shift on the electrostatic polarizability of the rare gas
atom. The results conform to the Onsager model, but on a microscopic
level (9).

Spectral shifts of phenylacetylene due to aggregation by argon atoms
are given in Table 1. Investigations with large-ringed systems have
generally shown an approximate additivity of spectral shifts based on
the number of rare gas atoms clustered on to the aromatic (7). In our own
work, we find that this additivity is apparently additive only up to the
clustering of two atoms per aromatic ring as seen from the data in
Table 1. Since the additivity is nearly exact for the two-atom case, the

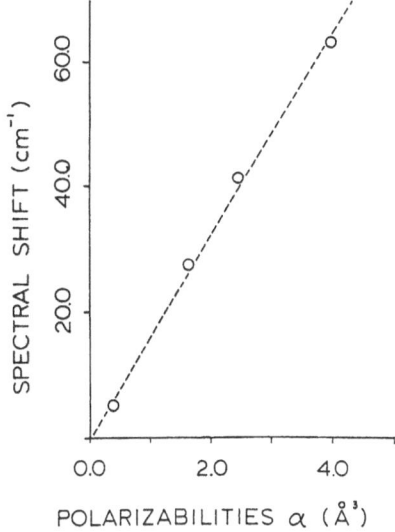

FIGURE 1. Spectral Shift of PA•R (relative to the nascent PA)
versus the polarizability of the rare-gas atom.
$\alpha = 0.40$ (Ne), 1.63 (Ar), 2.48 (Kr) and 4.01 Å^3 (Xe).

TABLE 1

Spectral shifts of the electronic origin of the S_1
excited state of the complex PA•R_n (R = rare gas atom).

Species	Spectral Shifts[a] $\delta\nu$ (cm^{-1})	Species	Spectral Shifts[a] $\delta\nu$ (cm^{-1})
PA•Ar	-27.8 ± 0.5	PA•Ar_8	-57.2 ± 0.8
PA•Ar_2	-53.3 ± 0.5 -25.4 ± 0.5[b]	PA•Ar_9	$-56 \quad \pm 1$
		PA•Ar_{10}	$-67 \quad \pm 1$
PA•Ar_3	-49.1 ± 0.5 -22.2 ± 0.5[b]	PA•Ar_{11-15}	(-54 to -47)
PA•Ar_4	-49.9 ± 0.5		
PA•Ar_5	-53.5 ± 0.5		
PA•Ar_6	-57.0 ± 0.5		
PA•Ar_7	-57.5 ± 0.5		

[a]Energy shift with respect to the unperturbed PA S_1
resonance. A negative value corresponds to a red-
shift.
[b]From stagnation pressure studies, we tentatively
assigned these two features to different conformers.

spectral shifts shown in Figure 1 are identical on a spectral shift per atom basis for the phenylacetylene system in the case of the two-atom containing rare gas complex.

Particularly interesting are the trends seen for larger clusters. Figure 2 shows a selected set of data for the spectral shifts relative to the S_1 electronic origin of phenylacetylene for clusters containing four to ten argon atoms. First it is interesting to note that the major feature asymptotically approaches a shift of approximately 50 cm^{-1}. Clearly the additivity rule does not apply. Secondly, the van der Waals modes to the right side of the main resonance begin to fill in for large cluster sizes. The spectra are broadening in analogy to those seen in the condensed phase and the features to the right resemble photon modes for a system of infinite lattice. The behavior of the spectral shift with cluster size in the case where phenylacetylene is attached to preexisting ammonia clusters (10) is illustrated in Figure 3.

DISSOCIATION DYNAMICS

The apparatus employed in these studies has been described in detail elsewhere (15) and only a brief description is given here. Clusters are formed via adiabatic cooling of a gas exiting a pulsed nozzle. The clusters are intercepted by a laser beam and undergo multiphoton ionization (MPI) in an electrostatic field of the time-of-flight mass spectrometer (TOFMS). The resulting ions are accelerated, typically to 1 or 2 keV, into a field-free region. At the end of this region, another electric field (the reflectron) reflects the ions before they are finally detected whereupon a time-of-flight (or arrival) spectrum is obtained.

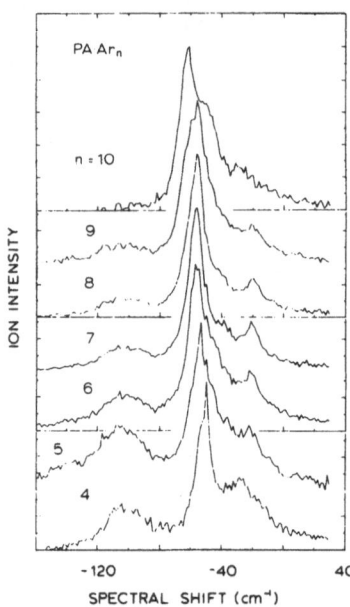

FIGURE 2.   R2PI current versus one-photon energy. The ion currents are recorded at the m/e ratios corresponding to PA•Ar_n (4≤n≤10). The energy scale is relative to the S_1 electronic origin of PA. The ion current scale is relative and different for each spectrum and p_0=300 Torr.

The reflectron is used in our laboratory to investigate dissociation in the field-free region, although this modification to TOFMS was originally developed to improve resolution (16). The separation of nondissociating parent ions and dissociated daughter ions occurs as a result of the loss of kinetic energy (due to mass loss) with essentially no change in the velocity of the ion upon dissociation. Species with greater kinetic energy penetrate deeper into the reflectron and hence arrive later at the detector. In addition to affording TOF separation, the reflectron may also be used as an energy analyzer whereby only daughter ions of kinetic energy less than the some threshold value are reflected and other ions pass through the reflectron. Typically, ions enter the field-free region about a microsecond after ionization and reach the reflectron 10 to 100 µsec later, so that dissociation processes on the order of 10^4 to 10^6 sec^{-1} are observable. The dissociation rate as the cluster enters the field-free region can be determined by changing the extraction voltages in such a way as to vary the acceleration time (the time interval from ionization to entry into the field-free region) but maintain a constant birth potential, U_o. Faster dissociation processes in the neighborhood of 10^6 to 10^8 sec^{-1} are often discernible in conventional TOFMS (i.e., no reflectron) because of the effect of dissociation in the accelerating fields on the peak shapes (17,18).

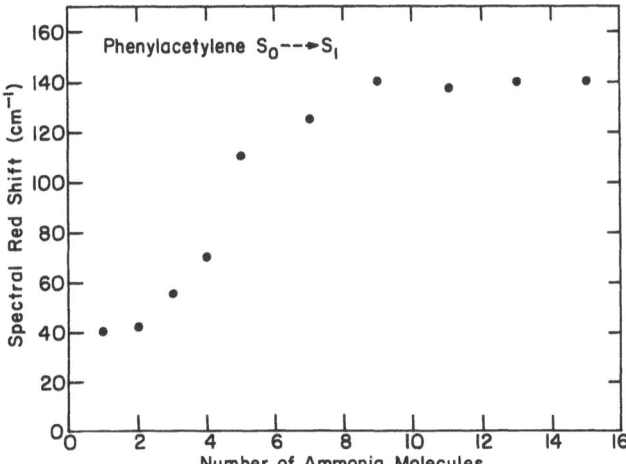

FIGURE 3.   Spectral shifts when phenylacetylene is adsorbed onto preexisting ammmonia clusters.

Ammonia Clusters

Interaction with the laser beam leads to ionization of one of the molecules in the cluster with ejection of an electron. Subsequently, the ionized molecule may react with a neighboring molecule in the cluster and molecules in the cluster will rearrange into an orientation dictated by the presence of the charge (due to ion-dipole interactions, for instance). The relaxation processes may include dissociation as well as radiative relaxation. An excited cluster loses internal energy via the energy

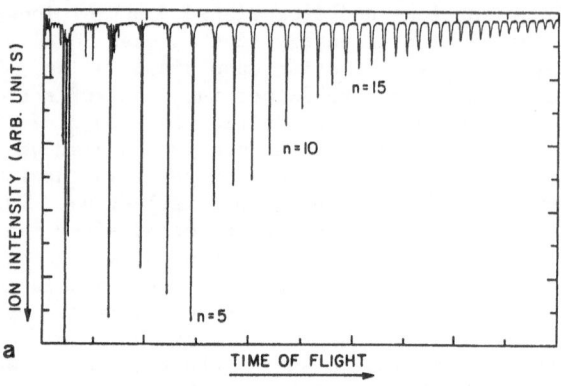

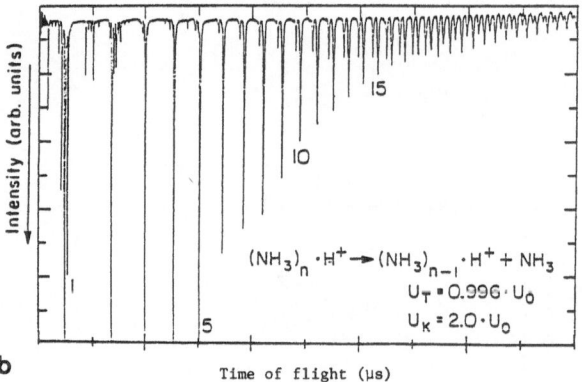

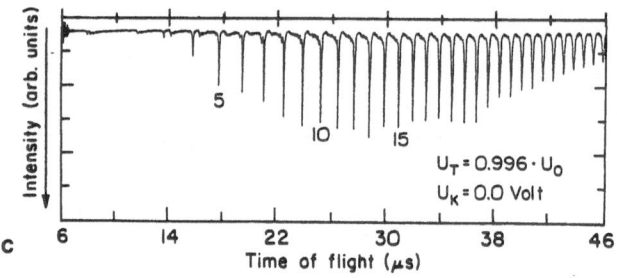

FIGURE 4. (A) A conventional TOF mass spectrum of $H^+(NH_3)_n$ clusters
from the multiphoton ionization of ammonia clusters at
266 nm. (B) Time-of-flight mass spectrum taken using the
reflectron. The potentials U_T and U_K are chosen such that
all ions are reflected but with the daughter ions arriving
at the particle detector earlier than their corresponding
parent ions. (C) Time-of-flight mass spectrum taken using
the reflectron. U_K is lowered to eliminate the parent ions.
Both spectra in Figures 4B and 4C are accumulated ion
signals over 2560 laser shots.

required to break a cluster bond along with any kinetic and internal energy that the departing molecule may carry. An example of these processes is shown below.

$$(NH_3)_m + nh\nu \quad \rightarrow \quad [NH_3^+(NH_3)_{m-1}]^* + e^- \qquad (1)$$

$$[NH_3^+(NH_3)_{m-1}]^* \quad \rightarrow \quad [NH_4^+(NH_3)_{m-2}]^* + NH_2 \qquad (2)$$

$$[NH_4^+(NH_3)_{m-2}]^* \quad \rightarrow \quad NH_4^+(NH_3)_{m-\ell-2} + \ell NH_3 \qquad (3)$$

The study of dissociation of ammonia clusters following MPI at 266 nm has been reported previously (15). Briefly, it was found that the $NH_4^+(NH_3)_n$ clusters are formed rapidly (in less than 10 ns) since the fragmentation of reaction (2) was not observed in the experimentally available time-scales. The extent of fragmentation is determined through use of the reflectron. Figure 4a shows a typical ammonia cluster ion experimental TOF without the reflectron while 4b shows the superimposed daughter distribution which arises through use of the reflectron. If the reflectron barrier is lowered sufficiently, only the lower kinetic energy products are reflected. The non-dissociating ions are thereby eliminated from the spectrum as shown in the lower part of Figure 4c. Further reduction of the reflecting potential improves the ability to discern small contributions from more extensive dissociation.

The contribution of unimolecular compared to collision induced dissociation is determined through experiments in which the pressure in the field-free region is varied. An extrapolation of the data to a zero of pressure can be made in a linear fashion under the thin collision approximation where, within the uncertainty of the true zero of pressure, the ordinate gives a direct measure of the fraction of dissociation by unimolecular (evaporative) loss. The component from unimolecular decay has been found to increase steadily with cluster size from n=4 to 25 all being of order 10^5 sec^{-1}. The implication of our results is that each larger cluster progressively has more excess internal energy due to factors such as dielectric relaxation after ionization, latent heat from the formation of the neutral cluster, and the lower ionization threshold of larger clusters. Confirmation of this picture follows from the observation that the rate for the loss of a second ammonia molecule decreases with increasing cluster size in accordance with expectations for clusters having little internal energy as a result of the loss of the first ammonia molecule.

Methanol Clusters

Similar studies to the foregoing were undertaken using 266 nm light to ionize $(CH_3OH)_n$ clusters (19). Several processes were observed, the first of which has a direct analogy to the ammonia studies. In particular, following ionization, methanol clusters undergo proton transfer with a concomitant loss of CH_3O. The exothermicity of the proton transfer reaction and subsequent relaxation of the alcohols around the protonated moiety also leads to excess internal energy and dissociation processes. The dissociation processes lead to unimolecular evaporation rates in a manner similar to the ammonia system and the rates are also in the order of 10^5 to 10^6 sec^{-1}.

In contrast to the ammonia system, a nonevaporative loss process was also observed, namely the elimination of H_2O from the protonated dimer.

This process, however, was not detected for the larger cluster ions. This observation is in direct correspondence with observations (20) of gas-phase ion-molecule reactions

$$CH_3OH_2^+ + CH_3OH \longrightarrow (CH_3)_2OH^+ + H_2O \qquad (4)$$

$$\xrightarrow{M} CH_3OH_2^+ \cdot CH_3OH \qquad (5)$$

and

$$CH_3OH_2^+ \cdot CH_3OH + CH_3OH \xrightarrow{\quad\times\quad} (CH_3)_2OH^+ \cdot CH_3OH + H_2O \qquad (6)$$

$$\xrightarrow{M} CH_3OH_2^+(CH_3OH)_2 \qquad (7)$$

The rate for the H_2O loss from $CH_3OH_2^+ \cdot CH_3OH$ has been measured to be 5.5×10^5 s^{-1}, in good agreement with the range of 10^5 to 10^6 s^{-1} estimated for the gas-phase ion-molecule process.

Based on measurements (21) and estimates for the thermodynamics of various dissociation channels in $H^+(CH_3OH)_3$ one would expect the loss of H_2O from this and larger methanol clusters to be at least as energetically favorable as that for the CH_3OH. Nibbering and coworkers (22) have suggested that an intermediate is important in the rearrangement process which leads to the loss of H_2O. Bowers and his colleagues (20) have suggested that the major product ion in the ion-molecule reaction is the formation of the symmetrical, proton-bound dimer. It may be that rearrangement from one conformation to the other is necessary for the water loss process and that this process can occur readily only in the protonated dimer, perhaps being impeded in higher order clusters due to structural hindrance. Rapid energy dissipation in larger clusters may also reduce the possibility of rearrangement. Hence, these data reflect the importance of structural rearrangements necessary for certain unimolecular processes to be operative, and thereby give further insight into differences which may be seen in the condensed phase compared to the gas for ion-molecule reactions.

Xenon Clusters

Recently, we conducted a multiphoton ionization study of xenon clusters (23). The relative cluster abundances of Xe_n^+ are virtually identical with mass spectra obtained by high-energy electron impact ionization (24). This observation indicates the importance of the energy liberated following ionization of the cluster in effecting the magic numbers and size distribution of weakly bound clusters as compared to the energy deposited to initiate ionization. From a study of the peak shapes of the Xe^+, Xe_2^+, and Xe_3^+ in the time-of-flight spectra under single field acceleration, the dissociation rates could be estimated. A tail was found toward longer times which is the result of rapid dissociation of larger clusters. The dissociation rates for Xe_5^+, Xe_4^+, and Xe_3^+ are found to lie in the neighborhood of 5×10^7 sec^{-1}. These are evidently the only species which are metastable in the time domain accessible by this method. The dissociation rates for larger clusters (ten or more atoms) have been reported to be much slower (25).

SHIFTS IN IONIZATION POTENTIAL WITH DEGREE OF AGGREGATION

The ionization potentials of p-xylene bound with argon ($PX \cdot Ar_n$) were determined through studies in which the energy of one photon was fixed at the L_b resonance and the wavelength of the second laser was scanned (11). Resonance-enhanced ionization with a single-color laser results in significant fragmentation due to the fact that the absorbed energy is substantially above the ionization threshold since the S_1 state lies more than halfway to the ionization continuum. Cluster fragmentation was found to be suppressed to a negligible amount in the two-color experiments, enabling a detailed investigation of the variation in ionization potential with degree of aggregation to be definitively established.

It is well known that the Stark effect leads to a shift in ionization potential when measured in an electric field and correction is necessary to account for shifts on the order of 50 cm^{-1}. The ionization potentials are found to vary with the square root of the electric field present in the region of ionization in accordance with expectations and findings of others (12). Extrapolation to zero field is readily accomplished in view of the linear dependence and the fact that various cluster systems display lines of identical slopes in these weakly perturbed rare-gas aggregates.

The shifts in ionization potential of p-xylene in the rare gas aggregates is shown in Figure 5 for clusters with one to six argon atoms. The shift in relative ionization potential is observed to display a broadly linear dependence on the number of argon atoms. The largest deviations from this trend are observed for the dimer and pentamer. The observed total shift of about 750 cm^{-1} for the hexamer is to be contrasted with the matrix isolated value which is about 6000 cm^{-1} for a similar molecule (benzene) in an argon matrix (13). Evidently, the "local environment" with which a molecule interacts is relatively large in such a matrix and the observed shift is far from the expected bulk value. This is in interesting contrast to the metal systems discussed in the next subsection. When corrected for ion image potential effects the ionization potential of aggregates comprised of only a few alkali metal atoms display nearly the bulk ionization potential of the polycrystalline metallic system (14).

Metastable Decay of Ar_3^+

In order to focus on the details of the metastable decay of a well-studied system, experiments were recently undertaken in our laboratory to investigate the decay of argon trimer to Ar_2^+ and Ar. The time window in the present work is much later and extends to a longer time than in previous studies; the metastable window starts at approximately 40 microseconds and extends to 120 microseconds. In these experiments argon clusters were produced in a supersonic expansion and ionization was accomplished through a crossed electron beam, the intersection position of which could be altered through a set of deflection plates. In this manner cluster ions, including the the one of interest, were produced either by ionization of preexisting neutral clusters by crossing the electron beam far downstream of the supersonic expansion or alternatively by ionizing the gas close to the nozzle exit and allowing cluster formation by a three-body association growth mechanism. It was found that the metastable content of argon trimer was much larger in the case where neutral clusters were ionized.

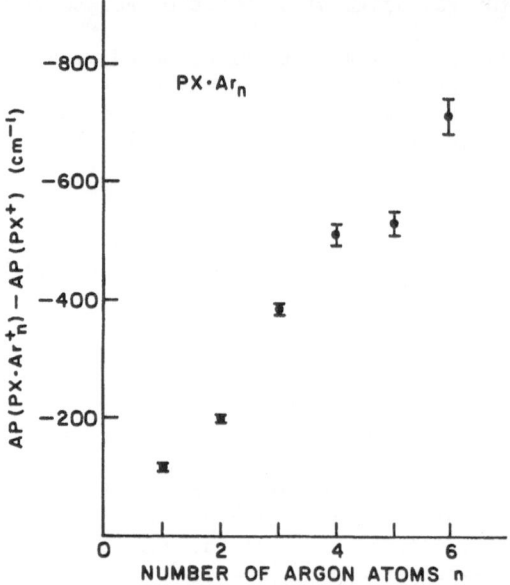

FIGURE 5. Relative appearance potentials. Field ionization of
PX·Ar$_n$ (n=0-6) in a 150V/cm dc field. AP(PX·Ar$_n^+$)
increases with the coordination number n.

The apparatus employed in these studies contained a Wien filter which
enabled the mass selection of the argon trimer ion, a differentially pumped
collision region separate from the ion source region, and a quadrupole mass
filter for product identification. Experiments made by increasing the
collision gas pressure revealed dissociation by a collision induced
process; linear extrapolation (thin collision approximation) to zero
pressure established a non-zero ordinate which is proportional to the
metastable decay component.

Very little metastable decay was observed in the case where the
cluster ions were born through the three-body growth mechanism, but decay
rates in the neighborhood of 10^2 to 10^3 were observed in those experiments
where the electron beam intersected the cluster beam more than about
one-half centimeter downstream from the cluster source. These are very
long metastable lifetimes and an interesting comparison is made with RRKM
calculations. These were made including correction for anharmonicity and
indicate that the minimum decay rate for Ar$_3^+$ → Ar$_2^+$ is 2×10^9 s^{-1} if
E_0 = 0.22 eV and 5×10^7 s^{-1} if E_0 = 0.75 eV. The first value of E_0 assumes
the enthalpy change for the dissociation reaction into Ar$_2^+$ and Ar while
the second value is obtained if it is assumed that the ground state of Ar$_3^+$
is linear and symmetric and that E_0 is the average bond energy.

The fraction of energized Ar$_3^+$ remaining after time t_a is given by
exp(-kt$_a$); in the present experiment metastability is observable 40 μsec
after ionization. Even assuming the minimum RRKM value for the unimolecular decay rate, the fraction of energized Ar$_3^+$ that could remain would be
essentially zero, i.e., all Ar$_3^+$ which could have decayed to Ar$_2^+$, even
assuming the minimum rate, would have done so before mass selection.
Hence, no metastable decay would have been observed if the unimolecular
dissociation into Ar$_2^+$ was occurring by the classical statistical
vibrational predissociation.

In view of the similarities in dissociation rates among many cluster systems and for a range of clusters in the argon system in particular, a common mechanism of dissociation is indicated. Obviously statistical predissociation is not possible and it is untenable to believe that so many different systems and cluster sizes could involve similar lifetimes as a result of crossing of electronic states. An alternative mechanism is that of tunneling through a rotational barrier. This would be a likely source even in clusters of heavy atoms or molecules due to the relative insensitivity of such tunneling to the mass of the escaping particle. This is a consequence of the way the one dimensional tunneling probability depends on the reduced mass μ and the effective potential V_{eff} (26),

$$P \propto \exp\left[\frac{-2}{h} \int_{r_1}^{r_2} [2\mu(V_{eff} - E)]^{1/2} dr \right] \qquad (8)$$

where r is the internuclear distance and E is the total internal energy of the ion. The effective potential energy is given by

$$V_{eff} = V(r) + \frac{h^2 L(L+1)}{2\mu r^2} \qquad (9)$$

where the second term on the right-hand side is the familiar rotational barrier, V(r) is the rotationless potential function, and L is the total angular momentum quantum number. At large r and L the second term becomes relatively more important and when it is dominant the reduced mass in the denominator tends to offset the reduced mass in Eq. (8). This offers a possibility for tunneling decay even for heavy masses.

In order to look at tunneling lifetimes for Ar_3^+, we have employed an effective potential of the form shown in Equation (9) to estimate the height and shape of barriers arising due to the angular momentum of the cluster ions. For the rotationless potential, V(r), the quasi-diatomic model is assumed, and the standard Morse potential function is used.

$$V(r) = D_o \{1 - \exp[- \alpha(r-r_e)]\}^2 \qquad (7)$$

All parameters employed in the calculation come from experiments and theoretical calculations (27-29). .

The actual distribution of angular momentum is unknown. For purposes of argument we assume a Boltzmann distribution of rotational levels

$$f_{rot} = \frac{1}{Q_R} (2L+1) \exp\left[\frac{-L(L+1)\theta_{rot}}{T} \right] \qquad (8)$$

where Q_r is the rotational partition function, and θ_{rot} is the characteristic rotational temperature. The Ar_3^+ ion has a very small rotational constant (0.03 cm^{-1}) (28), hence a very large range of L values are likely to be populated at very low temperatures. For example, even with rotational temperatures in the neighborhood of only 10 K, a typical value in molecular beams, a state with an L value of 30 would have a relative population of 0.5%. Rotational temperatures are likely to be even higher than this during the formation process. An additional important source of angular momentum of cluster ions also follows from the evaporative mechanism.

Calculations with a one-dimensional WKB approximation suggest that some metastable cluster ion decomposition can be observed in almost any time window for which measurements are made as found in ours and other studies. Since such a wide variety of clusters have been observed to have long lifetimes (milliseconds to microseconds) a very general explanation is indicated and we believe rotational tunneling must make a substantial contribution to these long lifetimes.

ACKNOWLEDGMENTS

Support by the the U. S. Army Research Office, Grant No. DAAG29-85-K-0215, and the U. S. Department of Energy, Grant No. DE-AC02-82ER60055, is gratefully acknowledged. The authors thank P.D. Dao, S. Morgan, J. J. Breen, E. E. Ferguson, C. R. Albertoni, R. Kuhn, and Z. Y. Chen for their assistance in performing the experiments.

REFERENCES

1. A. W. Castleman, Jr., in: Electronic and atomic collisions (J. Eichler, I.V. Hertel and N. Stolterfoht, Eds), Elsevier Science Publishers, Amsterdam, pp. 579-590 (1984).
2. A. W. Castleman, Jr. and R. G. Keesee, Chem. Rev. 86, 589 (1986).
3. A. W. Castleman, Jr. and R. G. Keesee, Ann. Rev. Phys. Chem. 37, 525 (1986).
4. A. W. Castleman, Jr. and R. G. Keesee, Accts. Chem. Res. 19, 413 (1986).
5. A. W. Castleman, Jr. and T. D. Mark, in: Gaseous Ion Chemistry/Mass Spectrometry (J. H. Futrell, Ed.) John Wiley and Sons, pp. 259-303 (1986).
6. M. F. Vernon, D. J. Krajnovich, H. S. Kwok, J. M. Lisy, Y. R. Shen, and Y. T. Lee, J. Chem. Phys. 77, 47 (1982); R. E. Miller, R. D. Watts and A. Ding, Chem. Phys. 83, 155 (1984); P. M. Dehmer and S. T. Pratt, J. Chem. Phys. 76, 843 (1982).
7. S. Leutwyler, U. Even and J. Jortner, J. Chem. Phys. 79, 5769 (1983).
8. S. Basu, Advan. Quantum Chem. 1, 145 (Eq. 46) (1964).
9. P. D. Dao, S. Morgan, and A. W. Castleman, Jr., Chem. Phys. Lett. 111, 38 (1984).
10. J. J. Breen, W.-B. Tzeng, S. Wei, and A. W. Castleman, Jr., to be published.
11. P. D. Dao, S. Morgan, and A. W. Castleman, Jr., Chem. Phys. Lett. 113, 219 (1985).
12. K. H. Fung, H. L. Selzle and E. W. Schlag, Z. Naturforsch 36a, 1257 (1981).
13. J. Jortner, in: Vacuum Ultraviolet Radiation Physics (E. E Koch, R. Haensel, and C. Kunz, Eds.) Pergamon Press, Oxford, p. 291 (1974).
14. K. I. Peterson, P. D. Dao, R. W. Farley, and A. W. Castleman, Jr. J. Chem. Phys. 80, 1780 (1984).
15. O. Echt, P. D. Dao, S. Morgan, and A. W. Castleman, Jr., J. Chem. Phys. 82, 4076 (1985). See also O. Echt, S. Morgan, P. D. Dao, R. J. Stanley, and A. W. Castleman, Jr., Ber. Bunsenges. Phys. Chem. 88, 217 (1984).
16. V. A. Mamyrin, V. I. Karataev, D. V. Shmikk, and V. A. Zauglin, Sov. Phys. JETP 37, 45 (1973).

17. J. L. Durant, D. M. Rider, S. L. Anderson, F. D. Proch, and R. N. Zare, J. Chem. Phys. 80, 1817 (1984).

18. H. Kuhlewind, U. Boesl, R. Weinkauf, H. J. Neusser, and E. W. Schlag, Laser Chem. 3, 3 (1983).

19. S. Morgan and A. W. Castleman, Jr., J. Am. Chem. Soc. 109, 2867 (1987).

20. L. M. Bass, R. D. Cates, M. F. Jarrold, N. J. Kirchner, and M. T. Bowers, J. Am. Chem. Soc. 105, 7024 (1983).

21. R. G. Keesee and A. W. Castleman, Jr., J. Phys. Chem. Ref. Data, 15, 1011 (1986).

22. J. C. Kleingeld and N. M. M. Nibbering, Org. Mass Spectrom. 17, 136 (1982).

23. O. Echt, M. Cook and A. W. Castleman, Jr., Chem. Phys. Lett. 135, 229 (1987).

24. O. Echt, K. Sattler, and E. Recknagel, Phys. Rev. Lett 47, 1121 (1981).

25. D. Kreisle, O. Echt, M. Knapp, and E. Recknagel, Phys. Rev. A 33, 768 (1986).

26. D. Bohm, Quantum Chemistry (New York: Prentice Hall, Inc., 1951) pp. 264-295.

27. D. L. Turner and D. C. Conway, J. Chem. Phys. 71(4), 1899 (1979).

28. H.-U. Bohmer and Sigrid D. Peyerimhoff, Z. Phys. D, Atoms, Molecules and Clusters 3, 195-205 (1986).

29. H. Eyring, J. Walter and G. E. Kimball, Quantum Chemistry (John Wiley and Sons, Inc.) p. 273 (1944).

PHOTODETACHMENT SPECTROSCOPY OF NEGATIVE CLUSTER IONS

Kit H. Bowen and Joseph G. Eaton
Department of Chemistry
The Johns Hopkins University
Baltimore, MD 21218, USA

ABSTRACT

Negative ion photoelectron spectra have been obtained for a variety of gas-phase cluster anions using visible photons. The negative cluster ions were generated by injecting relatively low energy electrons directly into the high density portion of an expanding supersonic jet. The spectra of $NO^-(N_2O)_{n=1,2}$, $H^-(NH_3)_{n=1,2}$, $NH_2^-(NH_3)_{n=1,2}$, $NO^-(Ar)_1$, $NO^-(Kr)_1$, $NO^-(Xe)_1$, $O_2^-(Ar)_1$, $(N_2O)_2^-$, and $(CS_2)_2^-$ reveal that they are simple ion-molecule complexes in which the excess negative charges are largely localized on sub-ions within the larger cluster anions. In addition to information on the bonding of cluster ions, the spectra also provide electron affinities and ion-solvent dissociation energies as a function of cluster size. The spectra of $(CO_2)_2^-$, $(SO_2)_2^-$, and $(NO)_2^-$, on the other hand, indicate that these species are more complicated cases, and that they are not well described as simple ion-molecule complexes. Also, in the case of NH_4^-, evidence is found not only for the ion-molecule complex, $H^-(NH_3)_1$, but also for a higher energy isomer of tetrahedral geometry. Other systems studied include negative cluster ions of water and alkali metal cluster anions. Even though H_2O^- is unstable, clusters of water are able to bind an electron to form $(H_2O)_n^-$. The spectra of $(H_2O)_{n=2,6,7,10-17,19}^-$ and $Ar(H_2O)_{n=2,6,7}^-$ provide the vertical detachment energies for these species. The alkali metals are the simplest of metals. The spectra of Na_{2-5}^-, K_{2-8}^-, Rb_{2-4}^-, and $Cs_{2,3}^-$ yield electron affinities as a function of cluster size as well as the electronic state splittings for neutral alkali metal clusters.

147

INTRODUCTION

The study of clusters affords an opportunity for a better understanding of the condensed phase at the microscopic level. Investigations
of cluster ions provide an avenue for exploring ion solvation on a
molecular scale. Negatively-charged cluster ions have analogs among
solvated anions and excess electron states in solution. Here, we report
the application of negative ion photoelectron (electron photodetachment)
spectroscopy to the study of size-selected negative cluster ions.

Our goals in studying the photodetachment of negative cluster ions
are (1) to explore their energetic properties as a function of cluster
size and (2) to develop a descriptive understanding of the bonding within
negative cluster ions. Important energetic properties include electron
affinities and step-wise ion-solvent dissociation (solvation) energies.
Clustering can be expected to stabilize the excess charge on negative
ions. One also expects that electron affinity values will increase
rapidly with cluster size for small clusters and then approach a limiting
value at some larger size as the mean number of solvent molecules interacting with the anion becomes constant. Step-wise solvation energies
should eventually decrease with increasing solvation numbers. The
experiments reported here map out both electron affinities and step-wise
solvation energies as a function of cluster size.

An important aspect of ion-neutral bonding concerns the distribution
of excess negative charge over the negative cluster ion. One can imagine
two extreme charge distribution categories where in one the excess charge
is localized on a single component of the cluster ion, and where in the
other there is a dispersal of the negative charge over part or all of the
cluster ion. The situation where the excess charge is localized on a
single component of the cluster ion is reminiscent of the usual notion of
a solvated anion in which a central negative ion is surrounded by a
sheath of neutral solvent molecules. There the central negative ion may
be thought of as remaining largely intact even though it is perturbed by
its solvents. In this case electrostatic interactions between the ion
and the solvent molecules presumably dominate the bonding. In other
cases, however, charge dispersal effects may also make significant
contributions to the bonding. These contributions may arise either
in the sense of covalency in ion-neutral bonds or in the sense of excess
electron delocalization via electron tunneling between energetically and
structurally equivalent sites within the cluster ion. In favorable cases
the photoelectron spectra of negative cluster ions can offer clues as to

the nature of the excess charge distribution in these species.

In this paper the results are organized into two parts;
(1) ion-molecule complexes where the excess electron is localized on one
component of the negative cluster ion, and (2) more complicated cluster
anions involving significant excess negative charge dispersal. The
spectra of $NO^-(N_2O)_{n=1,2}$, $H^-(NH_3)_{n=1,2}$, $NH_2^-(NH_3)_{n=1,2}$, $NO^-(Ar)_1$,
$NO^-(Kr)_1$, $NO^-(Xe)_1$, $O_2^-(Ar)_1$, $(N_2O)_2^-$, and $(CS_2)_2^-$ reveal that they are
simple ion-molecule complexes in which the excess negative charges are
largely localized on sub-ions within the larger cluster anions. The
spectra of $(CO_2)_2^-$, $(SO_2)_2^-$, and $(NO)_2^-$, on the other hand, indicate that
these species are more complicated cases, and that they are not well
described as simple ion-molecule complexes. Also, in the case of NH_4^-,
evidence is found not only for the ion-molecule complex, $H^-(NH_3)_1$, but
also for a higher energy isomer of tetrahedral geometry. Other systems
studied include negative cluster ions of water and alkali metal cluster
anions. The spectra of $(H_2O)_{n=2,6,7,10-17,19}^-$ and $Ar(H_2O)_{n=2,6,7}^-$ provide
the vertical detachment energies for these species. The spectra of
Na_{2-5}^-, K_{2-8}^-, Rb_{2-4}^-, and $Cs_{2,3}^-$ yield electron affinities as a function of
cluster size as well as the electronic state splittings for neutral
alkali metal clusters.

EXPERIMENTAL

Negative ion photoelectron spectroscopy is conducted by crossing a
mass-selected beam of negative ions with a fixed-frequency photon beam
and energy analyzing the resultant photodetached electrons. Subtraction
of the center-of-mass electron kinetic energy of an observed spectral
feature from the photon energy gives the transition energy (the electron
binding energy) from an occupied level in the negative ion to an
energetically accessible level in the corresponding neutral. Our negative ion photoelectron spectrometer has been described previously.[1]
It is comprised of three main component systems. These are (a) the beam
line along which negative ions are formed, transported, and mass-
selected, (b) the high-power argon ion laser operated intracavity in the
ion-photon interaction region, and (c) the doubly magnetically shielded,
high resolution hemispherical electron energy analyzer which is located
below the plane of the crossed ion and photon beams. The mass selector
is a cooled Colutron 600B Wien filter. This is a E x B velocity filter
with electrostatic shims which compensate for the focusing effects of
simple Wien filters. Mass selection allows us to "purify" our starting

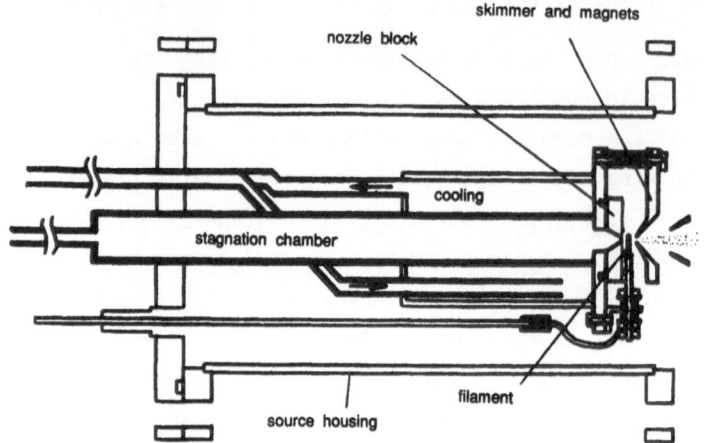

FIG. 1 Diagram of the supersonic expansion-
ion source used in most of these experiments.

sample of negative ions before photodetachment and thus to obtain
interference-free photoelectron spectra of specific negative ions. This
capability is well suited for studies of homologous series of ions such
as negative cluster ions.

Beams of negative cluster ions were generated in a supersonic
expansion-ion source similar in spirit to that developed by Haberland.[2]
Figure 1 presents a schematic of our version of this source. In this
source a biased filament located just outside the nozzle orifice injects
relatively low energy electrons into the supersonic expansion. Permanent
magnets placed near the expansion jet were found to enhance the
production of negative ions.

RESULTS AND DISCUSSION

**Ion-Molecule Complexes: Cluster anions with localized excess negative
charges**

$NO^-(N_2O)_{n=1,2}$ The photoelectron (photodetachment) spectra of the gas-
phase negative cluster ions, $NO^-(N_2O)_1$ and $NO^-(N_2O)_2$ were recorded using
2.540 eV photons.[3] Both spectra exhibit structured photoelectron
spectral patterns which strongly resemble that of free NO^-, but which
are shifted to successively lower electron kinetic energies with their
individual peaks broadened (see Figure 2). Each of these spectra is
interpreted in terms of a largely intact NO^- sub-ion which is solvated

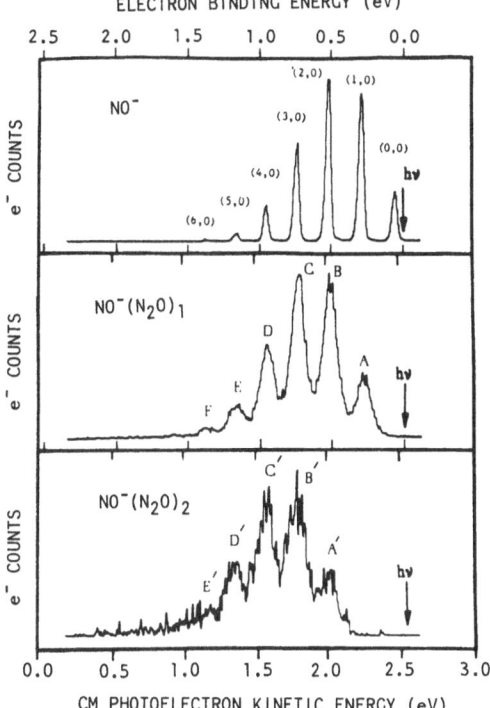

FIG. 2   The negative ion photoelectron spectra of
NO^-, $NO^-(N_2O)$, and $NO^-(N_2O)_2$ all presented
on the same center-of-mass electron kinetic
energy and electron binding energy scales.

and stabilized by nitrous oxide. For both $NO^-(N_2O)_1$ and $NO^-(N_2O)_2$,
the ion-solvent dissociation energies for the loss of single N_2O solvent
molecules were determined from origin peak shifts to be ~0.2 eV.
Electron affinities for $NO(N_2O)_1$ and $NO(N_2O)_2$ were determined to be
0.258 ± 0.009 eV and 0.513 ± 0.022 eV, respectively. The localization of
the cluster ion's excess negative charge onto its nitric oxide rather
than its nitrous oxide subunit was interpreted in terms of kinetic
factors and a possible barrier between the two forms of the solvated ion.

$H^-(NH_3)_{n=1,2}$ The gas-phase $H^-(NH_3)_1$ ion was first observed by
Nibbering[4] in a FT-ICR spectrometer. Theoretical calculations by Rosmus,
by Squires, by Schleyer, by Cremer, and by Ortiz all agree that the
hydride ion is bound at a relatively long distance to only one of
ammonia's hydrogens in the most stable configuration of the $H^-(NH_3)_1$ ion-
dipole complex, and that $H^-(NH_3)_1$ is more stable than $NH_2^-(H_2)_1$. These

calculations found global minima for $H^-(NH_3)_1$ in which the H^- ion lies almost in line with a N–H bond in ammonia. Most of them also found values for the dissociation energy of $H^-(NH_3)_1$ into $H^- + NH_3$ that ranged around about a third of an eV.

In this work, $H^-(NH_3)_{n=1,2}$ ions were generated from ammonia in a supersonic expansion-ion source and photodetached with 2.540 eV photons.[5,6] No homologous series in $NH_2^-(H_2)_n$ was observed. The photoelectron spectra of $H^-(NH_3)_1$ and $H^-(NH_3)_2$ are both dominated by large peaks which we have designated as peaks A and A', respectively, in Fig. 3. The $H^-(NH_3)_1$ spectrum also exhibits a smaller peak on the low electron kinetic energy side of peak A which we have labelled peak B. The shoulder on the low electron kinetic energy side of peak A' in the $H^-(NH_3)_2$ spectrum is marked in Fig. 3 as peak B'. A much smaller third peak (peak C) also exists in the $H^-(NH_3)_1$ spectrum. This will be discussed in a separate section below.

Our interpretation of peak A in the $H^-(NH_3)_1$ spectrum and of peak A' in the $H^-(NH_3)_2$ spectrum is that they contain the origins of their respective photodetachment transitions. Both peaks are due to the photodetachment of solvated hydride ion "chromophores" within $H^-(NH_3)_1$ and $H^-(NH_3)_2$. This results in the main features (peaks A and A') of the $H^-(NH_3)_1$ and $H^-(NH_3)_2$ photoelectron spectra resembling the photoelectron spectrum of free H^- (a single peak) except for being broadened and shifted to lower electron kinetic energies due to the stabilizing effect of solvation.

The electron binding energy of peak A is 1.11 eV. This is interpreted as an upper limit to the energy difference between the lower vibrational states of $H^-(NH_3)_1$ and the $H + NH_3 + e^-$ dissociation asymptote. This value is thus a reasonably close approximation to the dissociative detachment energy of $H^-(NH_3)_1$ [and to the electron affinity of $H(NH_3)$]. The electron binding energy of peak A' is 1.46 eV and it is similarly interpreted. An upper limit to the ion-solvent dissociation energy of $H^-(NH_3)_1$ into H^- and NH_3 (the gas-phase solvation energy) is given by subtracting the electron affinity of H (0.754 eV) from the value we have obtained for the upper limit to the dissociative detachment energy of $H^-(NH_3)_1$ ie. the origin peak shift. This value is 0.36 eV, and it is in very good agreement with theoretical calculations. Likewise, the ion-solvent dissociation energy of $H^-(NH_3)_2$ into $H^-(NH_3)_1$ and NH_3 is given by subtracting the dissociative detachment energy of $H^-(NH_3)_1$ from

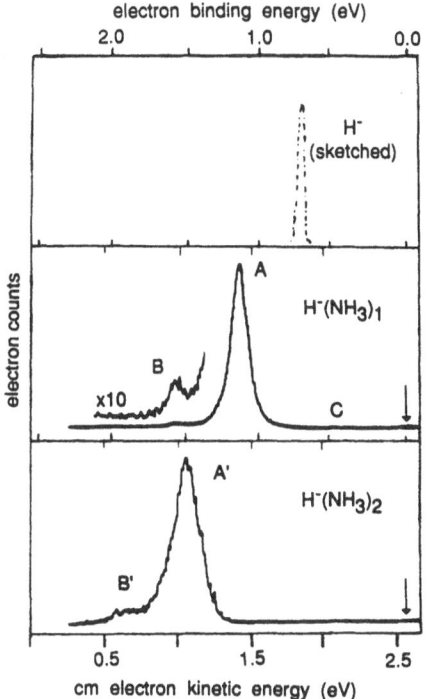

FIG. 3 The photoelectron spectra of
$H^-(NH_3)_1$ and $H^-(NH_3)_2$. The
spectrum of $H^-(NH_3)_1$ also shows
a x30 magnified trace.

that of $H^-(NH_3)_2$ <u>ie.</u> the shift between these origin peaks. This value is
0.35 eV.

Our interpretation of peak B is that it is primarily due to the
excitation of a stretching mode (or modes) in the ammonia solvent during
photodetachment. The small Franck-Condon factor observed suggests that
the ammonia solvent is only slightly distorted by its complexation with
H^-. The center of peak B is separated from that of peak A by 3480 ±
60 cm^{-1} in the $H^-(NH_3)_1$ spectrum, and in the $D^-(ND_3)_1$ spectrum, the
separation between the centers of peaks A and B is 2470 ± 80 cm^{-1}. These
peak separations are close to the observed stretching frequencies of NH_3
and ND_3, and they therefore support our interpretation. The B' shoulder
in the spectrum of $H^-(NH_3)_2$ is probably due to analogous transitions.
Taken together, the foregoing provides spectroscopic evidence for cluster
ions consisting of <u>intact</u> hydride ions which are perturbed and solvated
by ammonia.

$NH_2^-(NH_3)_{n=1,2}$ Figure 4 presents the photoelectron spectra of NH_2^-, $NH_2^-(NH_3)_1$, $NH_2^-(NH_3)_2$ all plotted on a common center-of-mass electron kinetic energy scale.[7] The well-known photoelectron spectrum of NH_2^- is dominated by a single peak, and it is presented in Figure 4 for comparative purposes. The spectra of the clustered ions are also dominated by large peaks (A and A'). These shift to lower electron kinetic energies with increasing solvation and are broadened. Our interpretation of peaks A and A' is that they both contain the origins of their respective photodetachment transitions. Both peaks arise due to the photodetachment of solvated amide ion "chromophores" within the $NH_2^-(NH_3)_1$ and $NH_2^-(NH_3)_2$ cluster ions. The spectral shifts are a consequence of the stablization of the NH_2^- sub-ion due to its interactions with the NH_3 "solvent" molecule(s) in the cluster ions.

The center of peak A in the $NH_2^-(NH_3)_1$ spectrum corresponds to an electron binding energy of 1.30 eV. This is interpreted as an upper limit to the energy difference between the lower vibrational states of $NH_2^-(NH_3)_1$ and the $NH_2 + NH_3 + e^-$ dissociation asymptote. This value is a close approximation to the dissociative detachment energy of $NH_2^-(NH_3)_1$, and to the electron affinity. The center of peak A' in the $NH_2^-(NH_3)_2$ spectrum corresponds to an electron binding energy of 1.78 eV, and it is similarly interpreted. An upper limit to the ion-solvent dissociation energy of $NH_2^-(NH_3)_1$ into NH_2^- and NH_3 (the gas-phase solvation energy) is given by the magnitude of the shift between the centers of the peak in the NH_2^- spectrum and peak A in the $NH_2^-(NH_3)_1$ spectrum. This is equivalent to subtracting our measured value for the electron affinity of NH_2 from the value of the upper limit to the dissociative detachment energy of $NH_2^-(NH_3)_1$. This value is 0.52 eV, and it is in good agreement with theoretical calculations by Squires. Likewise, the ion-solvent dissociation energy of $NH_2^-(NH_3)_2$ into $NH_2^-(NH_3)_1$ and NH_3 is given by the shift between the centers of peaks A and A' in the cluster ion spectra. This value is 0.48 eV, indicating an approximately equal stabilization of NH_2^- by both the first and the second NH_3 "solvent" molecules.

A less intense peak, designated as peak B, appears on the low electron kinetic energy side of peak A in the photoelectron spectrum of $NH_2^-(NH_3)_1$. The separation between the centers of peaks A and B is close to the observed values of the stretching frequencies of ammonia. Our interpretation of this peak is that it is primarily due to the excitation of a stretching mode (or modes) in the NH_3 solvent during photodetachment. The photoelectron spectrum of $ND_2^-(ND_3)_1$ offers further

154

support of this interpretation. Peak A occurs at essentially the same location in the spectra of $ND_2^-(ND_3)_1$ and $NH_2^-(NH_3)_1$. The spacing between peaks A and B in the spectrum of $ND_2^-(ND_3)_1$, however, has decreased to an energy which is equal to the values of the observed stretching frequencies of ND_3.

In previous photoelectron experiments on negative cluster ions, we have also studied the species, $H^-(NH_3)_1$ and $H^-(NH_3)_2$. Qualitatively the photoelectron spectra of $NH_2^-(NH_3)_{n=1,2}$ and $H^-(NH_3)_{n=1,2}$ are rather similar. Both sets of spectra exhibit large peaks (A and A' peaks). Both $H^-(NH_3)_1$ and $NH_2^-(NH_3)_1$ spectra have B peaks to the low electron energy side of their A peaks which are separated from them by energies corresponding to that of ammonia's stretching frequencies. Quantitatively, however, the $NH_2^-(NH_3)_{n=1,2}$ spectra exhibit larger shifts and more broadening than the $H^-(NH_3)_{n=1,2}$ spectra. We have found that the first and second ion-solvent dissociation energies for $NH_2^-(NH_3)_1$ and $NH_2^-(NH_3)_2$ are both ~0.5 eV, while those for $H^-(NH_3)_1$ and $H^-(NH_3)_2$ are both ~0.35 eV. Clearly, the interaction of NH_2^- with ammonia is stronger than that of H^- with ammonia. Calculations by Squires find that $H^-(NH_3)_1$ and $NH_2^-(NH_3)_1$ have similar gross structures. Using flowing afterglow techniques to study the $NH_2^- + H_2 \rightarrow H^- + NH_3$ reaction, Bohme has shown that NH_2^- is a stronger base than H^- in the gas-phase. It thus seems likely that the higher ion-solvent dissociation energy of $NH_2^-(NH_3)_1$ relative to that of $H^-(NH_3)_1$ is a consequence of NH_2^- being a stronger base than H^-.

The clustering of solvent molecules around a bare gas-phase anion stabilizes the excess negative charge on the ion, and this results in a decrease in the gas-phase basicity. It is often the case, in fact, that the ordering of basicities in the gas-phase is the reverse of their ordering in solution. Using our results to calculate the relative basicites of NH_2^- vs H^-, $NH_2^-(NH_3)_1$ vs $H^-(NH_3)_1$, and $NH_2^-(NH_3)_2$ vs $H^-(NH_3)_2$ shows that such a reversal in the ordering of basicities in these systems occurs by the addition of a second ammonia solvent to NH_2^- and to H^-. This illustrates the role that cluster ions can play in illuminating the regime between the gaseous and the condensed (solution) phase.

Both $NH_2^-(NH_3)_1$ and $H^-(NH_3)_1$ spectra show an A-B peak spacing that is indicative of an ammonia stretching frequency. The relative intensity of the B peak in each of these spectra is a measure of the degree to which the ammonia "solvent" molecule is distorted due to its complexation with the anion. This peak is larger in the $NH_2^-(NH_3)_1$ spectrum than in the

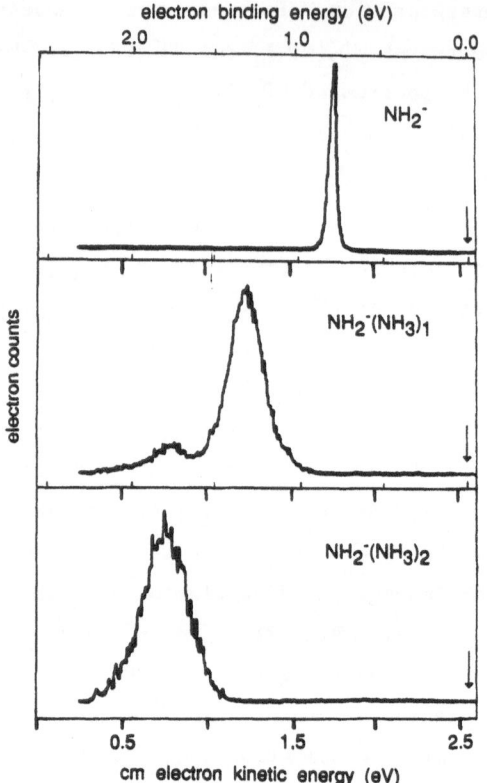

Fig. 4 The photoelectron spectra of NH_2^-, $NH_2^-(NH_3)_1$, and $NH_2^-(NH_3)_2$ all recorded with 2.540 eV photons.

$H^-(NH_3)_1$ spectrum, and this is consistent with the stronger interaction implied by the larger ion-solvent bond dissociation energy found for $NH_2^-(NH_3)_1$. These observations are also consistent with calculations by Squires. He finds that the ammonia N-H bond which interacts with the anion is more elongated in $NH_2^-(NH_3)_1$ than in $H^-(NH_3)_1$.

$NO^-(Ar)_1$, $NO^-(Kr)$, and $NO^-(Xe)_1$ The photoelectron spectra of $NO^-(Ar)_1$, $NO^-(Kr)_1$, and $NO^-(Xe)_1$ were recorded with 2.409 eV photons.[8] All of these rare gas (Rg) negative cluster ion spectra exhibit structured spectral patterns which strongly resemble that obtained for free NO^-, but which are shifted to lower electron energies with their individual peaks broadened (see Figure 5). Each of these spectra is interpreted in terms of a largely intact NO^- sub-ion which is solvated and stabilized by its rare gas solvent atom. The ion-solvent dissociation energy for a given $NO^-(Rg)_1$ cluster ion dissociating into NO^- and Rg is approximately given by the energy difference between the origin peak of the free NO^- spectrum and the origin peak of a given $NO^-(Rg)_1$ spectrum. The values of these shifts were found to be: 0.058 ± 0.011 eV, 0.099 ± 0.018 eV, 0.161 ± 0.024 eV for the argon, krypton, and xenon complexes, respectively. A plot of these energy shifts vs. the polarizabilities of the rare gas atoms gave a straight line. Values for the electron affinities of these complexes were found to be: 0.095 eV, 0.136 eV, and 0.204 eV for the argon, krypton, and xenon complexes, respectively.

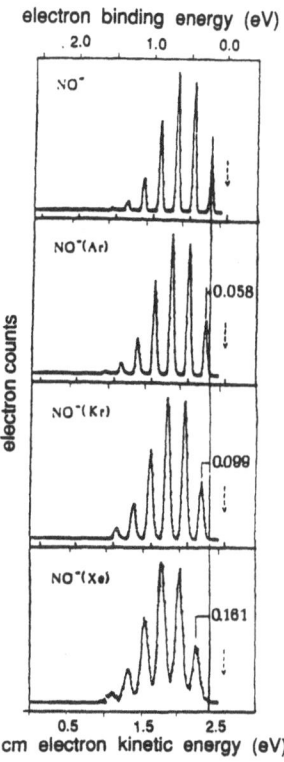

Fig. 5 The photoelectron spectra of NO^-, $NO^-(Ar)_1$, $NO^-(Kr)_1$, and $NO^-(Xe)_1$

$O_2^-(Ar)_1$ The photoelectron spectrum of $O_2^-(Ar)_1$ exhibits the highly structured photoelectron spectral pattern of free O_2^- shifted to lower electron kinetic energy by ~70 meV, the ion-atom dissociation energy of

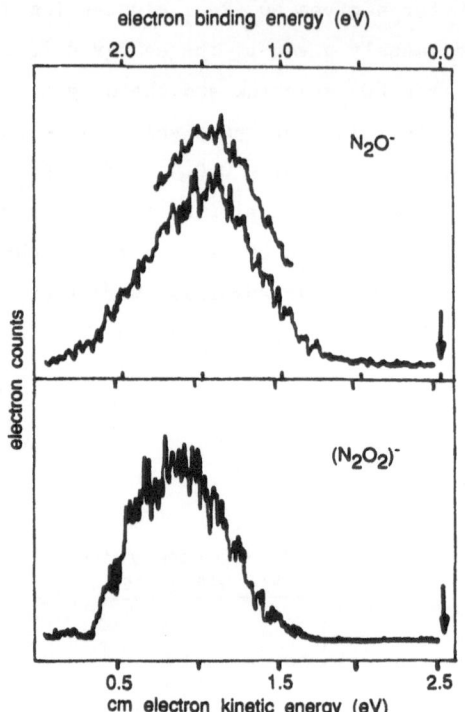

Fig. 6 The photoelectron spectrum of (a) N_2O^- and
(b) $(N_2O)_2^-$ presented in terms of center-of-mass
electron kinetic energies and electron binding energies.
Both spectra were recorded with 2.540 eV photons.
The limited range scan above the full spectrum of N_2O^-
has 2.5 times more signal.

this anion-complex. We are currently studying this system to see if an
$O_2^-(Ar)_1$ cluster ion with a vibrationally hot O_2^- component can survive the
$\sim 10^{-5}$ sec flight time between the ion source and the laser interaction
region without undergoing vibrational predissociation. If hot bands are
observed in the $O_2^-(Ar)_1$ spectrum, it will indicate that the vibrational
predissociation lifetime is $> 10^{-5}$ sec.

N_2O^- and $(N_2O)_2^-$ The photoelectron spectra of N_2O^- and $(N_2O)_2^-$ were
recorded with 2.540 eV photons[9] (see Figure 6). Because of the large
geometrical difference between N_2O (linear) and N_2O^- (bent), there is
little Franck-Condon overlap between the lowest-lying levels in the ion
and its neutral. We interpret the photoelectron spectrum of N_2O^- as

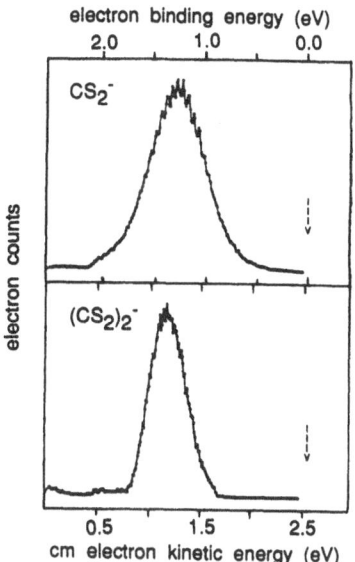

electron binding energy (eV)

CS_2^-

$(CS_2)_2^-$

cm electron kinetic energy (eV)

Fig. 7 The photoelectron spectra of the negative
ions of carbon disulfide monomer and dimer are shown.
The spectra were taken using 2.540 eV photons.

being largely due to an unresolved progression in the bending mode of
N_2O. The electron binding energy corresponding to the maximum in the
N_2O^- spectrum is ~1.5 eV and is a reasonable measure of the vertical
detachment energy of N_2O^-. The spectrum of $(N_2O)_2^-$ provides information
on the distribution of excess charge within the negative dimer ion. The
maximum in the $(N_2O)_2^-$ spectrum is shifted by 0.19 eV to lower electron
kinetic energy relative to the maximum in the N_2O^- spectrum. We
interpret the $(N_2O)_2^-$ spectrum as arising from the photodetachment of an
ionic species which is best described as a bent N_2O^- solvated by a
neutral linear N_2O, ie. as $N_2O^-(N_2O)_1$ and the ~ 0.2 eV shift between the
N_2O^- and the $(N_2O)_2^-$ spectra as a measure of the dimer anion's
dissociation energy into N_2O^- and N_2O.

$(CS_2)_2^-$ In addition to nitrous oxide, carbon disulfide and carbon
dioxide are also linear triatomic molecules with bent anions. The
photoelectron spectrum of $(CS_2)_2^-$ bears a substantial resemblance to that
of the monomeric ion, CS_2^- (ref. 10), except for the former being shifted
to lower electron energies relative to the latter (see Figure 7). Our
interpretation of the $(CS_2)_2^-$ spectrum leads to the conclusion that $(CS_2)_2^-$
is composed of a largely intact CS_2^- ion which is "solvated" by a neutral

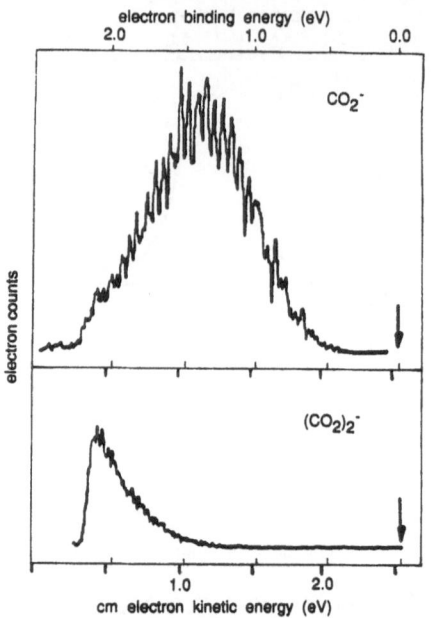

electron binding energy (eV)
2.0   1.0   0.0

electron counts

CO_2^-

$(CO_2)_2^-$

1.0   2.0

cm electron kinetic energy (eV)

Fig. 8 The photoelectron spectra of (a) CO_2^- and (b) $(CO_2)_2^-$.

CS_2, <u>ie</u>. the dimer ion is an ion-neutral complex. We have determined the
ion-solvent dissociation energy for the dimer ion dissociating into CS_2^-
and CS_2 to be 0.176 ± 0.025 eV.

More Complicated Cases: Cluster Anions with Excess Charge Dispersal

$(CO_2)_{n=1,2}^-$ The photoelectron spectra[11] of CO_2^- and $(CO_2)_2^-$, which were
both recorded with 2.540 eV photons, are presented in Figure 8. Unlike
N_2O and CS_2, CO_2 has a negative adiabatic electron affinity. The
negative ion, CO_2^-, is thus metastable and has an autodetachment lifetime
of ~10^{-4} sec. Because of the relative energies and large geometrical
differences between CO_2 and CO_2^-, there should be no Franck-Condon overlap
between the lowest lying levels of the ion and its neutral. We interpret
the photoelectron spectrum of CO_2^- as being largely due to a progression
in the bending mode of CO_2. The structure near the maximum of the CO_2^-
spectrum is real and the peak spacings probably correspond to the energy
differences between anharmonic bending levels in CO_2. The electron
binding energy corresponding to the maximum in our CO_2^- spectrum can be
associated with its vertical detachment energy. This value, 1.4 eV, is
in good agreement with Jordan's calculated value for the vertical
detachment energy of CO_2^-. The photoelectron spectrum of $(CO_2)_2^-$ is
presented in Fig. 8b. The downward turn on the low electron kinetic
energy side of this spectrum is an experimental artifact due to the rapid

160

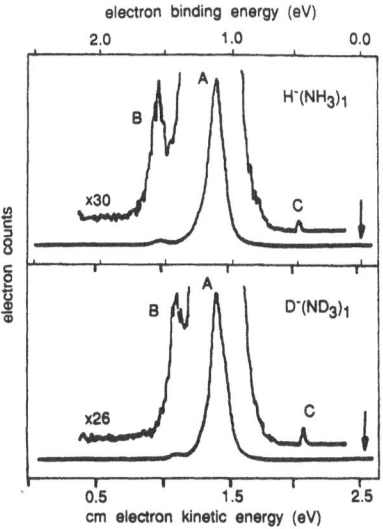

electron binding energy (eV)

2.0   1.0   0.0

$H^-(NH_3)_1$

B

A

C

x30

$D^-(ND_3)_1$

B

A

C

x26

electron counts

0.5   1.5   2.5

cm electron kinetic energy (eV)

Fig. 9 The photoelectron spectra of $H^-(NH_3)_1$, $D^-(ND_3)_1$, $NH_4^-(Td)$, and $ND_4^-(Td)$.

and unavoidable decrease in the transmission functions of electron energy analyzers at low electron kinetic energies. Spectra taken with 2.707 eV photons show that the photodetachment cross section is still increasing at the false maximum in the 2.540 eV spectrum. Thus, with visible photons the spectrum of $(CO_2)_2^-$ exhibits only the lower energy photodetachment transitions of $(CO_2)_2^-$. Assuming that there is a spectral maximum in the $(CO_2)_2^-$ spectrum, it must occur at an electron binding energy that is >2.4 eV, _ie_. the maximum in the $(CO_2)_2^-$ spectrum is shifted to lower electron kinetic energies by >1 eV with respect to the maximum in the CO_2^- spectrum. This suggests that there is a substantial difference between $(CO_2)_2^-$ and $(N_2O)_2^-$ and $(CS_2)_2^-$. Recent calculations by Jordan on the various possible structures for $(CO_2)_2^-$ shed substantial light on this problem. Upon reexamining the relative energies of the symmetrical D_{2d} form of $C_2O_4^-$ and the asymmetrical ion-molecule complex, Jordan found the former to be more stable by ~0.2 eV. More importantly, however, he also found that these two forms of the anion gave very different vertical detachment energies, and that the calculated VDE for the D_{2d} form is in agreement with our measurements. Thus, it appears that even though $(N_2O)_2^-$ and $(CS_2)_2^-$ are ion-molecule complexes, $(CO_2)_2^-$ is not. It is probably better described as $C_2O_4^-$.

NH_4^- As mentioned earlier, the photoelectron spectrum of NH_4^- is dominated by two peaks (A and B) which arise due to the photodetachment of electrons from the ion-molecule complex, $H^-(NH_3)_1$. In addition, however, there is also a much smaller third peak (C) in the spectrum, and this feature provides the first experimental evidence for a higher energy isomer of NH_4^- of tetrahedral geometry.[12] Fig. 9 shows magnified traces of peak C in both the $H^-(NH_3)_1$ and the $D^-(ND_3)_1$ spectra. Since negative ions in this experiment are carefully mass-selected before photodetachment, the existence of peak C in both of these spectra is good evidence that it is not due to an "impurity" ion. Peak C occurs at too high of an electron kinetic energy to be due to NH_2^- (or to OH^-). Since clustering is expected to stabilize the excess negative charge on an anion and to shift spectral features toward lower (rather than higher) electron energies, it is also unlikely that peak C is due to the presence of small amounts of $NH_2^-(H_2)_1$. The photodissociation of $H^-(NH_3)_1$ into H^- + NH_3 followed by the photodetachment of electrons from the nascent H^- is a two-step process which is energetically-accessible with 2.5 eV photons. If this process were to occur, however, it would result in electrons with kinetic energies substantially lower than that of peak C (possible kinematic effects having been carefully considered). In addition, the laser power dependence of peak C's intensity is linear, and while not proof in itself, this is consistent with a single photon process. Further insight into the possible origin of peak C derives from its intensity variation with source conditions and from its behavior in the spectrum of the deuterated cluster ion, $D^-(ND_3)_1$. While the relative intensities of peaks A and B are essentially constant as source conditions are varied, the relative intensity of peak C changes substantially from day to day. Such intensity variations are indicative of photodetachment transitions which originate from an excited state of the ion and they are often associated with vibrationally-excited negative ion states. Hot band peaks arising from such transitions, however, should shift with deuteration, and peak C does not. Peak C behaves as if it arises from the photodetachment of an electronically higher energy form of the negative ion. It seems unlikely that peak C is due to the photodetachment of an electronically-excited state of $H^-(NH_3)_1$. Our observations are consistent with it being due to the photodetachment of a higher energy isomer of an ion with molecular formula, NH_4^-. The width of peak C is rather narrow, much narrower than peaks A and B. This implies that the structure of the ion being photodetached and the equilibrium structure of its corresponding neutral are rather similar. Neutral NH_4 is known to have a tetrahedral configuration. This suggests that the form of NH_4^- that gives rise to peak C is also of tetrahedral geometry. Also, the united atom for NH_4 is Na. The electron affinity of Na is ~0.5 eV. The electron binding energy of the species that gives rise to peak C is ~0.5 eV. In addition, calculations by Schleyer, by Cremer, and by

Ortiz all find a higher energy isomer of NH_4^- of tetrahedral geometry. The energy of $NH_4^-(T_d)$ above the global minima of $H^-(NH_3)_1$ is also consistent with our spectra. It seems likely that NH_4^- (T_d) should be envisioned as an NH_4^+ core with two Rydberg-like electrons around it. Thus, NH_4^+, NH_4, and NH_4^- can all exist in tetrahedral forms. Moreover, they are all really the same thing ie. NH_4^+ cores with 0,1, and 2 loose electrons associated with them.

$(SO_2)_2^-$ and $(NO)_2^-$ The photoelectron spectra of $(SO_2)_2^-$ and of $(NO)_2^-$ do not show the shifted "fingerprint" spectral patterns of SO_2^- and of NO^- that one might expect of **simple**, localized excess charge ion-molecule complexes[13,14] ie. of $SO_2^-(SO_2)_1$ and of $NO^-(NO)$. The component parts of these dimer ions may resonantly share the excess electron. This appears likely in the case of $(SO_2)_2^-$ where the observed spectral shift between the origin peak of the SO_2^- spectrum and the suspected origin in the $(SO_2)_2^-$ spectrum implies an ion-solvent dissociation energy that is reasonable for an ion-molecule complex. The spectrum of $(NO)_2^-$, on the other hand, implies a lower limit to its electron affinity of ~2.1 eV. This implies that $(NO)_2^-$ enjoys a high degree of electron dispersal, and that it may be better described as the unclustered ion, $N_2O_2^-$. Studies by Johnson[15] at higher photon energies indicate the existence of other isomers of $N_2O_2^-$ as well.

$Na_{n=2-5}^-$, $K_{n=2-8}^-$, $Rb_{n=2-4}^-$, and $Cs_{n=2,3}^-$ The study of metal clusters provides an avenue for exploring the variation in the electronic properties of metals in the transition size regime between atoms and the solid state bulk. For metals, properties such as ionization potentials and electron affinities typically vary in magnitude by several eV from their atomic to their bulk (work function) values. Presumably, clusters of intermediate size have electronic properties of intermediate values. For phenomena where the properties of matter in the regime of small sizes are important, eg. surface reactivity and thin films, these variations in electronic properties can have pivotal effects. In principle, the study of the electronic properties of metal clusters as a function of cluster size could allow us to observe the evolution of the electronic states of metals from those of their atoms to those of band theory.

The alkali metals are the simplest of metals. We have recently recorded the photoelectron spectra of $Na_{n=2-5}^-$, $K_{n=2-8}^-$, $Rb_{n=2-4}^-$, and $Cs_{n=2,3}^-$ using 2.540 eV photons (see Fig. 10).[16] These highly structured spectra map out both the electron affinities vs cluster size for those cluster anions studied thus far and the electronic state splittings of

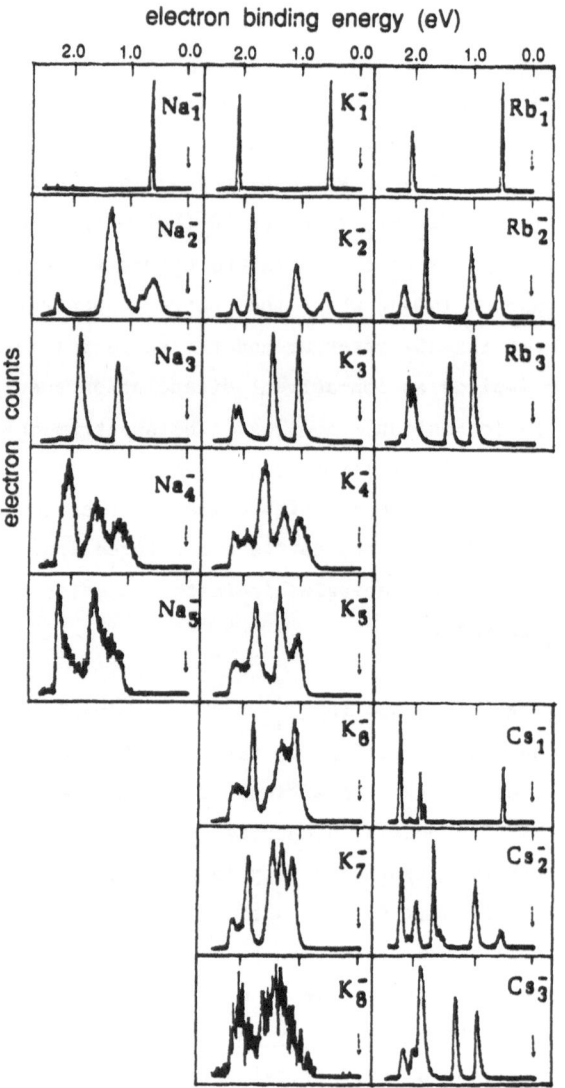

Fig. 10   The Photoelectron Spectra of
Alkali Metal Cluster Anions

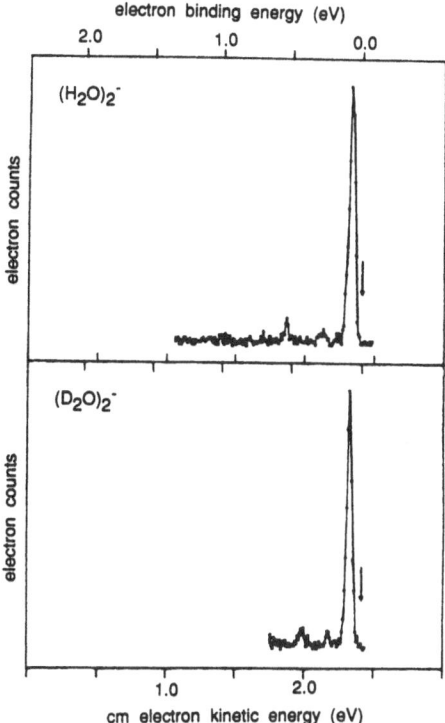

Fig. 11 The photoelectron spectra of $(H_2O)_2^-$ and $(D_2O)_2^-$

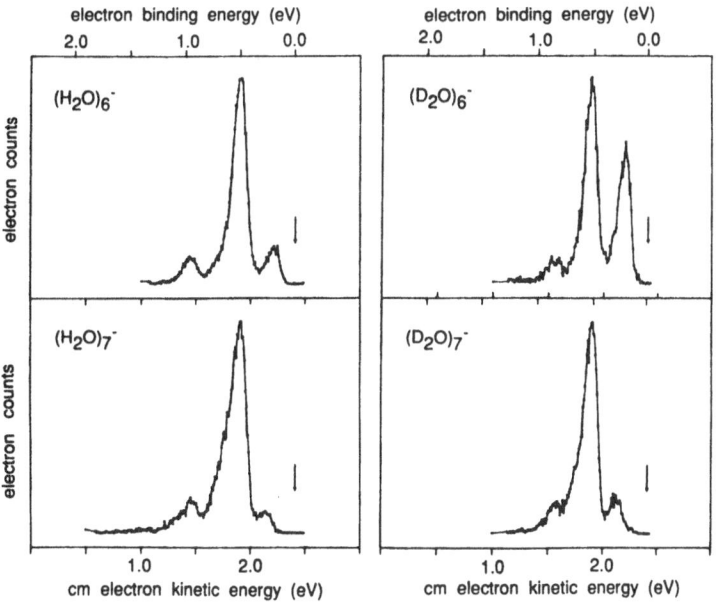

Fig. 12 The photoelectron spectra of $(H_2O)_{6,7}^-$ and $(D_2O)_{6,7}^-$

their corresponding <u>neutral</u> clusters (at the geometry of their cluster anions) vs cluster size. The dimer anion spectra have been completely assigned. These provide adiabatic electron affinities, vertical detachment energies, dimer anion dissociation energies, neutral dimer electronic state spacings, and bond lengths for the various excited electronic states of the neutral dimers. Thus far, we have obtained our most complete set of data on potassium cluster anions. Potassium ($[Ar]4s^1$) should be electronically analogous to copper ($[Ar]3d^{10}4s^1$). A compa.ison of the electron affinity vs cluster size trends for potassium clusters with those for copper clusters (measured by Lineberger[17] and by Smalley[18]) of the same size, shows quantitative differences (copper has a substantially larger work function) yet strikingly similar <u>qualitative</u> trends. Most of the structure in these spectra comes about due to the electronic states of the neutral clusters. One can see qualitative similarities between the UV photodetachment spectra of copper cluster anions and our visible spectra of potassium cluster anions. This correlation with the UV experiments is reasonable since the electronic states of neutral copper clusters might be expected to be more widely spaced in energy than those of neutral potassium clusters.

$(H_2O)^-_{n=2,6,7,10-17,19}$ and $Ar(H_2O)^-_{n=2,6,7}$ Over the years, it has often been suggested that gas-phase $(H_2O)^-_n$ cluster ions ought to exist, and that they might be gas-phase counterparts to condensed phase solvated (hydrated) electrons. A few years ago, these entities were observed for the first time in the gas-phase by Haberland.

Recently, we generated $(H_2O)^-_{n=11-21}$ from neat water expansions in a supersonic expansion ion source and recorded the photoelectron spectra of $(H_2O)^-_{n=11-15,19}$ using 2.409 eV photons.[19] Each of these spectra consists of a single broadened peak, and these are shifted to successively lower electron kinetic energies with increasing cluster ion size. We interpreted the electron binding energies of the fitted centers of these spectral peaks to correspond to the vertical detachment energies for each of the cluster anions. These vertical detachment energies vary smoothly from 0.75 eV for $(H_2O)^-_{11}$ to 1.12 eV for $(H_2O)^-_{19}$.

Currently, we are collaborating with H. Haberland and his student, C. Ludewigt to photodetach more water cluster anions. Thus far, we have looked at $(H_2O)^-_{2,6,7}$ and $(D_2O)^-_{2,6,7}$ [see Figure 11 and 12]; and $Ar(H_2O)^-_{2,6,7}$ and $Ar(D_2O)^-_{2,6,7}$. We plan to extend these studies to much larger cluster anions in the near future.

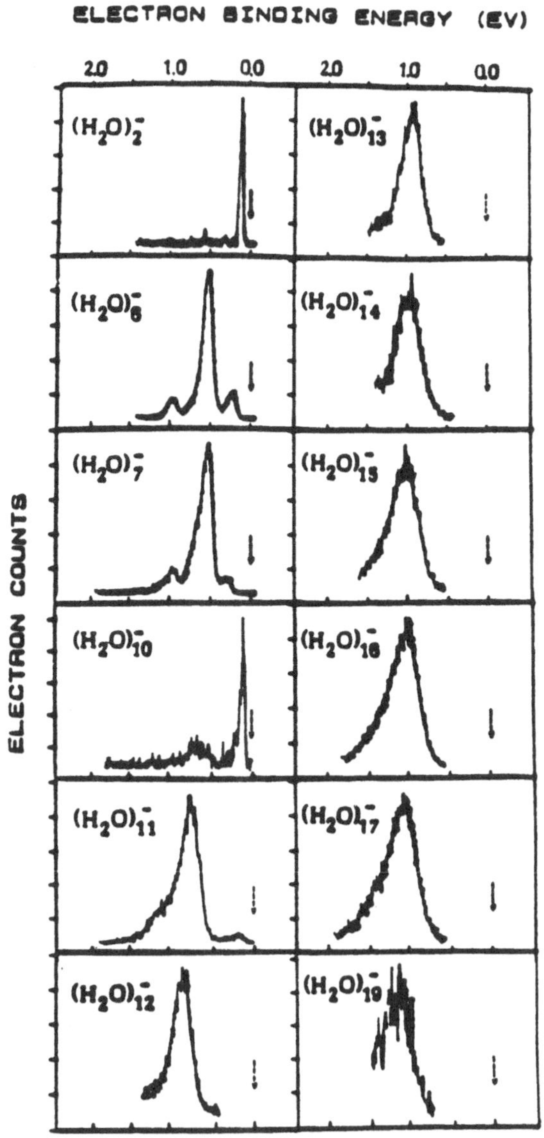

ELECTRON BINDING ENERGY (EV)

Fig. 13 The Photoelectric Spectra of
$(H_2O)^-_{n=2,6,7,10-17,19}$

The spectrum of the water dimer anion is particularly interesting. Calculations have generally predicted the structure of $(H_2O)_2^-$ to be the same as that of $(H_2O)_2$, ie., the excess electron was not found to distort the structure of neutral water dimer. In the spectrum of $(H_2O)_2^-$, however, we observe peak spacings which are characteristic of H_2O bending and H_2O stretching frequencies. In the $(D_2O)_2^-$ spectrum, we see these spacings shift appropriately for a D_2O bend and a D_2O stretch. Thus, it is reasonably clear that at least one water component within the water dimer anion is distorted! We also measure the vertical detachment energy to be ~34 meV for water dimer anion. Since there is at least some structural difference between the anion and its neutral, the adiabatic electron affinity of water dimer must be at least a little less than 34 meV. The field detachment experiments of Haberland determined the electron binding energy of $(H_2O)_2^-$ to be ~17 meV. Figure 13 presents the photoelectron spectra of all of the $(H_2O)_n^-$ species that we have studied thus far.

ACKNOWLEDGEMENTS

This research was supported by the National Science Foundation under Grant No. CHE-8511320. Some of the work on the negative cluster ions of water was performed in collaboration with H. Haberland, C. Ludewigt, and D. Worsnop, and was partially supported by a NATO Collaborative Research Grant (#86/307).

REFERENCES

1. J. V. Coe, J. T. Snodgrass, C. B. Freidhoff, K. M. McHugh, and K. H. Bowen, J. Chem. Phys. 84, 618 (1986).
2. H. Haberland, H.-G. Schindler, and D. R. Worsnop. Ber. Bunsenges. Phys. Chem. 88, 270 (1984).
3. J. V. Coe, J. T. Snodgrass, C. B. Freidhoff, K. M. McHugh, and K. H. Bowen, J. Chem. Phys. 87, 4302 (1987).
4. J. C. Kleinfeld, S. Ingemann, J. E. Jalonen, N. M. M. Nibbering, J. Am. Chem. Soc. 105, 2474 (1983).
5. J. V. Coe, J. T. Snodgrass, C. B. Freidhoff, K. M. McHugh, and K. H. Bowen, J. Chem. Phys. 83, 3169 (1985).
6. J. T. Snodgrass, J. V. Coe, C. B. Freidhoff, K. M. McHugh, and K. H. Bowen, to be published.
7. J. T. Snodgrass, J. V. Coe, C. B. Freidhoff, K. M. McHugh, and K. H. Bowen, to be published.
8. C. B. Freidhoff, J. T. Snodgrass, and K. H. Bowen, to be published.
9. J. V. Coe, J. T. Snodgrass, C. B. Freidhoff, K. M. McHugh, and K. H. Bowen, Chem. Phys. Lett. 124, 274 (1986).
10. J. M. Oakes and G. B. Ellison, Tetrahedron 42, 6263 (1986).
11. J. V. Coe, Ph.D. Thesis, The Johns Hopkins University, 1986.
12. J. T. Snodgrass, Ph.D. Thesis, The Johns Hopkins University, 1987.
13. J. T. Snodgrass, J. V. Coe, C. B. Freidhoff, K. M. McHugh, and K. H. Bowen, submitted.

14. C. B. Freidhoff, Ph.D. Thesis, The Johns Hopkins University, 1987.
15. M. Johnson (private communication).
16. K. M. McHugh, J. G. Eaton, G. H. Lee, H. Sarkas, L. Kidder, and K. H. Bowen, to be published.
17. D. G. Leopold, Joe Ho, and W. C. Lineberger, J. Chem. Phys. $\underline{86}$, 1715 (1987).
18. O. Cheshnovsky, P. F. Brucat, S. Yang, C. L. Pettiette, M. J. Craycraft, and R. E. Smalley, Proceedings of the International Symposium on the Physics and Chemistry of Small Clusters, Richmond, VA, Oct. 28-Nov. 1, 1986.
19. J. V. Coe, D. Worsnop, and K. H. Bowen, submitted.

THE STRUCTURE OF SMALL MOLECULES AND IONS AS DETERMINED BY PHOTOIONIZATION MASS SPECTROMETRY

J. Berkowitz[a], C. A. Mayhew[a] and B. Ruscic[a,b]

[a] Chemistry Division, Argonne National Laboratory, Argonne IL, U. S. A.
[b] Physical Chemistry Department, Rugjer Boskovic Institute, Zagreb, Yugoslavia

Photoionization mass spectrometry (PIMS) is an extremely versatile spectroscopic technique for the investigation of molecular ions and their parents. Whenever relative photoion yields and accurate adiabatic and appearance potentials are needed, PIMS is the method of choice. Judiciously used, the information obtained by this technique becomes a powerful basis for inferring the electronic and geometric structure of small molecules and ions. We shall corroborate this assertion by presenting a couple of recent studies from our laboratory.

THE VINYL RADICAL [1]

The heat of formation of the vinyl radical (or, almost equivalently, the C—H bond energy in ethylene) and the geometry of the ground state of the vinyl cation are two topics which are currently receiving considerable attention.

In order to address these questions, the photoionization spectrum of vinyl radical has been recorded, from its observed threshold to 1160 Å. Two methods of preparation have been employed: the abstraction reaction of F atoms with C_2H_4, and the pyrolysis of divinyl mercury at 1200 K. The observed ionization threshold for the ground state of vinyl cation is 8.59 ± 0.03 eV with the F atom reaction, and 8.43 ± 0.03 eV with the pyrolysis method. The lower value in the latter experiment is interpreted as a hot band. The relatively low value of the photoionization cross section near threshold implies a large geometry change between vinyl radical and ground state vinyl cation, favoring the view that the non-classical (bridged) structure is the most stable one for the vinyl cation. The photoionization spectra indicate a vibrational progression in the threshold region; it is quite possible that the 0 ← 0 transition lies one quantum lower than our detected limit. With this bracketed adiabatic ionization potential and the appearance potential of $C_2H_3^+$ from C_2H_4, a C—H bond energy in ethylene of 107 + 110 kcal/mole (at 0 K) is deduced.

Beyond the threshold region of the photoionization spectra, relatively sharp autoionization structure has been observed, and interpreted as a Rydberg series converging to the excited $^3A''$ state of vinyl cation. The analysis leads to an adiabatic ionization energy of

~10.7 eV for this state, with a structure similar to that of vinyl radical but with an increased C—C distance.

B_2H_6 AND BH_3 [2]

Some recent articles have focused attention on the dimerization energy of borane (the heat of reaction for $2BH_3 \rightarrow B_2H_6$) and its crucial role in determining the bond energies in hydroboranes.

This was the prime, although not the sole reason to undertake a photoionization study of B_2H_6 and BH_3. In this study, the photoion yield curves of $B_2H_n^+$ (n=2-6) and BH_n^+ (n=2,3) from B_2H_6, as well as BH_n^+ (n=1-3) from BH_3 (which was produced by pyrolysis of B_2H_6 at 800 K) have been obtained.

The results clearly show why the combination of appearance potential measurements for BH_3^+ from B_2H_6 and BH_3^+ from BH_3 yields a poor upper limit for $-\Delta_{dimeriz}H_{0K}(BH_3)$ of 52.7 kcal/mole. The combination of appearance potentials of BH_2^+ from B_2H_6 and BH_2^+ from BH_3 provides a somewhat better upper limit of 46.6 ± 0.6 kcal/mole for this quantity. However, the threshold for BH^+ from BH_3, combined with auxiliary data, provides the best current experimental value, (32.7 + 39.1) ± 2 kcal/mole, where the bracketed range depends on the particular choice for the value of $\Delta_{subl}H(B)$. The present data also allow the extraction of other relevant thermodynamical quantities, like the atomization energy of BH_3 and the upper limits on $\Delta_f H$ for the observed ionic species.

In conjunction with recent *ab initio* calculations, the experimental results make possible a meaningful discussion of plausible structures for the ionic species. Finally, it can be shown that the fragmentation behavior of photoions from diborane has a more facile explanation by quasi-equilibrium theory, than by a molecular orbital picture, with the probable exception of BH_3^+ from B_2H_6.

This research was supported by the U. S. Department of Energy (Office of Basic Energy Sciences) under contract W-31-109-ENG-38 and by the U. S. - Yugoslav Joint Board for Science and Technology through the U. S. Department of Energy Grant No. PN 561.

REFERENCES

1. J. Berkowitz, C. A. Mayhew and B. Ruscic, "A Photoionization Study of the Vinyl Radical", to appear in J. Chem. Phys.

2. B. Ruscic, C. A. Mayhew and J. Berkowitz, "Photoionization Studies of $(BH_3)_n$, (n=1,2)", to appear in J. Chem. Phys.

DYNAMIC ASPECT OF EXCITED-STATE PHOTOELECTRON SPECTROSCOPY:

SOME SMALL MOLECULES

Katsumi Kimura

Institute for Molecular Science
Okazaki 444
Japan

INTRODUCTION

During the last two decades, photoelectron spectroscopy with a HeI (58.4 nm) resonance lamp or other vacuum UV sources has been extensively applied to a variety of molecules in the gas phase. Such photoelectron spectroscopy gives us valuable data on the ionic states which are produced by single-photon ionization of molecules in their ground state.[1-3]

Combination of a nanosecond tunable UV/visible laser with a photo-electron spectroscopic technique has made it possible to observe excited-state photoelectron spectra according to resonant multiphoton ionization (MPI) or stepwise ionization. The author has recently published review articles on the situation of the development of excited-state photoelectron spectroscopy.[4] With this technique it is possible to study dynamic behavior of excited states. The dynamic aspect of the laser photoelectron spectroscopy is in striking contrast to the VUV photoelectron spectroscopy which is mostly concerned with static aspect of ground-state molecules. In this sense, it would be said that laser photoelectron spectroscopy provides a useful technique for studying the dynamic behavior of excited state molecules.

The present brief article will be confined mainly to the special subjects recently studied in my laboratory. The purpose of this paper is to show several examples in connection with the dynamic aspect of excited-state photoelectron spectroscopy.

EXCITED-STATE PHOTOIONIZATION

Excited-state photoelectron spectra provide the information which is analogous to fluorescence spectra. Ionization transition is always allowed transition for any excited states, and therefore ionization transition probabilities are the similar order of magnitude. However, fluorescence transition probability varies to a large extent of several orders of magnitude. Therefore, non-radiative excited states are also interesting subjects to be studied by laser photoelectron spectroscopy. Furthermore, ionic states with different electronic structures can be produced as the final states of photoionization. This situation also differs from fluorescence spectroscopy.

The photoionization at an excited state is in principle in competition with other deactivation processes such as relaxation or dissociation. The rate of photoionization at a specific excited state is given by the product of the laser intensity and the ionization cross-section, so that we can control the rate of photoionization by changing the laser intensity. This situation is shown schematically in Fig. 1.

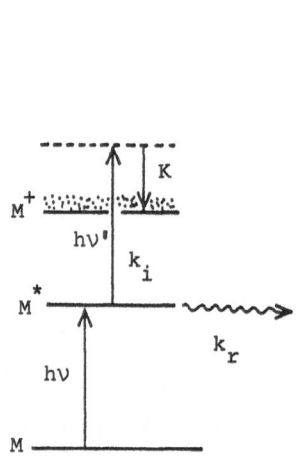

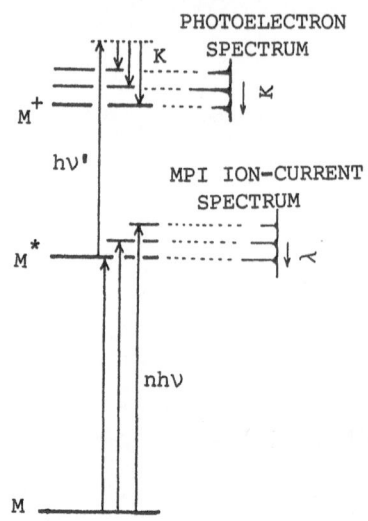

Fig. 1. Competition of ionization and deactivation processes at an excited state.

Fig. 2. Schematic energy diagram and (n+1') resonant ionization.

MPI ION-CURRENT AND PHOTOELECTRON SPECTRA

When molecules in the gas phase are excited by a pulsed UV/visible laser to their specific excited electronic states, resonant ionization is remarkably enhanced, ejecting photoelectrons by a total of two or more photons. The resonant ionization or stepwise ionization of molecules has been studied mainly by three techniques; (1) ion-current measurements as a function of laser wavelength, (2) mass spectroscopic measurements, and (3) photoelectron energy measurements as a function of electron kinetic energy at each ion-current peak.

Scanning the laser wavelength in the UV/visible region, one can obtain an MPI ion-current spectrum which consists of many peaks. Each ion-current peak corresponds to each resonant excited level (vibrationally and rotationally resolved). All possible (n+m)-type resonant ionizations are included in MPI ion-current spectroscopy.[5]

When a single UV/visible laser is used, the case of m=1 in the (n+m) resonant ionization is most important in photoelectron measurements, otherwise the resulting photoelectron spectrum might be attributable to a higher unknown excited state. Therefore, the ideal photoelectron experiment is to use two kinds of lasers, one is for excitation and the other for ionization. In other words, the (n+1')-type resonant ionization with a set of two lasers is the most desirable from a photoelectron point of view. Such two-color photoelectron spectroscopy is especially important for studying excited state dynamics. An energy level diagram relevant to the resonant or stepwise ionization is schematically shown in Fig. 2.

Let us compare a HeI photoelectron spectrum and a laser photoelectron spectrum in the case of NO molecule. A laser photoelectron spectrum

obtained by (1+1) resonant ionization of NO via Rydberg A state (v'= 0) is
shown in Fig. 3(a), while a HeI spectrum is shown in Fig. 3(b) for
comparison. This comparison demonstrates a dramatic difference in spectral
pattern.[6] The spectrum of the Rydberg A state shows only a single
vibrational peak due to the $v^+=0$ level of the $NO^+(X)$ ion, since the
equilibrium bond distance is essentially unchanged during photoionization.
On the other hand, the equilibrium bond distance of the neutral ground
state is considerably different from that of the NO^+ ion. In the case of
(1+1) resonant ionization of NO via the Rydberg A state, photoelectron
spectra have also been reported by other workers.[7,8]

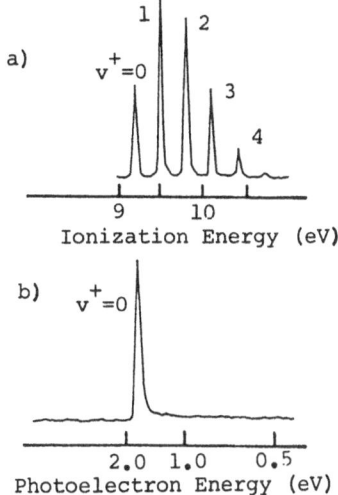

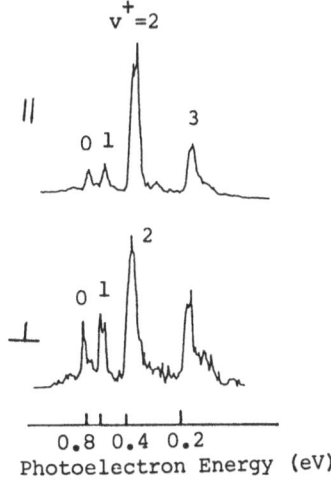

Fig. 3. Photoelectron spectra of
NO, obtained (a) by (1+1)
resonant ionization via
the A v=0 state and (b) by
a HeI source (Ref. 6).

Fig. 4. Photoelectron spectra of
O_2 obtained at 287.9 nm
in the direction parallel
and perpendicular to the
polarization vector
(Ref. 16).

In general, removal of a Rydberg electron in (n+1)-type resonant
ionization through an excited Rydberg state (especially in its low-lying
vibrational levels) dominantly gives rise to a vibrational band due to $\Delta v=0$
ionization transition in a photoelectron spectrum. Such $\Delta v=0$ vibrational
bands have been observed for the Rydberg states of H_2,[9] N_2,[10] NO,[6-8,11,12]
and NH_3.[12-14]

ONE-PHOTON FORBIDDEN EXCITED STATES

The Rydberg C state ($^3\Pi_g$) of molecular oxygen, which is one-photon
forbidden (two-photon allowed) from the ground electronic state, has been
studied by (2+1) MPI ion-current and photoelectron spectra in the laser
wavelength region 287-289 nm.[15,16] The resulting photoelectron spectra are
shown in Fig. 4. The following points have been pointed out.[17] (1) All
the energetically possible vibrational bands ($v^+=0-3$) of O_2^+ are clearly
observed. (2) In each spectrum, the $v^+=2$ band is strongest among the four
observed vibrational bands. (3) The $v^+=2$ photoelectron intensity largely
depends on the laser polarization vector, but the other peaks ($v^+=0,1,3$)
show no significant angular dependence. The photoelectron spectra of the
O_2 C state observed at different rotational levels have shown the non-
Franck-Condon distribution that considerably deviates the $\Delta v=0$ pattern.[16]

Furthermore, the $v^+=2$ band shows the photoelectron angular dependence
different from the other vibrational bands ($v^+=0,1,3$).[16] From the angular

dependences, two kinds of ionization processes have been suggested; one is the process associated with $\Delta v = 0$ ionization producing the $v^+=2$ bands, and the other is the ionization producing the vibrational distribution of $v^+=0$-3.[16] In order to explain the photoelectron vibrational distribution, it is necessary to know electronic characters of both the resonant and the ionic states. In connection of the width of MPI ion-current rotational lines of O_2, Sur et al.[15b] have discussed the interaction between the Rydberg C state and a repulsive valence state. Mixing of nearby valence electronic states with the C state would cause more or less non-Franck-Condon patterns in photoelectron spectra.

Deviation from expected Franck-Condon factors in (n+1) resonant ionization would be attributed to autoionization or configuration mixing of the resonant state. In (n+2)-type resonant ionization, there is a possibility of accidental resonance, and in fact a non-Franck-Condon situation has been demonstrated for (3+2) and (3+3) resonant ionization of CO via the Rydberg A state.[17]

AUTOIONIZATION

The autoionization pathway through dissociative super-excited valence states of NO molecule has been studied by (2+1) resonant ionization via the valence excited B state at many rotational levels of the $v'=9$ vibrational state.[18] The reason of the selection of the B-9 state was to remove a possibility of direct ionization of producing the ground-state NO^+ ion.[18] The direct ionization of NO from the B state to the ground ionic state is forbidden in the sense of one-electron transition.

As a result, it has been indicated that an ion-current spectrum for the B-9 region shows several intensity-anomalous rotational lines as well as normal Q-branch rotational lines. The photoelectron spectrum obtained for each intenisty-normal rotational line shows three energetically accessible vibrational bands ($v^+=0,1,2$), whereas that obtained for each intensity-anomalous rotational line indicates a relatively high yield of the $v^+=0$ ion in addition to the $v^+=0,1,2$ bands, as shown in Fig. 5[18]

On the basis of these results of the (2+1) resonance ionization of NO through the valence B state, two autoionization processes have been proposed:[18] One is the process in which the third photon state is the super-excited valence I state coupling with the ground state of the ion, and the other is the process in which the third photon state is the Rydberg N ($v'=6$) state which couples with the super-excited valence B' state and the ground state of the ion.

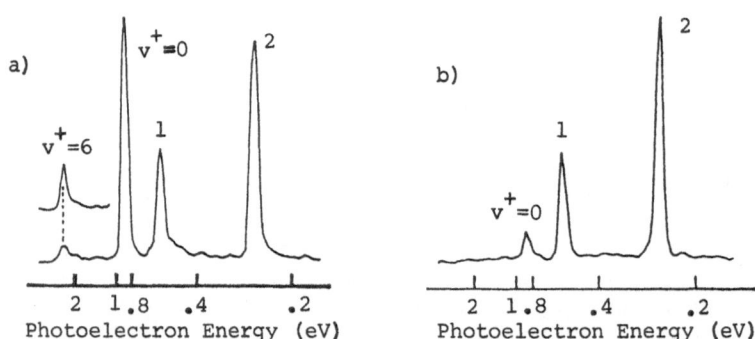

Fig. 5. Photoelectron spectra obtained by (2+1) resonant
        ionization via the B-9 state of NO, obtained at
        (a) normal intensity line and (b) anomalous
        intensity line (Ref. 18).

A double-resonance technique using a two-color laser system is especially powerful for studying super-excited states at which autoionization occurs. Super-excited states can be produced selectively by the second laser from a specific excited state initially chosen by the first laser. One of important subjects is to determine the branching ratio of producing different vibrational states of ions in autoionization processes. Such two-color photoelectron spectroscopy can distinguish autoionization from direct ionization.

For example, single-photon excitation from the v'=4 levels of Rydberg C state (3pπ) gives rise to the v'=4 levels of super-excited Rydberg states (nsσ; ndσ,π,δ), according to Δv=+1 selection rule.[6] Three super-excited states can be seen from an ion-current spectrum by scanning the second laser in the laser wavelength region 560-580 nm, corresponding to 6sσ, 5dδ, and 5dσ,π.[6] The photoelectron spectra obtained from these super-excited states have indicated that the transitions of Δv=-1 as well as other transitions take place together in the autoionization, as shown in Fig. 6.[6]

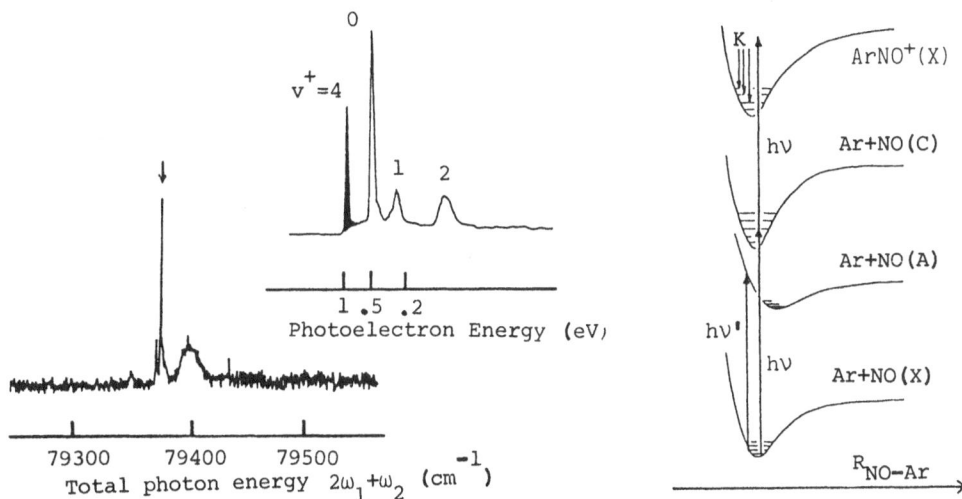

Fig. 6. (a) A two-color MPI ion-current spectrum of NO, associated with autoionization, and (b) the resulting photoelectron spectrum (Ref. 6).

Fig. 7. Schematic potential curves of ArNO, showing (1+1) resonant ionization and photodissociation.

SMALL VAN DER WAALS MOLECULES AND PHOTODISSOCIATION

Some laser photoelectron spectroscopic studies of small van der Waals molecules have been demonstrated, with ArNO and $(NO)_2$, which are produced in supersonic jets.[19-21] In the (2+1) resonant ionization of ArNO via the Rydberg C state, the $ArNO^+$ ion and the corresponding photoelectron spectra have been observed.[19] In the excitation of ArNO to the A state, dissociation takes place, giving rise to the excited A state of the NO component.[20] Schematic potential curves of ArNO are shown in Fig. 7. The excited A state of NO has been also observed in the UV excitation of $(NO)_2$.[21]

ArNO[19,20]

For ArNO, the total ion-current spectrum has been measured in the 380-385 nm wavelength region, and photoelectron spectra have been obtained at several ion-current peaks.[19] The resulting ion-current spectrum shows a new vibrational progression consisting of four peaks on the longer wavelength side of the (2+1) peaks of free NO molecule; that is, (2+1) resonant ionization via the Rydberg C state (v"=0).[19] A similar result has been reported by another worker.[22]

The following results have been obtained for ArNO.[19] The new ion-current peaks have been attributed to the transitions of the ArNO complex from its ground state to the two-photon resonant state (C v'=0) in which the NO component is in the 3p Rydberg state. The final state is $ArNO^+(X)$. Each ion-current peak separation is about 50 cm^{-1}, corresponding to the Ar-NO intermolecular stretching frequency, but showing strong anharmonicity. The dissociation energy of the ArNO C state has been found to be $D_o = 0.055 \pm 0.001$ eV. From the photoelectron spectra, it has been indicated that the adiabatic ionization energy of ArNO is $I_a = 9.148 \pm 0.005$ eV and the dissociation energy of $ArNO^+(X)$ is $D_o = 0.129 \pm 0.005$ eV.

The dissociation energy of the $ArNO^+$ ion is considerably larger than that of ArNO C state.[19] This indicates that the intermolecular potential of the ArNO C state is different from that of the $ArNO^+$ ion. Therefore, a vibrational progression of the Ar-NO stretching vibration is expected to appear in the photoelectron spectra. If the O-N-Ar angle of the ionic state is considerably different from the resonant state, a progression of the intermolecular bending vibration is expected to appear.

According to a LIF study,[23] the electronic transition of ArNO from the ground state to the dissociative continuum (correlated to the ground state of Ar and the excited A state of NO) is Franck-Condon allowed. The rotational distribution of the Rydberg A state of the NO component produced by photodissociation of the ArNO complex has been measured by using a (1+1) resonant ionization technique.[20] Two eminent maxima have been observed in the distributions at all the laser energies studied. It has been found that the dependences of these maxima on the excess energy and the reduced mass are the same as those of the rotational rainbow peaks in rotationally inelastic scattering.[20] A theoretical study of partial cross sections by a quantum mechanical close-coupling method has strongly supported the experimentally observed anomalous rotational-state distribution is a manifestation of the rotational rainbow effect.[20]

NO Dimer[21]

Laser photoelectron energy measurements in a supersonic NO jet have been carried out at the far-UV laser wavelengths 193, 199.8, 204.2, 208.8 and 217.8 nm. Direct spectroscopic evidence for formation of the excited A state of NO from $(NO)_2$ has been obtained at the various far-UV wave-lengths. If the two-photon resonant ionizations of $(NO)_2$ take place before fragmentation (mechanism 1), the photoelectron kinetic energy K is given by

$$K = 2h\nu - I' - E', \qquad (1)$$

where $h\nu$ is the one-photon energy, I' the ionization potential of $(NO)_2$, and E' the internal energy of $(NO)_2^+$. One the other hand, if one-photon ionization takes place from the NO excited state (mechanism 2), the photoelectron energy is given by

$$K = h\nu - I + T + E - E^+, \qquad (2)$$

where $h\nu$ is the photon energy, I is the first ionization potential of NO, T and E are the electronic and the vibrational energy of the excited NO, respectively, and E^+ is the vibrational energy of the NO^+ ground state.

From the experimental results, it has been concluded that the precursor of the NO^+ production is the NO A state which is produced from $(NO)_2$ by one-photon absorption. The mechanism of the NO^+ formation in the far-UV laser irradiation is schematically shown in Fig. 8. The observed photoelectron spectra of the excited A state of NO at several laser wavelengths are shown in Fig. 9.

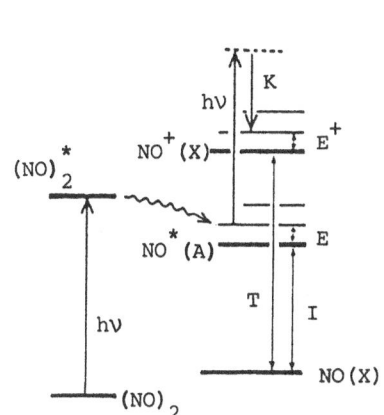

Fig. 8. Schematic energy diagram of the NO dimer and NO, showing photodissociation followed by ionization.

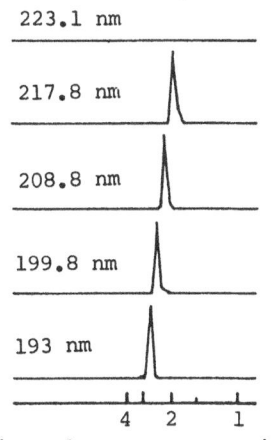

Photoelectron Energy (eV)

Fig. 9. Photoelectron spectra of the NO A state obtained by photodissociation of the NO dimer (Ref. 21).

CONCLUDING REMARKS

In studies of resonant multiphoton ionization, ion-current and mass spectroscopic informations have been widely measured with UV/visible lasers, while there are still only a limited number of photoelectron studies. Photoelectron measurements are very useful for the characterization or diagnosis of resonant excited states and ionic states. In addition to photoelectron kinetic-energy distribution, the photoelectron branching ratios of producing different ionic states are especially powerful for studying autoionization mechanisms and the dynamic behavior of excited states.

REFERENCES

1. D. W. Turner, A. D. Baker, C. Baker, and C. R. Brundle, "Molecular Photoelectron Spectroscopy", Interscience, London (1970).
2. K. Siegbahn, C. Nordling, G. Johansson, J. Hedman, P. F. Heden, K. Hamrin, U. Gelius, T. Bergmark, L. O. Werme, R. Manne, and Y. Baer, "ESCA Applied to Free Molecules", North-Holland, Amsterdam (1969).
3. K. Kimura, S. Katsumata, Y. Achiba, T. Yamazaki, and S. Iwata, "Handbook of HeI Photoelectron Spectra of Fundamental Organic Molecules", Halsted, New York, (1981).
4. (a) K. Kimura, Adv. Chem. Phys., **60**, 161. (1985). (b) K. Kimura, Inter. Rev. Phys. Chem., **6**, 195 (1987).
5. P. M. Johnson, Acc. Chem. Res., **13**, 20 (1980).
6. Y. Achiba and K. Kimura, J. Chem. Soc. Japan (in Japanese), 1529 (1984).
7. J. C. Miller and R. N. Compton, J. Chem. Phys., **84**, 675 (1986).
8. K. S. Viswanathan, E. Sekreta, E. Davidson, and J. P. Reilly, J. Phys. Chem., **90**, 5658 (1986).
9. S. T. Pratt, J. L. Dehmer, and P. M. Dehmer, Chem. Phys. Lett., **105**, 28 (1984).

10. S. T. Pratt, J. L. Dehmer, and P. M. Dehmer, J. Chem. Phys., **80**, 1706 (1984).
11. J. C. Miller and R. N. Compton, J. Chem. Phys., **75**, 22 (1981).
12. Y. Achiba, K. Sato, K. Shobatake, and K. Kimura, J. Chem. Phys., **78**, 5474 (1983).
13. J. H. Glownia, S. J. Riley, S. D. Colson, J. C. Miller, and R. N. Compton, J. Chem. Phys., **77**, 68 (1982).
14. W. E. Conaway, R. J. S. Morrison, R. N. Zare, Chem. Phys. Lett., **113**, 429 (1985).
15. (a) A. Sur, C. V. Ramana, and S. D. Colson, J. Chem. Phys., **83**, 904 (1985). (b) A. Sur, C. V. Ramana, W. A. Chupka, and S. D. Colson, J. Chem. Phys., **84**, 69 (1986).
16. S. Katsumata, K. Sato, Y. Achiba, and K. Kimura, J. Elect. Spectrosc., Rel. Phenom., **41**, 325 (1986).
17. S. R. Pratt, E. D. Poliakoff, P. M. Dehmer, P. M. and J. L. Dehmer, J. Chem. Phys., **78**, 65 (1983).
18. Y. Achiba, K. Sato, and K. Kimura, J. Chem. Phys., **82**, 3959 (1985).
19. K. Sato, Y. Achiba, and K. Kimura, J. Chem. Phys., **80**, 57 (1984).
20. K. Sato, Y. Achiba, H. Nakamura, and K. Kimura, J. Chem. Phys., **85**, 1418 (1986).
21. K. Sato, Y. Achiba, and K. Kimura, Chem. Phys. Lett,, **130**, 231 (1986).
22. J. C. Miller, J. Chem. Phys., **86**, 3166 (1987).
23. P. R. R. Langridge-Smith, E. Carrasquillo M., D. H. Levy, J. Chem. Phys., **74**, 6513 (1981).

PROBING MOLECULAR DYNAMICS WITH LASER

PHOTOELECTRON SPECTROSCOPY

Ellen Sekreta, Charles W. Wilkerson, Meiying Hou
and James P. Reilly

Department of Chemistry
Indiana University
Bloomington, Indiana, 47401

INTRODUCTION

Nonradiative relaxation processes such as internal conversion and intersystem crossing have been of interest to photochemists for a number of years. Although the decay of an initially prepared state can be studied by monitoring fluorescence lifetimes or quantum yields, observation of the state that is populated by the radiationless transition is usually difficult. In principal, this can be accomplished by absorption experiments, but in practice such signals are usually weak. In recent years the method of multiphoton ionization spectroscopy has grown in popularity.[1-3] It has often been pointed out that one of the advantages of this technique is that it can be used to detect "dark" or non-fluorescing excited states such as those populated during radiationless relaxation. Smalley, Johnson and coworkers[4,5] have shown that triplet states of benzene populated following intersystem crossing from the initially excited $^1B_{2u}$ state are ionizable and that kinetic information about the triplet state populated can be measured. Yet even in these experiments, it was impossible to distinguish ions produced by ionizing triplet states from those generated by ionizing singlet states. To overcome this fundamental problem, Pallix and Colson[6] have demonstrated that by recording the kinetic energy distribution of the photoelectrons generated during the two-step ionization of sym-triazine they could readily distinguish signals associated with singlet and triplet ionization from each other. Along similar lines, we have been monitoring the photoelectrons ejected in the two-step ionization of benzene following its excitation to the $^1B_{2u}$ and $^1B_{1u}$ electronic states. It is well-known that the relaxation rate of the latter electronic state is extremely rapid. Both single color and two-color experiments have been performed using light pulses of nanosecond and picosecond duration. Laser ionization mass and photoelectron spectra have been recorded and will be discussed.

EXPERIMENTAL

The time-of-flight mass and photoelectron spectrometers
used in these experiments have been previously described.[7-9]
The low-energy electron transmission cutoff is between 50 and
70 meV.   In order to observe electrons slower than this,
acceleration grids are mounted around the ionization region,
and a few volts are applied to them.   The light used to
excite molecules and induce ionization was produced by
nanosecond and picosecond pulse duration Nd:YAG lasers and
a nanosecond-pulse excimer laser-pumped dye laser. A pair of
particularly interesting wavelengths, 208nm and 192 nm, can
be generated as the 7th and 8th anti-stokes Raman-shifted
lines from a frequency-doubled Nd:YAG laser.   Tunable
radiation between 202 and 212 nm was produced by frequency
doubling and Raman shifting the output of a Nd:YAG laser
pumped dye laser. In some cases with sufficiently energetic
laser light an electron signal was produced without
introducing any sample molecules, and it was necessary to
subtract this background from our photoelectron spectra. Gas
phase benzene was introduced into the spectrometer through
both effusive and pulsed supersonic valve sources. Data were
collected and stored with a microcomputer.

RESULTS AND DISCUSSION

In order to demonstrate that intersystem crossing
between the $^1B_{2u}$ and the triplet state manifold can be
directly observed, we excited benzene's 6^1_0 vibronic band at
259 nm, and, after a variable time delay, irradiated the
excited molecules with a pulse of 192 nm light.   The
resulting electron kinetic energy distributions are displayed
in Figure 1. Note that two bands appear in each of these

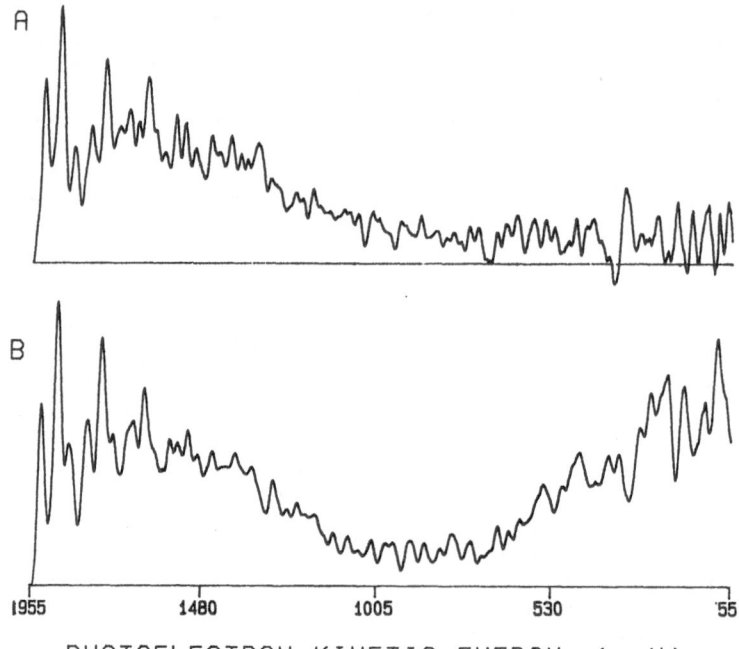

PHOTOELECTRON KINETIC ENERGY (meV)

Fig. 1. Two-color laser photoelectron spectra of room
        temperature benzene recorded with 259 and 208 nm
        light pulses. (a) 0 nsec delay (b) 50 nsec delay

spectra. The first, which results from ionizing $^1B_{2u}$ benzene with 192 nm light, has vibronic structure similar to that which we previously recorded with 259 nm radiation alone.[8] Its intensity decreases as the time delay between laser pulses increases. The second band is broad and structureless, and its relative intensity increases as the time delay increases. We attribute this peak to the ionization of $^3B_{1u}$ benzene that was produced by intersystem crossing. One of the reasons for recording this spectrum was to determine whether there is any selectivity associated with the triplet state production, but the lack of structure in the spectrum suggests that there is not.

Two experiments were undertaken to excite and ionize $^1B_{1u}$ benzene using 208 and 192 nm laser radiation. The former wavelength is near the origin of this state, the

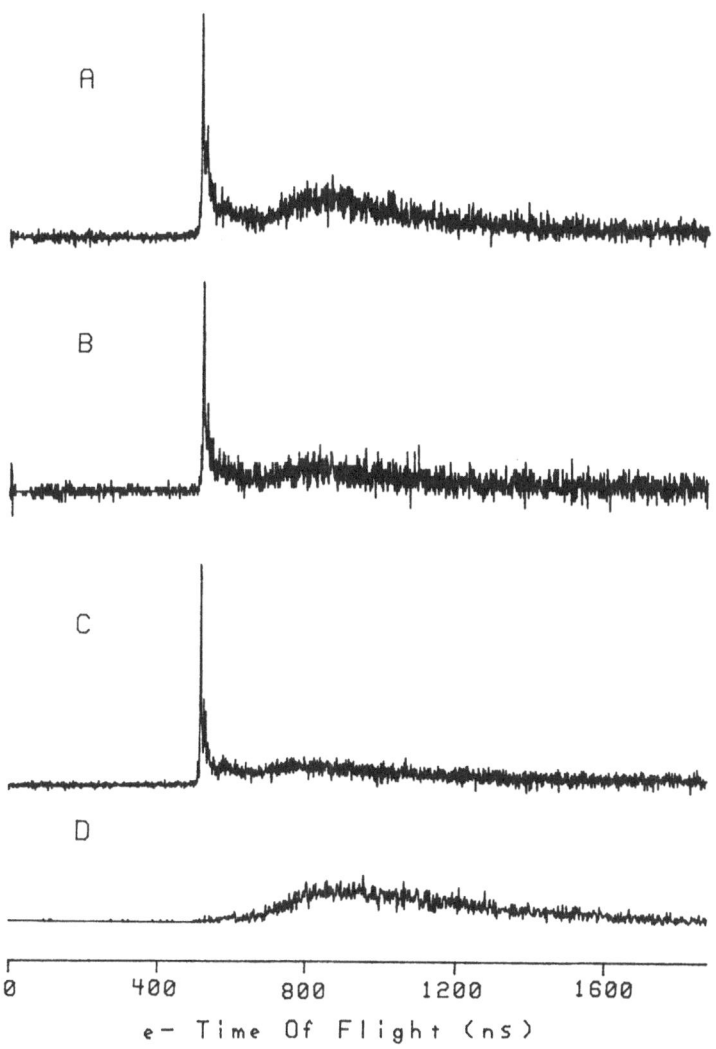

Fig. 2. Laser photoelectron spectra of benzene recorded with
208 nm light. (a) room temperature, 2 nsec pulses
(b) room temperature, 15 psec pulses (c) same as
(a) but 50 K (d) same as (a) but less than 5 K

latter is approximately 0.5 eV above the origin. Figure 2A
displays data recorded with room temperature benzene at the
former wavelength. It consists of two bands. The first is
attributed to ionization of $^1B_{1u}$ benzene, the second to
ionization following relaxation to a lower singlet or triplet
state. The first peak in the former band is located at
exactly the right electron kinetic energy to correspond to
production of ions with near zero vibrational excitation.
Based on this interpretation of the two bands, we anticipated
that if a picosecond pulse of light were used, the relative
intensities of the first and second peaks in the spectrum
would change. In particular we anticipated that the first
band would grow relative to the second. This is seen to be
the case in Figure 2B, although the effect is not
particularly dramatic. When the benzene is modestly cooled
to about 50 K in a pulsed expansion using a low stagnation
pressure, there is still little change. (Figure 2C) However
when the benzene is cooled to just a few K in a high pressure
supersonic expansion, the effect is remarkable. The
resulting photoelectron spectrum is displayed in Figure 2D.
The sharp peak that appeared at room temperature is clearly
missing, suggesting a change in the mechanism by which
benzene is ionized. Pertinent to this mechanism are laser
ionization mass spectra generated with light at this same
wavelength. Figures 3A and 3B display time-of-flight mass
spectra of room temperature and very cold benzene. The

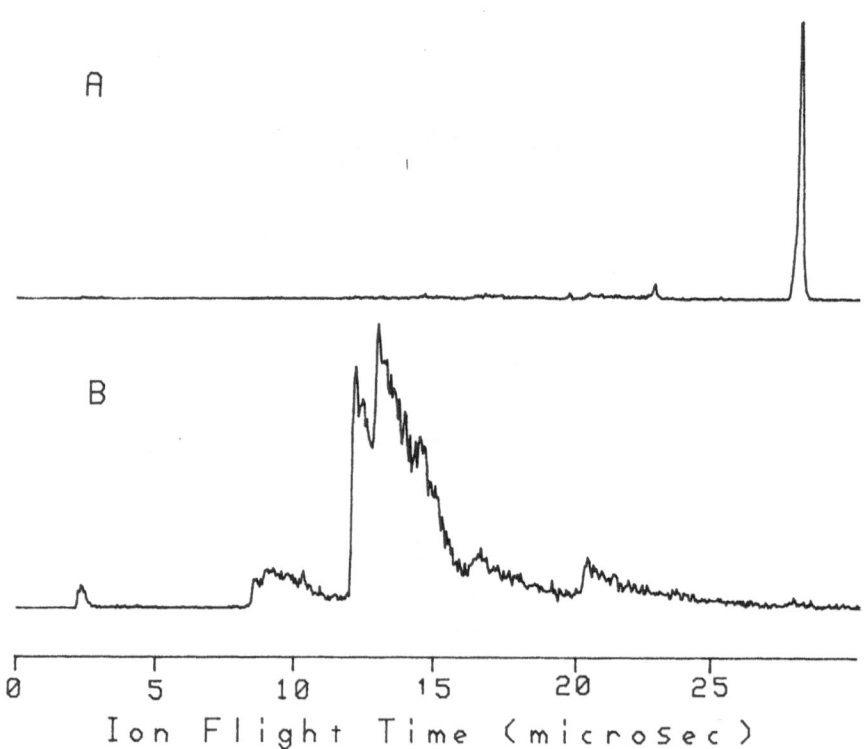

Fig. 3. Laser ionization mass spectra of benzene recorded at
208 nm. (a) room temperature (b) jet-cooled

differences between the two mass spectra are even more
dramatic than the differences in the respective
photoelectron spectra. We interpret the broadened mass
spectral peaks obtained with cold benzene as metastable ion
fragments. Although we have no evidence to decide whether
they follow the absorption of two or three photons,
appearance potential data strongly suggest that at least
three photons are required to produce these fragments.[10] A
consistent interpretation of the photoelectron and mass
spectra is as follows: For some unknown reason when <u>jet-
cooled</u> benzene is excited at this wavelength, it rapidly
relaxes to a state that is highly vibrationally excited.
When this state is ionized, the ions are produced in a
similar state of high vibrational excitation. Consequently,
the photoelectron spectrum shows no evidence of vibrationless
ions and the mass spectrum exhibits only ion fragments. The
most likely candidate for the electronic state to which $^1B_{1u}$
benzene relaxes is the $^3B_{1u}$ state which is 2.4 eV below it.

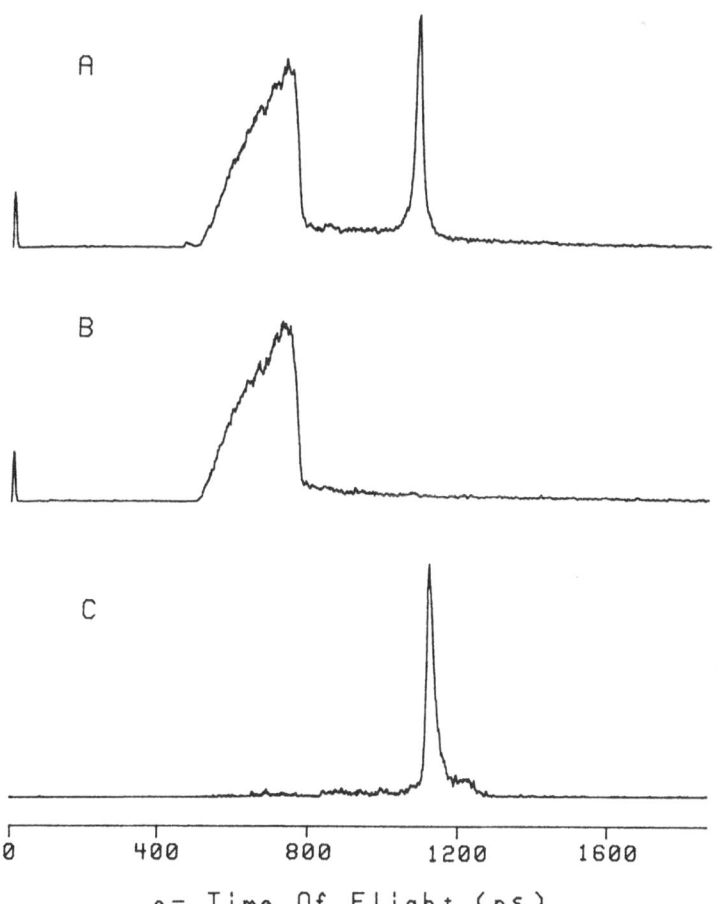

Fig. 4. Laser photoelectron spectra of benzene recorded with
        208 nm light and 1.5 volt accelerating potential
        (a) room temperature sample
        (b) background only; no sample introduced
        (c) jet-cooled sample

We have previously found that in order for our
spectrometer to detect very slow (less than about 50 meV)
electrons we must accelerate them from the ionization region
toward the detector.[9]   Figure 4A displays the laser
photoelectron spectrum of room temperature benzene recorded
under these conditions with 1.5 volts applied to the
acceleration grids.   Figure 4B is a background spectrum
recorded under similar conditions but with no benzene in the
spectrometer.   The central peak in Figure 4A is almost
entirely due to graphite photoemission.   The largest peak,
however, corresponds to electrons with kinetic energies less
than 20 meV.   Figure 4C is similar to Figure 4A except that
the benzene is jet-cooled.   The high energy electron peak is
once again not evident in the Figure 4C spectrum and the
dominating peak corresponds to low energy electrons.   The

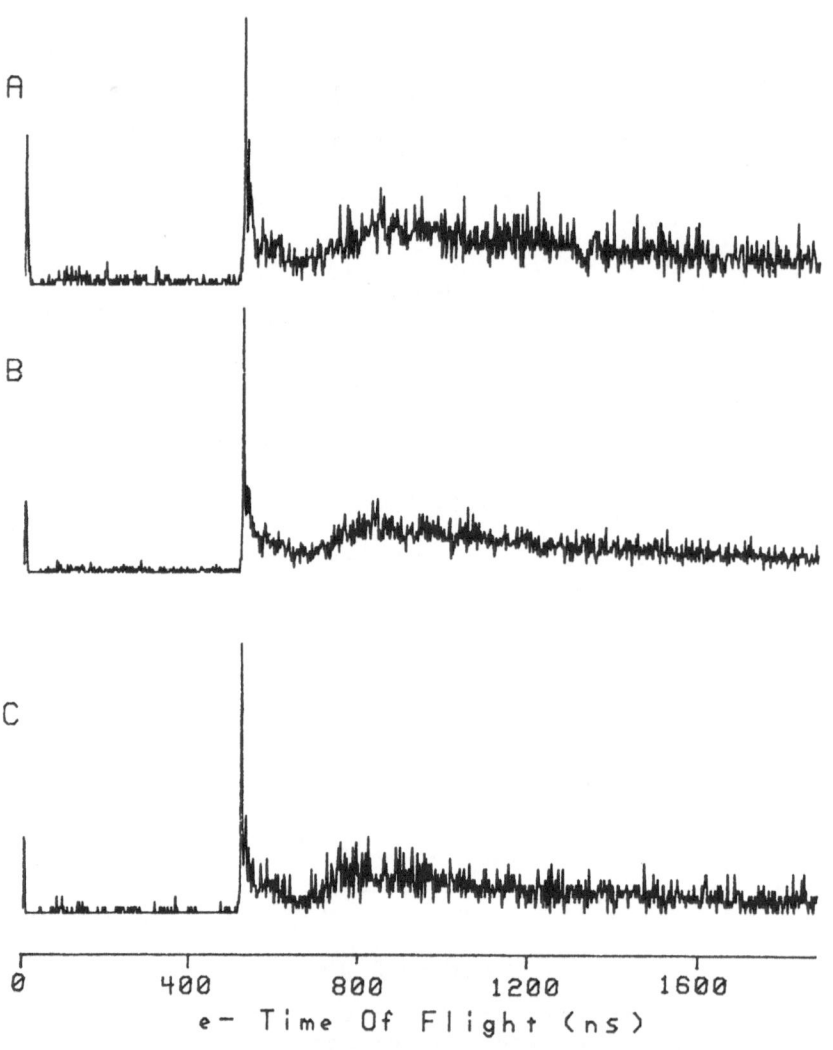

Fig. 5. Laser photoelectron spectra of room temperature
         benzene recorded with different tunable light
         wavelengths. (a) 208.3 nm (b) 208.0 nm (c) 207.7 nm

energy of these electrons is just what would be expected for ionization from the lowest triplet state. We hypothesize that the $^1B_{1u}$ state of benzene is initially populated by our laser pulse, that it relaxes to the $^3B_{1u}$ state and that because of the Franck-Condon principle ions are produced with comparable vibrational excitation.

In order to ascertain whether the difference between the behavior of room temperature and jet-cooled benzene is simply the result of band-sharpening of the spectrum of jet-cooled benzene, we recorded laser photoelectron spectra of room temperature benzene using several ultraviolet wavelengths between 207.7 and 208.8 nm. These are presented in Figure 5. It is quite evident from this data that the wavelength used is not highly critical and that the spectra are all qualitatively similar. We conclude that the appearance of the sharp, high-energy electron band in the spectra of Figure 4 does not result from a narrow, resonant excitation, but rather that it reflects the behavior of the short-lived $^1B_{1u}$ electronic state. Further work on the wavelength and intensity dependence of particular individual peaks in the photoelectron and mass spectra is currently in progress in our laboratory.

This work has been supported by the Environmental Protection Agency and the National Science Foundation. JPR is a Camille and Henry Dreyfus Teacher-Scholar.

REFERENCES

1. P. Johnson, Accounts Chem. Res. 13,20 (1980).
2. H. Reisler and C. Wittig, Adv. Chem. Phys. 60, 1, (1980).
3. D.H. Parker in "Ultrasensitive Laser Spectroscopy", ed. by S.S. Kliger, Academic Press:New York, (1983).
4. M.A. Duncan, T.G. Dietz, M.G. Liverman and R.E. Smalley, J. Phys. Chem. 85, 7, (1981).
5. C.E. Otis, J.L. Knee and P.M. Johnson, J. Phys. Chem. 87 2232, (1983).
6. J.B. Pallix and S.D. Colson, Chem. Phys. Lett. 119,38, (1985).
7. R.B. Opsal, K. Owens and J.P. Reilly, Anal. Chem. 57, 1884, (1985).
8. S.R. Long, J.T. Meek and J.P. Reilly, J. Chem. Phys. 79, 3206, (1983).
9. K.S. Viswanathan, E. Sekreta and J.P. Reilly, J. Phys. Chem. 90, 5658, (1986).
10. B.O. Jonsson and E. Lindholm, Arkiv fur Physik 39, 65 (1968).

A PHOTOION-PHOTOELECTRON COINCIDENCE STUDY OF $(CO)_2$

K. Norwood, J.-H. Guo, G. Luo and C. Y. Ng

Ames Laboratory, U.S. Department of Energy
and
Department of Chemistry
Iowa State University
Ames, Iowa 50011

INTRODUCTION

Molecular beam photoionization[1] and equilibrium[2] mass spectrometric measurements provide valuable energetic information about dimer and cluster ions in their ground state. In spite of recent intense research activities[3-9] in cluster ion chemistry, little is known about the interaction energies of an excited state ion with neutral molecules. A systematic method for determining the binding energies of a ground state as well as an excited state ion with neutral species is to measure the adiabatic ionization energies (IE) of the appropriate clusters by photoelectron spectroscopic techniques. The concentrations of clusters produced in a supersonic beam are usually much lower than that of the monomers. This, together with the fact that photoelectron bands of monomers and clusters often overlap in energy, makes the measurement of the photoelectron spectrum (PES) of a specific cluster difficult.

The photoion-photoelectron coincidence (PIPECO) technique,[10-14] which utilizes the flight time correlation of an ion-electron pair, is most promising for measuring the PES of a size-selected cluster. Attempts to obtain PIPECO spectra of simple van der Waals clusters have met with some success. The PIPECO spectra for Kr and Xe cluster ions[15,16] at photon energies below the IEs of the corresponding atoms have been reported. Photoelectrons originating from monomers at photon energies above their IEs give rise to a high false coincidence rate and a poor signal-to-noise ratio. As a result of the high false coincidence rate, the commonly used coincidence circuits, which execute single electron start-single (or multi) ion stop cycles,[15,16] are paralyzed[17] and artificially suppress the true coincidence rate. We have overcome this problem by using a multi-channel scaler to measure the ion time-of-flight (TOF) distribution after the triggering by an electron pulse. In this paper, we present preliminary results on the PIPECO study of CO clusters using the above method.

EXPERIMENTAL

The PIPECO apparatus used in this experiment was modified from the molecular beam photoionization apparatus[18,19] described previously. A zero-kinetic energy electron analyzer similar to the design used by Stockbauer[12] was installed opposite to the vertical quadrupole mass spectrometer (QMS).[19] The electron energy resolution of the analyzer used in this study was ~50 meV. The CO molecular beam was produced by supersonic expansion of pure CO at a stagnation pressure (P_0) of 200 or 350 Torr through a nozzle at a nozzle temperature (T) of 120K. The high-intensity central portion of the supersonic jet was collimated into the ionization chamber by a conical skimmer before intersecting at 90° with the dispersed vacuum ultraviolet (VUV) light beam emitted from the exit slit of the VUV monochromator. The electrons and ions formed in the photoionization region were guided by electrostatic lenses toward the electron and ion detectors

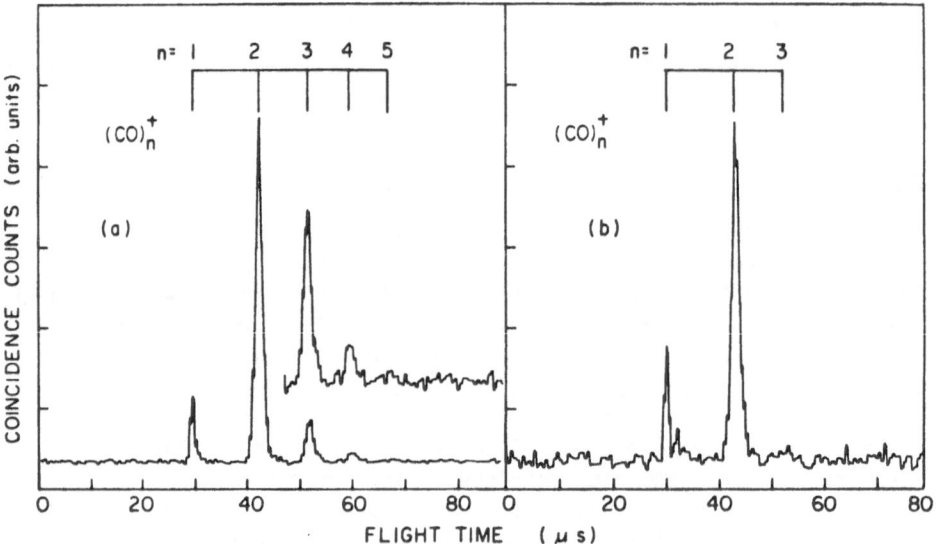

Figure 1. PIPECO TOF mass spectra observed at 910Å.
(a) P_0 = 350 Torr, T = 120K; (b) P_0 = 200 Torr, T = 120K.

RESULTS AND DISCUSSION

Figures 1(a) and 1(b) show the coincidence TOF spectra recorded at 910Å and P_0 = 350 and 200 Torr, respectively. These spectra were obtained when the QMS was operated as a radio frequency ion guide such that all ions were transmitted to the ion detector. The photon energy (13.62 eV) corresponding to 910Å is below the IE of CO. The formation of CO^+ is believed to be caused by scattered light. Similar coincidence TOF spectra were observed at 940Å. The spectra show that the difference in flight time of the correlated $e^- - (CO)_2^+$ pair is 42 μs. At P_0 = 350 Torr, the relative intensities of $(CO)_2^+$, $(CO)_3^+$ and $(CO)_4^+$ were found to be 1.00:0.13:0.03. The TOF peaks for $(CO)_n^+$, n≥3 are indiscernible in the spectrum shown in Fig. 1(b), indicating that the concentrations of $(CO)_n$, n≥3, formed in the CO beam at P_0 = 200 Torr is negligible compared to that of $(CO)_2$. The intensity of $(CO)_2^+$ measured at 900Å and P_0 = 200 Torr was more than a factor of 5 lower than that at P_0 = 350 Torr.

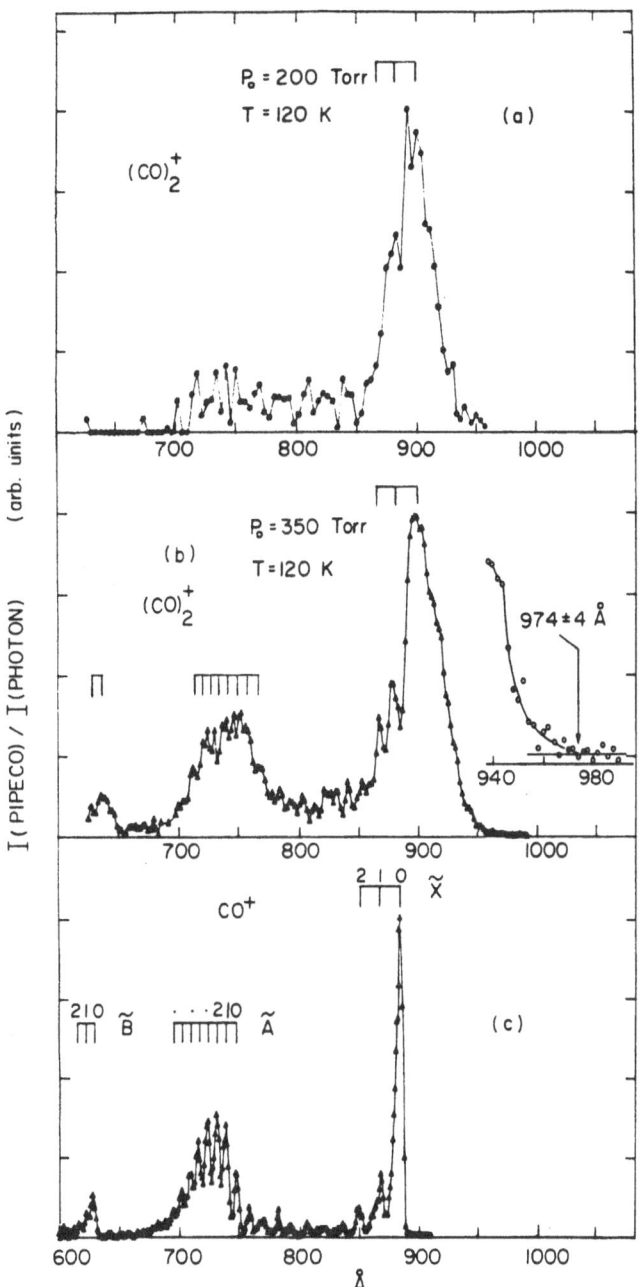

Figure 2. (a) PIPECO spectrum for $(CO)_2^+$ obtained at P_0 = 200 Torr and T = 120K; (b) PIPECO spectrum for $(CO)_2^+$ obtained at P_0 = 350 Torr and T = 120K; (c) PIPECO spectrum for CO^+.

The PIPECO spectra for $(CO)_2^+$ in the wavelength region of 625-990Å obtained at P_0 = 200 and 350 Torr are compared to the PIPECO spectrum for CO^+ in Figs. 2(a)-(c). The three electronic bands resolved in the CO^+ spectrum have been assigned to the $\tilde{X}^2\Sigma^+$, $\tilde{A}^2\Pi$, and $\tilde{B}^2\Sigma^+$ state of CO^+.[20]

The positions of the vibronic states observed in the HeI PES of CO are indicated by tic marks in Fig. 2(c). Vibrational peaks corresponding to the formation of $CO^+(\tilde{X}, v = 3\text{-}7)$ via autoionization are also discernible in the CO^+ spectrum.

Since photoionization involves a vertical transition, $CO^+(\tilde{X}, A, or\tilde{B})$ in $CO^+(\tilde{X}, \tilde{A}, or\tilde{B})\cdot CO$ initially formed at the equilibrium nuclear configuration of $(CO)_2$ can be viewed as a perturbed monomer ion.[1] If $CO^+(\tilde{X}, \tilde{A}, or\tilde{B})CO$ is stable within the flight time (42 μs) of the dimer ion, we expect to find three electronic bands in the PIPECO spectrum of $(CO)_2^+$ which can be associated with the interactions of the $CO^+(\tilde{X})CO$, $CO^+(\tilde{A})\cdot CO$, and $CO(\tilde{B})\cdot CO$ pairs. The $(CO)_2^+$ spectrum measured at $P_0 = 200$ Torr reveals a strong $CO^+(\tilde{X})\cdot CO$ electronic band peaked at 900Å and a weak $CO^+(\tilde{A})\cdot CO$ band at 750Å. Within the sensitivity of this experiment, the $CO^+(\tilde{B})\cdot CO$ band could not be found. The weakness of the $CO^+(\tilde{A})\cdot CO$ and $CO^+(\tilde{B})\cdot CO$ electronic bands suggest that $CO^+(\tilde{A}or\tilde{B})\cdot CO$ formed by photoionization of $(CO)_2$ is dissociative in a time scale shorter than the flight time of $(CO)_2^+$. The $CO^+(\tilde{A}or\tilde{B})\cdot CO$ dimer ions can be stabilized by radiative decay. The radiative lifetimes of $CO^+(\tilde{A})$ and $CO^+(\tilde{B})$ were determined to be ~4 μs[21,22] and ~50 ns,[23] respectively. Assuming the radiative lifetimes of CO^+ $(\tilde{A}, \tilde{B})$ and $CO^+(\tilde{A}, \tilde{B})\cdot CO$ are identical, we estimate that the dissociation lifetimes of $CO^+(\tilde{A})\cdot CO$ and $CO^+(\tilde{B})\cdot CO$ are shorter than ~4 μs and 50 ns, respectively.

The rapid dissociation of $CO^+(\tilde{A}or\tilde{B}), v')\cdot CO$ (X,v=0) initially formed by photoionization of $(CO)_2$ may be rationalized by the mechanism:

$$CO^+(\tilde{A}or\tilde{B}), v')\cdot CO(X,v=0) \xrightarrow{(a)} CO(x,v)\cdot CO^+(\tilde{X},v') \xrightarrow{(b)} CO(X,v) + CO^+(\tilde{X},v') \quad (1)$$

Step 1(a) involves a near resonance intramolecular charge transfer to produce a highly vibrationally excited complex, $CO(X,v)\cdot CO^+(\tilde{X},v')$. This vibrationally excited dimer ion is likely to fall apart efficiently via a vibrational predissociation process (Step 1(b)) because its internal energy content is greater than the dissociation energy of $CO^+(\tilde{X},v'=0)\cdot CO(X,v=0)$. The absence of the $CO^+(\tilde{A})\cdot CO$ and $CO^+(\tilde{B})\cdot CO$ bands is also consistent with repulsive interactions of the $CO^+(\tilde{A}or\tilde{B})\cdot CO$ pairs.

All three electronic bands were observed in the $(CO)_2^+$ spectrum shown in Fig. 2(b). It is interesting to note that the relative intensities of these three bands are similar to those of the CO^+ spectrum. Vibrational structures similar to those resolved in the CO^+ spectrum can also be identified in the dimer ion spectrum. The first electronic band centered at 900Å closely resembles that found in Fig. 2(a). The second and third bands, peaked at 750 and 635Å, respectively, necessarily arise from fragmentation of higher CO cluster ions. Based on the coincidence TOF spectrum observed at $P_0 = 350$ Torr (Fig. 1(a)), it is logical to conclude that $(CO)_2^+$ ions detected at the second and third bands are predominantly produced by fragmentation of $CO^+(\tilde{A}or\tilde{B},v')\cdot (CO)_n$, n = 2 and 3, initially formed by photoionization of $(CO)_3$ and $(CO)_4$. Rapid intramolecular charge transfer processes followed by ejection of vibrationally excited CO molecules can be a very effective mechanism in disposing the excess internal energy of an excited CO cluster ion. For example, a stable $CO^+(\tilde{X},v=0)\cdot CO$ dimer may be formed by the decomposition of a

$CO^+(\tilde{A} \text{ or } \tilde{B}, v') \cdot (CO)_3$ ion via the mechanism:

$$CO^+(\tilde{A} \text{ or } \tilde{B}) \cdot (CO)_3 \xrightarrow{(a)} CO(X,v) \cdot CO^+(\tilde{X},v') \cdot (CO)_2$$

$$\xrightarrow{(b)} CO(X,v) + CO^+(\tilde{X},v') \cdot (CO)_2 \qquad (2)$$

$$CO^+(\tilde{X},v') \cdot (CO)_2 \xrightarrow{(a)} CO(X,v) \cdot CO^+(\tilde{X},v'=0) \cdot CO$$

$$\xrightarrow{(b)} CO(X,v) + CO^+(\tilde{X},v'=0) \cdot CO \qquad (3)$$

Steps 2(a) and 2(b) are similar to Steps 1(a) and 1(b). If $CO^+(\tilde{X},v') \cdot (CO)_2$ produced in Step 2(b) is unstable, it may further eject a $CO(X,v)$ as shown in Steps 3(a) and 3(b) to form a stable $CO^+(\tilde{X},v'=0) \cdot CO$ dimer ion. Since the vibrational spacings in $CO^+(\tilde{X})$ and $CO(X)$ are nearly the same, the interconversion between $CO^+(\tilde{X},v') \cdot (CO)_2$ and $CO(X,v) \cdot CO^+(\tilde{X},v'=0) \cdot CO$ by a resonance charge transfer process (Step 3(a)) should be highly efficient.

The threshold energy for the dissociative photoionization process

$$(CO)_3 + h\nu \longrightarrow (CO)_2^+ + CO \qquad (4)$$

is expected to be higher than the IE for $CO^+(\tilde{X},v'=0) \cdot CO(X,v=0)$. Therefore the IE determination for $(CO)_2$ should not be affected by fragmentation of higher CO cluster ions. The adiabatic IE for $CO^+(\tilde{X},v'=0) \cdot CO(X,v=0)$ is determined to be 12.73±0.05 eV (974±4Å), lower by 0.32 eV than the value obtained from photoion measurement.[24] We find that the PIPECO technique is more sensitive than photoion and photoelectron measurements in determining IE because background ionization events due to scattered light are greatly reduced in coincidence studies. Using the IEs for $CO^+(\tilde{X},v'=0)$[25] and $CO^+(\tilde{X},v'=0) \cdot CO(X,v=0)$ and the estimated binding energy for $CO \cdot CO$,[26] we calculate a value of 29.8±1.0 kcal/mol for the bond dissociation energy of $CO^+(\tilde{X},v'=0) \cdot CO(X,v=0)$.

Following the argument presented above, the second and third electronic bands of the $(CO)_2$ spectrum in Fig. 2(b) are assigned to the formation of $CO^+(\tilde{A} \text{ or } \tilde{B}, v') \cdot (CO)_n$, n = 2 and 3 from photoionization of $(CO)_n$, n = 3 and 4. The fact that these electronic bands are red-shifted by ~0.5 eV compared to the corresponding bands in the CO^+ spectrum indicates that the $CO(\tilde{A} \text{ or } \tilde{B}) \cdot (CO)_n$, n = 2 and 3 complexes are bound. The upper limits for the adiabatic IEs of the $\tilde{A}$ and $\tilde{B}$ bands are estimated to be 15.79 and 18.99 eV, respectively, suggesting that interaction energies of $CO^+(\tilde{A} \text{ or } \tilde{B})$ with 2 or 3 CO are $\geq$ 17 kcal/mol. The previous measurements on the energetics of simple cluster ions[1] show that the stabilization of a cluster ion such as $(CO)_n^+$ is dominated by the dimeric interaction between CO^+ and a CO in the cluster. The observation of bound $CO^+(\tilde{A} \text{ or } \tilde{B}) \cdot (CO)_n$, n = 2 and 3, ions can be taken as a strong support for the conclusion that $CO^+(\tilde{A}) \cdot CO$ and $CO^+(\tilde{B}) \cdot CO$ are bound by more than 0.5 eV.

The ab initio calculation of the isoelectronic $(N_2)_2^+$ systems[27] indicates that potential energy surfaces correlated to $N_2^+(\tilde{A}) + N_2$ are not bound. Recent high level ab initio calculations[28,29] predict a symmetric, trans planar O-C-C-O structure for the ground (^2B_u) and excited (^2B_g) states of $(CO)_2^+$. The study of Blair et al.[28] suggests that the ^2B_u

ground state potential has a well of ~60 kcal/mol at a C-C distance (r_{cc}) of 1.3Å. The ^2B_g excited state is also found to have a potential well at $r_{cc} \sim 1.3$ Å. As r_{cc} increases, the ^2B_g potential surface rises to a barrier at $r_{cc} \sim 1.7$ Å and descends steeply toward the $CO^+(\tilde{A})$ + CO asymptote. Since the equilibrium bond distance of $(CO)_2$ is large (~3.5Å),[30] based on the Franck-Condon consideration it is unlikely that the experiment can sample the inner potential wells at $r_{cc} < 1.7$ Å. The evidence that the $CO^+(\tilde{A}) \cdot CO$ complex may be bound by more than 12 kcal/mol seems to contradict the theoretical prediction.[28]

The experiment to measure the PIPECO spectra of higher CO cluster ions is in progress. The detailed report of this work will be published elsewhere.

Acknowledgement

This work was supported by the Director Office of Basic Energy Sciences of the Departament of Energy under Contract No. W-7405-Eng-82.

References

1. C. Y. Ng, Adv. Chem. Phys. _52_, 263 (1983).

2. P. Kebarle, Ann. Rev. Phys. Chem. _28_, 445 (1977).

3. T. D. Märk and A. W. Castleman, Jr., Adv. At. Mol. Phys. _20_, 65 (1984).

4. A. W. Castleman, Jr. and R. G. Keesee, Ann. Rev. Phys. Chem. _37_, 525 (1986).

5. J. C. Phillips, Chem. Rev. _86_, 619 (1986).

6. A. W. Castleman, Jr. and R. G. Keesee, Chem. Rev. _86_, 589 (1986).

7. "The Physics and Chemistry of Small Clusters", P. Jena, Ed. (Plenum, New York, 1987) (Proceedings of the International Symposium on the Physics and Chemistry of Small Clusters, Richard, VA, 1986).

8. M. F. Jarrold, A. J. Illies, and M. T. Bowers, J. Chem. Phys. _81_, 222 (1984); _79_, 6086 (1983); _81_, 214 (1984); A. J. Illies, M. F. Jarrold, W. Wagner-Redeker and M. T. Bowers, J. Phys. Chem. _88_, 5204 (1984).

9. S. C. Ostrander and J. C. Weisshaar, Chem. Phys. Lett. _129_, 220 (1986); S. C. Ostrander, L. Sanders and J. C. Weisshaar, J. Chem. Phys. _84_, 529 (1986).

10. B. Brehm and E. von Puttkamer, Z. Naturforsch. Teil _A22_, 8 (1967).

11. J.H.D. Eland, Int. J. Mass Spectrom. Ion Phys. **8**, 143 (1972).

12. R. Stockbauer, J. Chem. Phys. **58**, 3800 (1973).

13. M. E. Gellender and A. D. Baker, in C. R. Brundle and A. D. Baker, Eds., "Electron Spectroscopy", (Academic Press, New York, 1977), Vol. 1, p. 435.

14. T. Baer, in M. T. Bowers, Ed., "Gas Phase Ion Chemistry", (Academic Press, New York, 1979), Vol. 1, p. 153.

15. E. D. Poliakoff, P. M. Dehmer, J. L. Dehmer and R. Stockbauer, J. Chem. Phys. **75**, 5214 (1982).

16. L. Cordis, G. Ganteför, J. Beßlich and A. Ding, Z. Phys. **D3**, 323 (1986).

17. C. Holzapfel, Rev. Sci. Instrum. **45**, 894 (1974).

18. Y. Ono, S. H. Linn, H. F. Prest, M. E. Gress and C. Y. Ng, J. Chem. Phys. **73**, 2523 (1980).

19. C.-L. Liao, J.-D. Shao, R. Xu, G. D. Flesch, Y.-G. Li and C. Y. Ng, J. Chem. Phys. **85**, 3874 (1986).

20. D. W. Turner, C. Baker, A. D. Baker and C. R. Brundle, "M9lecular Photoelectron Spectroscopy", (Wiley, New York, 1970).

21. J. Danon, G. Mauclaire, T. R. Govers and R. Marx, J. Chem. Phys. **76**, 1255 (1982).

22. V. E. Bondybey and T. A. Miller, J. Chem. Phys. **69**, 3597 (1978).

23. M. Bloch and D. W. Turner, Chem. Phys. Lett. **30**, 344 (1975).

24. S. A. Linn, Y. Ono and C. Y. Ng, J. Chem. Phys. **74**, 3342 (1981).

25. J. H. Fock, P. Gürtler and E. E. Koch, Chem. Phys. **47**, 87 (1980).

26. J. O. Hirschfelder, C. F. Curtiss and R. B. Bird, "Molecular Theory of Gases and Liquids", (Wiley, New York, 1964), p. 111.

27. S. C. deCastro, H. F. Schaefer III, and R. M. Pitzer, J. Chem. Phys. **74**, 550 (1981).

28. J. T. Blair, J. C. Weisshaar, J. E. Carpenter and Frank Weinhold, J. Chem. Phys. **87**, 392 (1987).

29. L. B. Knight, J. Steadman, P. K. Miller, D. E. Bowman, E. R. Davidson D. Feller, J. Chem. Phys. **80**, 4593 (1984).

30. R. M. Berns and A. van der Avoird, J. Chem. Phys. **72**, 6107 (1980).

DIRECT CALCULATION OF MOLECULAR TRANSITION ENERGIES

BY THE OPEN-SHELL COUPLED-CLUSTER METHOD[*]

Uzi Kaldor and Sigalit Ben-Shlomo

School of Chemistry
Tel Aviv University
69 978 Tel Aviv, Israel

INTRODUCTION

The exp(S) or coupled-cluster (CC) method[1-5] has been used widely in recent years for ab initio electronic structure calculations in closed-shell, non-degenerate systems, with highly satisfactory results.[6] The CCSD approximation,[7] in which single and double excitations are included to all orders, is usually employed; a few calculations including the effect of triple excitations (CCSDT) have appeared recently.[8] The theory becomes considerably more complicated when the system of interest cannot be described in terms of a closed-shell structure. A variety of multireference, open-shell (OSCC) formulations, designed to handle such situations, have been described.[9-23]

Recently we reported[24-31] the application of OSCC to the direct calculation of electron affinities (EA), ionization potentials (IP) and excitation energies (EE). These applications are summarized below, and new results are described. The method used largely follows Lindgren's normal-ordered formalism.[14] Single and double virtual excitations are included to all orders, while triple excitations with one or more electrons excited out of the valence shell are calculated to lowest order, using the converged CCSD amplitudes (the CCSD+T approximation[25]). Excitations of three closed-shell electrons do not contribute to electronic transition energies in this approximation, and are therefore ignored.

---
[*]Supported in part by the U.S.-Israel Binational Science Foundation.

METHOD

The Hamiltonian of the system is separated in the usual way into a zero-order operator H_0, with known eigenfunctions, and a perturbation V,

$$H = H_0 + V \tag{1}$$

$$H_0|\alpha\rangle = E_0{}^\alpha|\alpha\rangle \; . \tag{2}$$

A d-dimensional model space P and its complement Q are defined by projection operators,

$$P = \sum_{\alpha \in P} |\alpha\rangle \langle \alpha| \; , \quad Q = 1 - P \; . \tag{3}$$

There will usually be d eigenfunctions of H with major components in the model space,

$$H\Psi^a = E^a\Psi^a \; , \tag{4}$$

$$P\Psi^a = \Psi_0{}^a \; , \quad a=1,2,...,d \tag{5}$$

where $\Psi_0{}^a$ are linear combinations of $|\alpha\rangle$ $\alpha \in P$. The wave operator Ω transforms the model functions into exact ones,

$$\Omega\Psi_0{}^a = \Psi^a \; , \quad a=1,2,...,d \; . \tag{6}$$

Intermediate normalization is assumed,

$$\langle\Psi^a|\Psi_0{}^a\rangle = \langle\Psi_0{}^a|\Psi_0{}^a\rangle = 1 \; . \tag{7}$$

The key equation in Lindgren's derivation[14] is the generalized Bloch equation

$$[\Omega,H_0]P = V\Omega P - \Omega PWP, \tag{8}$$

where W is the effective interaction

$$W = V\Omega \; . \tag{9}$$

An alternative form of (8) is

$$[\chi,H_0]P = QWP - \chi PWP \; , \tag{10}$$

where the correlation operator χ is defined by

$$\Omega = 1 + \chi . \tag{11}$$

The energies of interest are obtained by diagonalizing the effective Hamiltonian in the model space,

$$H_{eff} \Psi_0{}^a = E^a \Psi_0{}^a , \tag{12}$$
where

$$H_{eff} = PH\Omega P = P(H_0 + W)P . \tag{13}$$

The correlation operator χ includes single, double, ... , virtual excitations and may be written as

$$\chi = C_1 + C_2 + ... = \sum_{ij} \{a_i^\dagger a_j\} \, t_j^i + \frac{1}{2} \sum_{ijkl} \{a_i^\dagger a_j^\dagger a_l a_k\} \, t_{kl}^{ij} + ... \tag{14}$$

t_j^i, t_{kl}^{ij}, ... , are excitation amplitudes, and the curly brackets denote normal order with respect to a reference (core) determinant. <u>All</u> terms, connected as well as disconnected, are included in (14). The operator used in CC is the excitation operator T, related to Ω by

$$\Omega = \{\exp(T)\} = 1 + T + \frac{1}{2}\{T^2\} + ... \tag{15}$$

T is given by summing the rhs of (14) over <u>connected</u> terms only. Perturbative or non-perturbative schemes for calculating the excitation operator and correlation energies may be derived from either of the following two equations, which include connected terms only[14]

$$[T,H_0] = (QV\Omega - \chi PV\Omega)_{conn} , \tag{16}$$
or

$$[T,H_0] = W_{op,conn} - (\chi W_{cl})_{conn} . \tag{17}$$

$W_{op,conn}$ describes all connected diagrams which have some open (non-valence) lines, corresponding to P→Q transitions. W_{cl} diagrams, with no external non-valence lines, describe P→P transitions. The latter also appear in the effective Hamiltonian, which may be written as

$$H_{eff} = PH_0 P + W_{cl} . \tag{18}$$

The second term in equation (16) or (17) gives rise to the so-called folded diagrams.

The T operator for an open-shell system may be partitioned according to the number of valence orbitals excited,

$$T = T^{(0)} + T^{(1)} + T^{(2)} + ... \quad . \tag{19}$$

Haque and Mukherjee[21] have shown that partial decoupling of the equations is then possible, as the equations for $T^{(n)}$ involve only $T^{(m)}$ elements with m≤n. This decoupling is helpful in reducing the computational effort required, and has been used in our calculations.[24-31]

The H_{eff} or W_{cl} diagrams (see Eq. (18)) may be separated into core and valence parts,

$$H_{eff} = H_{eff}^{core} + H_{eff}^{val} , \qquad (20)$$

where the first term on the rhs consists of diagrams without any external lines. The eigenvalues of H_{eff}^{val} will then give directly the transition energies from the core, with correlation effects included for both core and valence electrons. The physical significance of these energies depends on the nature of the model space. Thus, electron affinities may be calculated by constructing a model space with valence particles only,[24,25,29] ionization potentials are given using valence holes,[26] and both types are included for the purpose of getting excitations out of a closed-shell system.[27,28,30] The expansion (19) is replaced in the latter case by

$$T = T^{(0,0)} + T^{(1,0)} + T^{(0,1)} + T^{(1,1)} + .. \qquad (21)$$

where the two superscripts give the numbers of valence holes and particles, respectively. The same decoupling holds as for Eq. (19).

Model Spaces

Model (P) spaces in OSCC or in many-body perturbation theory (MBPT) are usually constructed by first partitioning the orbitals into core, valence and particles, and including in P determinants with all possible combinations of valence orbitals (core orbitals are always occupied). This choice, called[32] a "complete" model space, is assumed in the well-known derivations of linked-diagram theorems.[33,14] The first MBPT method capable of handling more general model spaces was described and applied by Hose and Kaldor,[32,34] and a similar OSCC method was later given by Jeziorski and Monkhorst.[18] A complete model space presents a reasonable choice when all valence orbitals are close in energy and belong to the same type (holes or particles). Molecular excited states will usually involve both valence holes and particles with respect to a closed-shell reference determinant; complete model spaces would then be very large, with determinants describing multiple ionizations and additions of electrons. Such a model space will have a very broad energy span and involve many types of T elements, making the calculation unwieldy at best and unlikely to converge in most cases, due to the appearance of intruder states.[35] A reasonable model space for n-excited states would include only determinants with exactly n valence holes and n valence particles. This so-called "quasicomplete" space, proposed by Lindgren,[15] will lead in general to unlinked diagrams.[15,36,37] It has however been shown[37] that the usual coupled-cluster expansion remains linked for a quasicomplete space with one valence hole and one valence particle.

Such spaces are used for singly-excited atomic and molecular states.[27,28,30] A more general method, applicable to any model spaces, has been derived recently by Mukherjee.[36] A key step in Mukherjee's derivation is abandoning the intermediate normalization (7). An arbitrary model space with m valence orbitals is defined, and operators of valence rank k ($0 \leq k \leq m$) are called m-open if they can lead to $P \rightarrow Q$ transitions in the m-valence sector. The resulting equations for Ω defined by (15) are[37]

$$[T,H_0]_{m-op}^{(k)} = \{V\Omega - \Omega W\}_{m-op}^{(k)} \tag{22}$$

$$\{\Omega W\}_{cl}^{(k)} = \{V\Omega\}_{cl}^{(k)} . \tag{23}$$

Equation (22) is identical with (16), except for the different structure of the model space. Equation (23) is somewhat more complicated than its complete-model-space counterpart (9), leading to coupled equations for the elements of the effective interaction W. These equations have low dimensionality and are simpler than the equations for the T elements, (16) or (22), and the added complications are not severe. An application of the method to several states of LiH is described below. A slightly different scheme has been recently applied to H_2 by Koch and Mukherjee.[38]

DIAGRAMS

The application of the coupled-cluster method is most easily described in terms of Goldstone-type diagrams. The atomic and molecular computations described below employ the CCSD approximation, including single and double excitations,

$$T \simeq T_1 + T_2 . \tag{24}$$

T_3 is calculated approximately[25] in some cases. T elements with no more than two valence lines (hole or particle) are calculated. This requires diagrams for $T_k^{(n,m)}$, with k=1,2 and $0 \leq n+m \leq 2$, where n and m are the numbers of incoming hole- and particle-valence lines, respectively, and k is the total number of incoming lines. The rules for drawing and calculating the diagrams are well-known (see e.g. Lindgren and Morrison,[14] Chap. 15). A modification of the program generating many-body perturbation theory diagrams[39] can be used. The Ω expansion under the approximation (24) is finite,

$$\Omega = 1 + T_1 + \frac{1}{2}\{T_1^2\} + T_2 + \frac{1}{3!}\{T_1^3\} + \{T_1 T_2\}$$
$$+ \frac{1}{4!}\{T_1^4\} + \frac{1}{2}\{T_1^2 T_2\} + \frac{1}{2}\{T_2^2\} . \tag{25}$$

The closed-shell diagrams (n=m=0, k=1,2) have been listed by Cullen and Zerner.[7] More recent publications include full descriptions of one- and two-body diagrams with a single valence particle[29] (n=0, m=1, k=1,2), from which one-valence-hole diagrams are easily obtained by reversing all line directions, and of $T_k^{(1,1)}$ diagrams, with k=1,2.[30] Lowest-order $T_3^{(0,1)}$ diagrams have also been listed.[25]

APPLICATIONS

Several applications implementing the methods described above have already been reported, and are briefly summarized below. These include the ionization potentials and several excitation energies (about ten per system) of Be,[27] Ne,[27] Mg,[28] Ar,[28] and H_2O,[30] as well as ionization potentials and electron affinities of the alkali atoms Li, Na, K, Rb, and Cs.[31] Highly satisfactory results were obtained in all cases; approximate inclusion of T_3 was necessary for some systems. Calculations of transition energies in other systems, namely O_2, N_2 and LiH, have not been reported before, and preliminary results for them are given here.

Table 1. O_2 Levels.

| | R_e(A) | | | ω_e(cm^{-1}) | | | T_e(eV) | |
	$^3\Sigma_g^-$	$^1\Delta_g$	$^1\Sigma_g^+$	$^3\Sigma_g^-$	$^1\Delta_g$	$^1\Sigma_g^+$	$^1\Delta_g$	$^1\Sigma_g^+$
Exp[42]	1.208	1.216	1.227	1580	1509	1433	0.982	1.636
4s2p1d basis								
CI[43]	1.24	1.25	1.26	1693	1595	1505	1.09	1.69
CCSD	1.23	1.24	1.25	1543	1485	1429	1.25	2.15
CCSD+T	1.23	1.23	1.24	1602	1583	1561	1.13	1.85
4s3p2d basis								
CCSD	1.23	1.23	1.24	1550	1530	1515	1.22	2.13
CCSD+T	1.22	1.22	1.23	1560	1554	1572	1.09	1.79

The Three Lowest States of the Oxygen Molecule

The energies of the three lowest states of O_2, resulting from the $1\pi_u^2$ electron configuration, were calculated by the open-shell CCSD method. Two basis sets were used,[40,41] the DZP set with 4 s, 2 p and 1 d orbital on each atom, and the somewhat larger 4s3p2d set. O_2^{--} served as the closed-shell reference state, and the states of interest were obtaned by ionizing two electrons from the $1\pi_u$ orbital. The results are summarized and compared with experiment[42] and previous calculations[43] in Table 1. The calculation with the larger basis and the lowest-order inclusion of triple excitations reproduces the experimental equilibrium separations R_e to 0.01A, the vibrational frequency ω_e to 20-140cm^{-1}, and the adiabatic excitation energies T_e to 0.15eV.

N_2 has a rich excitation spectrum, and is therefore a good test case for the method discussed here. Its IP's and EE's were computed with two basis sets, the 4s2p1d (double zeta plus polarization) set used by Bauschlicher and Langhoff,[44] and the 6311G**+ set developed by Pople's group,[45] a 5s4p1d basis which includes a set of sp Rydberg orbitals. The valence space for the smaller basis included only the $3\sigma_g$ and $1\pi_u$ holes and the $1\pi_g$ particles. For the larger basis, the $2\sigma_u$ also served as a valence hole, while the valence

Table 2. Ionization Potentials and Excitation Energies of N_2 (eV).

		Exp[47]	4s2p1d CCSD	5s4p1d CCSD	CI[46]
IP	$3\sigma_g$	15.58	15.18	15.44	14.91
	$1\pi_u$	16.70	16.94	17.11	16.64
	$2\sigma_u$	18.75	--	18.66	18.08
$3\sigma_g \to 1\pi_g$	B $^3\Pi_g$	8.12	7.98	8.05	8.28
	a $^1\Pi_g$	9.39	9.31	9.27	9.63
$1\pi_u \to 1\pi_g$	A $^3\Sigma_u^+$	7.78	7.68	7.55	7.93
	W $^3\Delta_u$	8.9	9.04	8.92	9.39
	B' $^3\Sigma_u^-$	9.7	9.31	9.86	9.98
	a' $^1\Sigma_u^-$	9.9	10.02	10.08	10.56
	w $^1\Delta_u$	10.3	10.23	10.53	10.73
	b' $^1\Sigma_u^+$	14.4	16.44	14.89	14.74
$1\pi_u \to 3s$	C $^3\Pi_u$	11.1	--	11.19	11.36
	b $^1\Pi_u$	13.4	--	13.62	13.77
Average error		--	0.19[a]	0.19	0.37

[a]Not including the b' state.

particle space included two σ_g, two σ_u, one π_u and two π_g orbitals. The calculated energies are collected in Table 2 and compared with the results of large CI[46] and with experiment.[47] Very good agreement with experiment is obtained even with the small basis, except for the b'$^1\Sigma_u^+$ state which has a strong Rydberg character. The larger basis gives three IP's and 10 EE's with an average error of 0.19eV. This may be compared with an average error of 0.37eV for extensive CI calculation with an even larger basis.[46]

Excitation Energies of Closed-Shell Atoms and H_2O

Excitation energies and ionization potentials were calculated for the closed-shell atoms Be,[27] Ne,[27] Mg[28] and Ar,[28] and for the water molecule.[30] About 10 transition energies of each system were studied, and T_3 effects were included to lowest order.[25] All the results were within 0.1-0.2eV of experiment. This includes rather sensitive levels, such as the lowest D states of Mg, which appeared in reverse order (singlet below triplet) as they should. T_3 effects proved very important for Ne and isoelectronic H_2O, contributing 0.2-0.4eV to excitation energies and 0.5-0.7eV to ionization potentials. Be and Mg, essentially two-electron systems, showed as expected small effects of triple excitations, up to 0.15eV. Somewhat surprising were the small (≤ 0.1eV) T_3 corrections for Ar. A possible explanation is the greater spatial extension of the Ar valence shell, as compared with Ne. Another interesting result involves the use of Hartree-Fock orbitals for the singly-ionized system rather than the neutral atom or molecule. Contrary to expectations, such orbitals gave poorer CCSD results; these were largely offset by bigger T_3 corrections, making the final CCSD+T values of the two orbital sets very close.

Electron Affinities and Ionization Potentials of the Alkali Atoms

A preliminary study[29] of Li and Na, with small basis sets, gave very good electron affinities (errors of 0.015eV or 3%). It should be noted that the Hartree-Fock EA's are negative (the energy of the anion is higher than that of the neutral atom), and correlation is crucial in getting correct values. The HF ionization potentials, on the other hand, are in fair agreement with experiment (the error is 0.2eV for Na), and very little improvement was obtained in the small-basis CCSD calculation.[29] Previous studies[48] using the configuration interaction (CI) method found that important contributions to the IP come from the core-valence correlation, i.e. the correlation between the ns and (n-1)p electrons. A serious problem encountered in the CI work is that inclusion of core-core correlation, in addition to core-valence, gives much poorer IP's. This effect results from the size-inconsistency of the CI method,[49] manifested by the appearance of unlinked terms in the expansion.

The CCSD method was applied to the alkali atoms using moderately large bases, from a set of 5 s, 5 p and one d contracted Gaussian-type orbitals for Li, to 11 s, 9 p and 5 d orbitals for Cs.[31] The M^+ ion served as the reference closed-shell system. Two sets of calculations were performed, correlating the ns electrons with or without the (n-1)sp shell. The main contribution to the electron affinities comes from the $(ns)^2$ correlation, which by itself brings the results to within 0.02eV of experiment. The inclusion of (n-1)s and p correlation has small effect (<0.03eV) and gives some improvement, cutting the error to 0.01eV. The correlated ^2S ionization potentials are within 0.02eV (Li) to 0.22eV (Cs) of experiment; when relativistic corrections, estimated from numerical Hartree-Fock and Dirac-Fock calculations, are added, the agreement with experiment is better than 0.1eV for all atoms. The ^2P ionization potentials are even better, with errors of 0.02-0.04eV.[31] It should be noted that the inclusion of both core-core and core-valence correlation causes no problems.

Convergence problems appear very often in multi-reference many-body perturbation theory (MBPT) and coupled-cluster calculations. They usually result from the use of "complete"[32] model spaces, comprising determinants constructed from all possible combinations of valence orbitals. The LiH molecule is a case in point. We tried to construct potential functions for the ground $X^1\Sigma^+$ state, with the configuration $1\sigma^2 2\sigma^2$, and the four lowest excited states, the $A^1\Sigma^+$ and $a^3\Sigma^+$ obtained by the $2\sigma \rightarrow 3\sigma$ transition and the $B^1\Pi$ and $b^3\Pi$ resulting from $2\sigma \rightarrow 1\pi$. Starting from the LiH^{++} determinant and adding the 2σ, 3σ and 1π orbitals as valence particles, the CCSD iterations could not be converged. The Hartree-Fock orbital energies are (in Hartree atomic units) -0.82, -0.39, -0.38, -0.30, and -0.26 for the 2σ, 1π, 3σ, 4σ and 5σ orbitals, respectively. A complete model space including the determinants $2\sigma^2$ and $3\sigma^2$, among others, will obviously have many Q-space determinants (e.g. $2\sigma 4\sigma$ and $2\sigma 5\sigma$) with energies inside the P-space range, a recipe for divergence. Freed and his coworkers[42] solved the problem within the MBPT framework by assigning equal energies to all valence orbitals, and including these energy shifts in the perturbation. The additional perturbation shows up in higher order only, and good results have been obtained in this way for a variety of atoms and molecules by second and third-order MBPT.[50] This stratagem cannot be used in CC work, as the infinite-order summation inherent in the method will bring the orbital energies back to their original values. Divergence can be avoided only by using an incomplete P space without the $1\sigma^2 3\sigma^2$ and $1\sigma^2 1\pi^2$ determinants, which are anyhow not expected to make major contributions to the molecular states of interest.

Table 3. LiH Results.

		D_e(eV)	R_e(A)	T_e(eV)	ω_e(cm^{-1})
$X^1\Sigma^+$	exp	2.52	1.5957	--	1406
	CCSD	2.42	1.588	--	1380
$A^1\Sigma^+$	exp	1.08	2.60	3.287	281
	CCSD	1.05	2.45	3.243	299
$B^1\Pi$	exp	0.03	2.378	4.328	130
	CCSD	0.01	2.50	4.299	121
$a^3\Sigma^+$	CCSD	repulsive			
$b^3\Pi$	CCSD	0.22	1.98	4.084	623

The concept of incomplete model spaces was first proposed by Hose and Kaldor,[32] who derived a general-model-space MBPT. Both the Hose-Kaldor formalism and the Jeziorski-Monkhorst[18] CC method include disconnected diagrams, which however do not spoil the size-consistence of the method, as no individual part of the disconnected diagram is a legitimate term by itself. Mukherjee[36] has recently described an incomplete-model-space CC scheme without disconnected diagrams, and this method is applied to LiH, with the P space including the determinants $1\sigma^2 2\sigma^2$, $1\sigma^2 2\sigma 3\sigma$, and $1\sigma^2 2\sigma 1\pi$. Results of calculations using the 37-function basis set of Liu et al.[51] are reported in Table 3. Excitation and dissociation energies in very good agreement with experiment are obtained.

SUMMARY AND CONCLUSION

The open-shell coupled cluster method has been applied to the direct calculation of electronic transition energies, including excitation energies, electron affinities and ionization potentials, of several atomic and molecular systems. Virtual single and double excitations are summed to infinite order (the CCSD approximation); triple excitations are included in the lowest order they appear (CCSD+T) in certain cases. Good agreement with experiment is achieved. The errors in the calculated results are 0.01eV for the alkali atoms electron affinities, 0.02-0.1eV for their ionization potentials (when corrected for relativity), and 0.1-0.2eV for the excitation energies of Ne, Mg, Ar, N_2, O_2 and H_2O. These errors are due to deficiencies in the basis sets and to approximations made in the excitation operator T. Of the latter, we believe the inclusion of T_3 to lowest-order only is more serious than the neglect of connected four- and higher-electron excitations.

REFERENCES

1. J. Hubbard, Proc. Roy. Soc. A240:539 (1957); ibid. A243:336 (1958).
2. F. Coester, Nucl. Phys. 7:421 (1958); F. Coester and H. Kümmel, Nucl. Phys. 17:477 (1960); H. Kümmel, K. H. Lührmann and J. G. Zabolitzky, Phys. Rept. 36:1 (1978).
3. O. Sinanoglu, Adv. Chem. Phys. 6:315 (1964).
4. J. Cizek, J. Chem. Phys. 45:4256 (1966); Adv. Chem. Phys. 14:35 (1969).
5. J. Paldus, J. Cizek and I. Shavitt, Phys. Rev. A 5:50 (1972); J. Paldus, J. Chem. Phys. 67:303 (1977); B. G. Adams and J. Paldus, Phys. Rev. A 20:1 (1979).
6. For a review see R. J. Bartlett, Ann. Rev. Phys. Chem. 32:359 (1981).
7. G. D. Purvis and R. J. Bartlett, J. Chem. Phys. 76:1910 (1982); J. M. Cullen and M. C. Zerner, J. Chem. Phys. 77:4088 (1982). Both references give the explicit CCSD equations, and the second one also shows the CCSD diagrams.
8. Y. S. Lee and R. J. Bartlett, J. Chem. Phys. 80:4371 (1984); Y. S. Lee, S. A. Kucharsky, and R. J. Bartlett, J. Chem. Phys. 81:5906 (1984); M. Urban, J. Noga, S. J. Cole, and R. J. Bartlett, J. Chem. Phys. 83:4041 (1985); J. Noga and R. J. Bartlett, J. Chem. Phys. 86:7041 (1987).
9. F. E. Harris, Intern. J. Quantum Chem. S11:403 (1977).
10. H. J. Monkhorst, Intern. J. Quantum Chem. S11:421 (1977).
11. J. Paldus, J. Cizek, M. Saute and A. Laforgue, Phys. Rev. A 17:805 (1978); M. Saute, J. Paldus and J. Cizek, Intern. J. Quantum Chem. 15:463 (1979).

12. D. Mukherjee, R. K. Moitra and A. Mukhopadhyay, Pramana 4:247 (1975); Mol. Phys. 30:1861 (1975); A. Mukhopadhyay, R. K. Moitra and D. Mukherjee, J. Phys. B 12:1 (1979); D. Mukherjee and P. K. Mukherjee, Chem. Phys. 39:325 (1979); S. S. Adnan, S. Bhattacharyya and D. Mukherjee, Mol. Phys. 39:519 (1980); Chem. Phys. Lett. 85:204 (1981).

13. R. Offerman, W. Ey and H. Kümmel, Nucl. Phys. A273:349 (1976); R. Offerman, Nucl. Phys. A273:368 (1976); W. Ey, Nucl. Phys. A296:189 (1978).

14. I. Lindgren, Intern. J. Quantum Chem. S12:33 (1978); S. Salomonson, I. Lindgren and A. M. Martensson, Phys. Scr. 21:351 (1980); I. Lindgren and J. Morrison, "Atomic Many-Body Theory", Springer, Berlin, (1982).

15. I. Lindgren, Phys. Scr. 32:291, 32:611 (1985).

16. H. Nakatsuji, Chem. Phys. Lett. 59:362 (1978); ibid. 67:329 (1979); Chem. Phys. 75:425 (1983); ibid. 76:283 (1983); J. Chem. Phys. 80:3703 (1984).

17. H. Reitz and W. Kutzelnigg, Chem. Phys. Lett. 66:111 (1979); W. Kutzelnigg, J. Chem. Phys. 77:3081 (1981); ibid. 80:822 (1984).

18. B. Jeziorski and H. J. Monkhorst, Phys. Rev. A 24:1668 (1981); L. Z. Stolarczyk and H. J. Monkhorst, Phys. Rev. A 32:725, 32:743 (1985).

19. A. Banerjee and J. Simons, Intern. J. Quantum Chem. 19:207 (1981).

20. V. Kvasnicka, Chem. Phys. Lett. 79:89 (1981).

21. A. Haque and D. Mukherjee, J. Chem. Phys. 80:5058 (1984); Pramana 23:651 (1984).

22. J. Arponen, Ann. Phys. (NY) 151:311 (1983).

23. K. Tanaka and H. Terashima, Chem. Phys. Lett. 106:558 (1984).

24. A. Haque and U. Kaldor, Chem. Phys. Lett. 117:347 (1985).

25. A. Haque and U. Kaldor, Chem. Phys. Lett. 120:261 (1985).

26. A. Haque and U. Kaldor, Intern. J. Quantum Chem. 29:425 (1986).

27. U. Kaldor and A. Haque, Chem. Phys. Lett. 128:45 (1986).

28. U. Kaldor, Intern. J. Quantum Chem. S20:445 (1986).

29. U. Kaldor, J. Comput. Chem. 8:448 (1987).

30. U. Kaldor, J. Chem. Phys. 87:467 (1987).

31. U. Kaldor, J. Chem. Phys. 87:4693 (1987).

32. G. Hose and U. Kaldor, J. Phys. B 12:3827 (1979).

33. B. H. Brandow, Rev. Mod. Phys. 39:771 (1967).

34. G. Hose and U. Kaldor, Phys. Scr. 21:357 (1980); Chem. Phys. 63:165 (1981); J. Phys. Chem. 86:2133 (1982); Phys. Rev. A 30:2932 (1984); U. Kaldor, J. Chem. Phys. 81:2406 (1984).

35. T. H. Schucan and H. A. Weidenmuller, Ann. Phys. (NY) 73:108 (1972); ibid. 76:483 (1973).

36. D. Mukherjee, Chem. Phys. Lett. 125:207 (1986); Intern. J. Quantum Chem. S20:409 (1986).

37. D. Sinha, S. Mukhopadhyay, and D. Mukherjee, Chem. Phys. Lett. 129:369 (1986).

38. S. Koch and D. Mukherjee, preprint (1987).

39. U. Kaldor, J. Comput. Phys. 20:432 (1976).

40. S. Huzinaga, J. Chem. Phys. 42:1293 (1965).

41. T. H. Dunning, J. Chem. Phys. 53:2823 (1970). The d exponents were 0.9 for the 4s2p1d set, 0.62 and 2.09 for the 4s3p2d set.

42. P. Krupenie, J. Phys. Chem. Ref. Data 1:423 (1972).

43. B. J. Moss and W. A. Goddard III, J. Chem. Phys. 63:3523 (1975).

44. C. W. Bauschlicher and S. R. Langhoff, J. Chem. Phys. 86:5595 (1987).

45. R. Krishnan, J. S. Brinkley, R. Seeger, and J. A. Pople, J. Chem. Phys. 72:650 (1980); M. J. Frisch, J. A. Pople, and J. S. Binkley, J. Chem. Phys. 80:3265 (1984).

46. S. K. Shih, W. Butscher, R. J. Buenker, and S. D. Peyerimhoff, Chem. Phys. 29:241 (1978).

47. K. P. Huber and G. Herzberg, "Constants of Diatomic Molecules", Van Nostrand Reinhold, New York (1979); J. Oddershede, Adv. Chem. Phys. 69:201 (1987); A. Lofthus and P. H. Krupenie, J. Phys. Chem. Ref. Data 6:113 (1977).

48. G. H. Jeung, J. P. Daudey, and J. P. Malrieu, J. Chem. Phys. 77:3571 (1982); Chem. Phys. Lett. 94:300 (1983); J. Phys. B 16:699 (1983); S. P. Walch, C. W. Bauschlicher, P. E. M. Siegbahn, and H. Partridge, Chem. Phys. Lett. 92:54 (1982); H. Partridge, D. A. Dixon, S. P. Walch, C. W. Bauschlicher, and J. L. Gole, J. Chem. Phys. 79:1859 (1983); H. Partridge, C. W. Bauschlicher, S. P. Walch, and B. Liu, J. Chem. Phys. 79:1866 (1983); W. Müller, J. Flesch, and W. Meyer, J. Chem. Phys. 80:3297 (1984).

49. H. Primas, in "Modern Quantum Chemistry", O. Sinanoglu, ed., Academic, New-York (1965), Vol 2.

50. S. Iwata and K. F. Freed, J. Chem. Phys. 61:1500 (1974); K. F. Freed, in "Modern Thoretical Chemistry", G. A. Segal, ed., Plenum, New York (1977); H. Sung, K. F. Freed, M. F. Herman, and D. L. Yeager, J. Chem. Phys. 72:4158 (1980); M. G. Sheppard and K. F. Freed, J. Chem. Phys. 75:4525 (1981).

51. S. Liu, M. F. Daskalakis, and C. E. Dykstra, J. Chem. Phys. 85:5877 (1986). The basis denoted L was used, with five orbitals in each d set.

TOPOLOGY OF THE GROUND-STATE

SURFACE OF CH_4^+

Regina F. Frey and Ernest R. Davidson

Department of Chemistry
Indiana University
Bloomington, IN 47405

INTRODUCTION

Potential-energy surfaces (PES) are used to explain the equilibrium
structures and spectra of molecules, as well as chemical reactions between
molecules. Improvements in both experimental and theoretical techniques
have led to increased interest in detailed studies of polyatomic systems.
This in turn has led to a need for understanding the potential-energy
surfaces of polyatomic systems.

In a previous paper,[1] the potential-energy surfaces of the three
lowest electronic states of the methane radical cation, CH_4^+, were
calculated. The tetrahedral configuration of CH_4^+ has a triply degenerate
$^2T_2$ state that is subject to a Jahn-Teller (JT) splitting. The CI energies
of the three states were fit to the eigenvalues of a 3x3 matrix whose
elements are a power-series expansion in the normal-mode displacements
around the SCF tetrahedral geometry. Various stationary points were
located on the fitted potential and the pseudo-rotation between the ground-
state C_{2v} minima was studied.

In this paper, the topology of the ground-state PES of CH_4^+ is studied
in greater detail. In particular, attention is paid to the regions around
the intersection of two or more states. In addition, the existence of an

asymptotic bifurcation in the Fukui reaction path between adjacent C_{2v} minima is discussed.

INTERSECTION OF POTENTIAL SURFACES

In the neighborhood of the intersection of two or more molecular states, the structure of the surface and the related wavefunction have been described by Von Neumann and Wigner,[2] Jahn and Teller,[3,4] and Herzberg and Longuet-Higgins.[5,6] Von Neumann and Wigner[2] described the case of degeneracy due to crossing of surfaces without regard to symmetry. They showed that two potential surfaces obtained from a Hamiltonian that depends on K parameters will almost always intersect in a surface of dimensionality K-2. Similarly, the dimensionality for triple degeneracy was shown to be K-5. Teller[3] applied this theorem to derive the "noncrossing" rule for diatomic molecules which says that the potential-energy curves only intersect if the states differ in symmetry or spin and Jahn and Teller[4] used the theorem to show that degeneracy induced by point-group symmetry can always be removed by a symmetry-lowering distortion that leads to a lower energy. As Herzberg and Longuet-Higgins[5] showed, the extension of the noncrossing rule to polyatomic systems is <u>not</u> true; i.e., in polyatomic molecules, two potential surfaces can intersect even if they belong to states of the same symmetry and spin-multiplicity. This implies the existence of various geometries where the degeneracy persists. Herzberg and Longuet-Higgins further showed that the electronic wavefunction for a given state could not be made a continuous function of nuclear coordinates on a closed loop around the intersection because it must change sign. This Longuet-Higgins theorem[6] was proven for a doubly degenerate case; there are no theorems concerning the sign reversal of the wavefunction for the intersection of three states.

A number of workers have studied this problem of degenerate non-symmetric geometries. Dixon[7] examined the Jahn-Teller (JT) distortions from the tetrahedral geometry of CH_4^+ that lower the energy by producing strong coupling between the $^2T_2$ components. Stone[8] discussed the distortions of ReF_6 that preserve the degeneracy of the $^2T_{2g}$ ground state. Davidson[9] described the positions of degenerate energies in triatomic molecules. Katriel and Davidson[10] determined various non-symmetric geometries for CH_4^+ that were triply degenerate.

Near T_d geometries, the 9 internal coordinates of CH_4^+ split into 4 sets: a_1 (symmetric stretch), s (t_2 stretching mode), b (t_2 bending mode), and e (bending mode). The s and b t_2 modes enter the effective 3x3 Hamiltonian in a fixed linear combination $t = \lambda s_i + \mu b_i$ (i=x,y,z) that defines the Jahn-Teller active t_2 mode. Near the T_d geometry, the

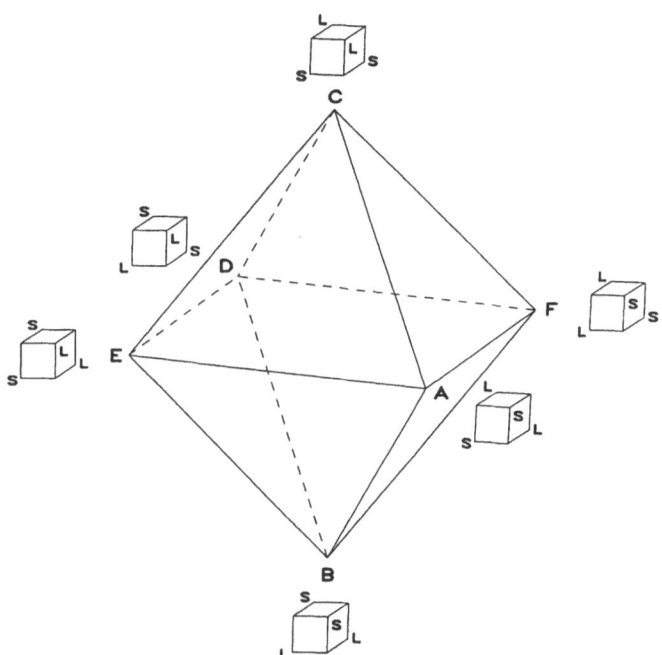

Fig. 1. A diagram of the six equivalent permutations of the C_{2v} minimum projected onto the JT active t_2 mode. The pure t_2 distortions of T_d are labelled as t_x, t_y, and t_z. The center of the faces of the octahedron contain a C_{3v} structure; four of the C_{3v} structures have a nondegenerate ground state and the other four structures have a degenerate ground state. In addition to the t_2 distortions, the optimum C_{2v} structures have large e distortions while the optimum C_{3v} structures have no e distortions. The optimum structure with no t_2 distortion has a large e distortion and is a D_{2d} configuration rather than the T_d structure.

orthogonal combination, $-\mu s_i + \lambda b_i$, defines an inactive t_2 mode which has no effect on the energy even though it may completely destroy all symmetry.

In this paper, the application of the Longuet-Higgins theorem in the regions surrounding a D_{2d} and two C_{3v} structures is considered. For CH_4^+, there are 12 equivalent hydrogen permutations of the C_{2v} minimum structure. As shown in Ref. 11, these 12 permutations may be divided into two distinct sets, each containing 6 elements, where the elements of the first set are enantiomers of the elements of the second set. This study is limited to the pseudo-rotation between the C_{2v} minima in one of the sets, however the results are valid for either set. Within each set (shown in Fig. 1), there are 6 accessible hydrogen arrangements of the C_{2v} minimum structure. Connecting the minima in this set, there are 12 equivalent transition state (TS) structures. In addition, there are 3 permutations of the D_{2d} structure, 4 permutations of the C_{3v} structure having a nondegenerate ground state, and 4 permutations of the C_{3v} structure having a degenerate

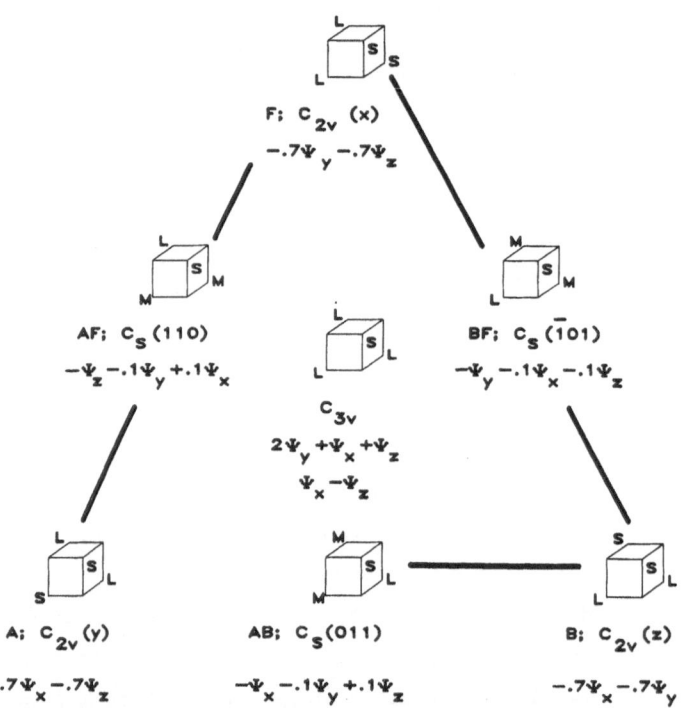

Fig. 2. A schematic diagram showing the behavior of a
wavefunction on a closed loop around a C_{3v} structure
having a degenerate ground state.

ground state. The geometries and energetics of these five structures are given in Ref. 1.

CH$_4^+$ has 9 internal degrees of freedom. Within this 9-dimensional space, there exists a 7-dimensional subspace where the ground state of CH$_4^+$ is doubly degenerate. For example, the C$_{3v}$ structure at the center of the ABF face in Fig. 1 has a degenerate ground state and is one configuration that is in this 7-dimensional subspace. When projected onto an appropriate two-dimensional subspace, this C$_{3v}$ geometry appears as the apex of a double cone in the potential-energy surface with three surrounding C$_{2v}$ minima. As seen in Fig. 2, the wavefunction at Permutation A is $.7\Psi_x - .7\Psi_z$. Following along the path to the AF permutation of the transition state, the wavefunction becomes $.1\Psi_x - .1\Psi_y - \Psi_z$ and, at Permutation F, the wavefunction is $-.7\Psi_y - .7\Psi_z$. Proceeding along the loop A → F → B → A, the wavefunction at A should finally be $-.7\Psi_x - .7\Psi_y$ to make the wavefunction remain continuous along this loop. However, this is the negative of the

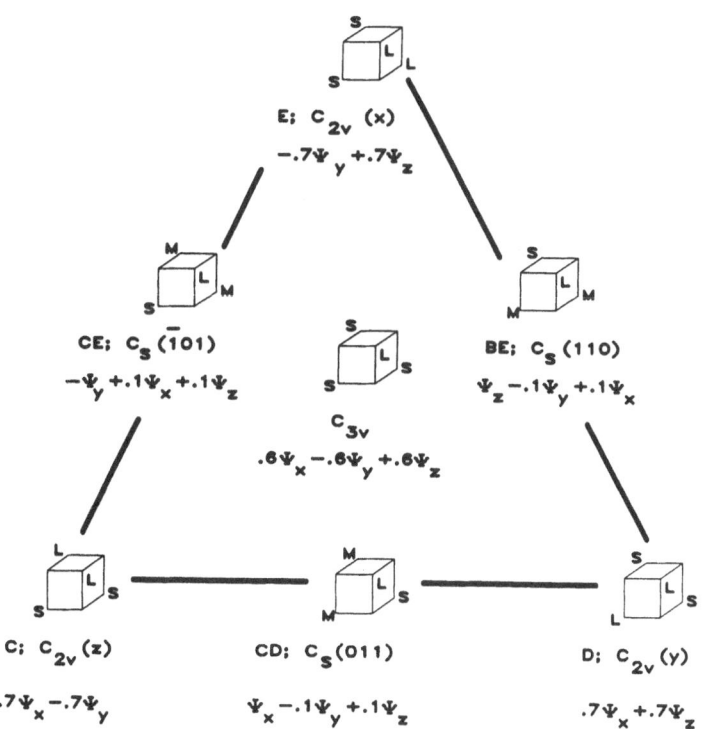

Fig. 3. A schematic diagram showing the behavior of a
wavefunction on a closed loop around a C$_{3v}$ structure
having a nondegenerate ground state.

original wavefunction; i.e., there has been a sign change in the wavefunction while going around the C_{3v} structure having a degenerate ground state. This is in agreement with the Longuet-Higgins theorem.

In Fig. 3, a loop encircles the C_{3v} structure at the center of the CDE face in Fig. 1 which has a nondegenerate ground state. Starting at Permutation C, the wavefunction is $.7\Psi_x - .7\Psi_y$. In going to Permutation E, the wavefunction changes from $-\Psi_y + .1\Psi_x + .1\Psi_z$ at the CE transition state to $-.7\Psi_y + .7\Psi_z$. Continuing along the path (C → E → D → C), the wavefunction at Permutation C finally becomes $.7\Psi_x - .7\Psi_y$ which is the same as the original wavefunction. That is, the wavefunction did not go through a phase change during the loop; therefore, the enclosed C_{3v} structure does not have a degenerate ground state. Again, this is in agreement with the Louguet-Higgins theorem.

Fig. 4 contains a schematic of the loop surrounding the D_{2d} structure. The wavefunction has a similar behavior here as it does around the nondegenerate C_{3v} structure; i.e., the wavefunction is a continuous function of nuclear coordinates on this closed loop.

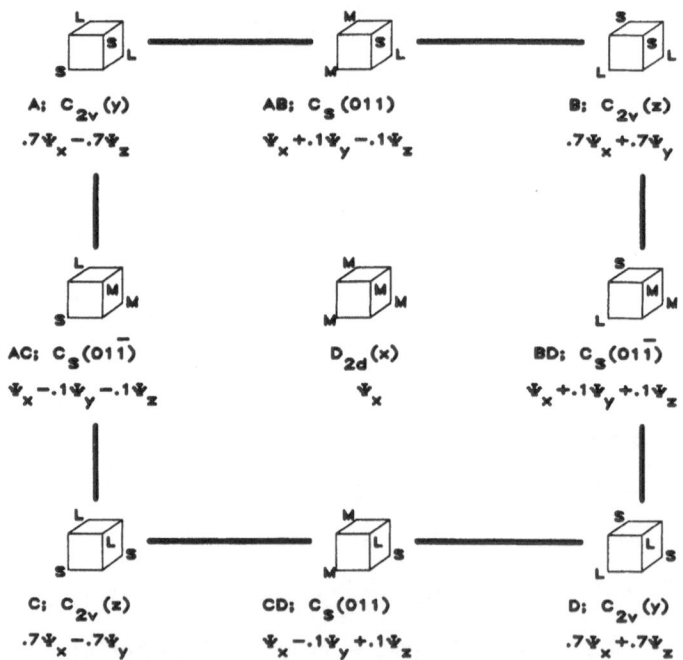

Fig. 4. A schematic diagram showing the behavior of a
wavefunction on a closed loop around a D_{2d} structure.

It should be noted that the dimension of the degenerate subspace is 7. It is not at all clear what it means to have a 1-dimensional loop encircle a 7-dimensional subspace. At each point in this 7-dimensional subspace, there are 4 modes that do not affect the energy, 2 Jahn-Teller active modes, and 1 mode that retains the doubly degeneracy of the ground state, but shifts the energy. Project a loop onto the local 2-dimensional space defined by the active JT modes. If the projection encircles such a degenerate point, the wavefunction will change sign when going around that

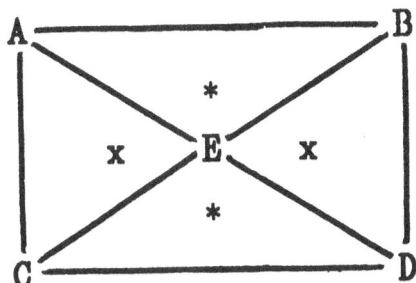

Fig. 5. A schematic diagram of the A → B → C → D → A loop where Permutation E is considered the center of the loop. C_{3v} structures having a degenerate ground state are denoted by the symbol *. C_{3v} structures having a nondegenerate ground state are denoted by the symbol x.

projected loop. In addition, Davidson[9] pointed out that a loop encircling 2 disjoint degenerate subspaces results in no sign change of the wavefunction. Therefore, if the degeneracy of the state within the "encircled" subspace is unknown and the wavefunction does not change sign, it is not clear whether the loop encircles no degenerate subspace or the loop encircles 2 degenerate subspaces. As shown in Fig. 5, the loop that "encircles" the nondegenerate D_{2d} structure in Fig. 4 can also be thought of as "encircling" two C_{3v} structures that have degenerate ground states if Permutation E is regarded as the center of the A → B → C → D → A loop. Since there is no sign change, these two C_{3v} structures must lie in disjoint 7-dimensional subspaces.

In the previous paper,[1] the Fukui reaction path[12-15] that connected adjacent minima was calculated. The path that was studied connected permutation A of the C_{2v} minimum to permutation C through the transition state AC. According to the definition of the Fukui path,[12] the reaction path must initially follow the normal mode of lowest frequency. In the case of CH_4^+, this normal mode is a twisting motion which preserves C_2 symmetry. Since the Fukui reaction path follows the gradient, the uphill path from the minimum along this mode would retain C_2 symmetry. Recall, however, that the adjacent minima and the transition state connecting those minima have no symmetry element in common. Therefore, according to more global theorems,[16,17] all symmetry must be lost along the path. These two seemingly contradictory statements imply that a bifurcation must exist at some point along the path that causes the loss of C_2 symmetry in the reaction path.

Fig. 6 is a schematic diagram of this bifurcation which occurs for each permutation of the C_{2v} minimum. Each C_{2v} minimum can be reached by four other minimum; for example (Fig. 6), permutation A is surrounded by permutations C, B, E, and F. As determined from a frequency analysis,[1] the normal mode of the lowest frequency for the C_{2v} minimum is an e mode of T_d; the mode that causes the bifurcation in the path is a t mode of T_d. To go from Permutation A to Permutation E, the distortion is in the t_x mode;

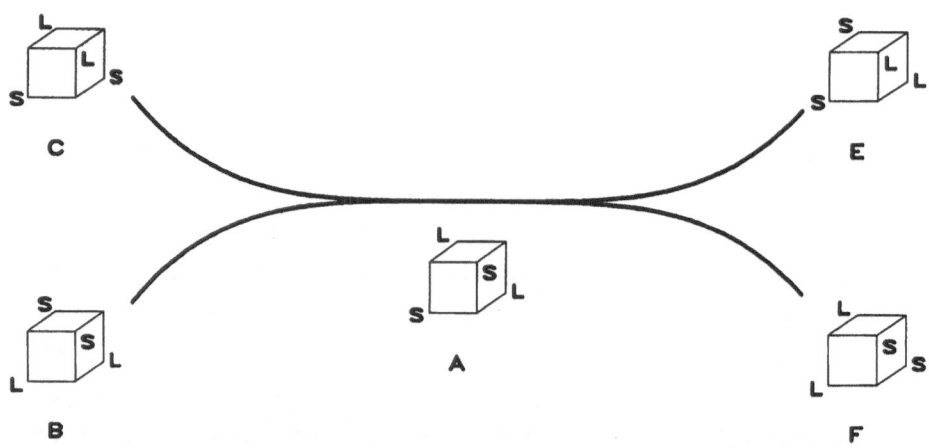

Fig. 6. A diagram of the four minima that are accessible to
Permutation A via the Fukui reaction path. The axes of
the diagram are t mode/e mode versus path length.

whereas, to go from A to F, the bifurcating distortion is in the t_x mode, but in the opposite direction. Similarly, to go from A to C or from A to B, the t_z mode is distorted in opposite directions. Calculations confirm that the Fukui definition and the global theorem can only be simultaneously true if there is an asymptotic bifurcation in the reaction path near C_{2v}. That is, the ratio of t distortion to e distortion vanishes exponentially as the reaction path approaches C_{2v} from the transition state. The Fukui definition of the reaction path does not allow discontinuous bifurcations in the path.

CONCLUSIONS

In this paper, the topology of the ground-state surface of CH_4^+ was studied. The change in the wavefunction for the pseudo-rotation between the equivalent C_{2v} minima was determined and related to the Longuet-Higgins theorem. In addition, the existence of an asymptotic bifurcation in the Fukui reaction path was discussed.

ACKNOWLEDGMENTS

The authors gratefully acknowledge financial support of this research by the National Science Foundation, Grant No. CHE-86-03416. Calculations were performed at the National Center for Supercomputing Applications using the computational time allocated by the National Science Foundation. Calculations were also performed at the Indiana University Chemistry Computer Facility, which was established in part by grants from the National Science Foundation, Grant Nos. CHE-83-09446 and CHE-84-05851.

REFERENCES

1. R. F. Frey and E. R. Davidson, accepted in J. Chem. Phys.

2. J. V. Neumann and E. P. Wigner, Z. Phys. 30, 467 (1929).

3. E. Teller, J. Phys. Chem. 41, 209 (1937).

4. H. A. Jahn and E. Teller, Proc. R. Soc. Lond. A, 161, 220 (1937).

5. G. Herzberg and H. C. Longuet-Higgins, Discuss. Faraday Soc. 35, 77 (1963).

6. H. C. Longuet-Higgins, FRS, Proc. R. Soc. Lond. A, 344, 147 (1975).

7. R. N. Dixon, Mol. Phys. 20, 113 (1971).

8. A. J. Stone, Proc. R. Soc. Lond. A, 351, 141 (1976).

9. E. R. Davidson, J. Am. Chem. Soc. 99, 397 (1977).

10. J. Katriel and E. R. Davidson, Chem. Phys. Lett. 76, 259 (1980).

11. M. N. Paddon-Row, D. J. Fox, J. A. Pople, K. N. Houk, and D. W. Pratt, J. Am. Chem. Soc. 107, 7696 (1985).

12. K. Fukui, Acc. Chem. Res. 14, 363 (1981).

13. K. Yamashita, T. Yamabe, and K. Fukui, Chem. Phys. Lett. 84, 123 (1981).

14. K. Yamashita, T. Yamabe, and K. Fukui, Theoret. Chim. Acta. 60, 523 (1982).

15. K. Yamashita, T. Yamabe, and K. Fukui, J. Am. Chem. Soc. 106, 2255 (1984).

16. J. W. McIver, Jr. and R. E. Stanton, J. Am. Chem. Soc. 94, 8618 (1972).

17. J. W. McIver, Jr., Acc. Chem. Res. 7, 72 (1974).

FRAGMENTATION MECHANISMS FOR MULTIPLY-CHARGED CATIONS

Leo Radom, Peter M.W. Gill and Ming Wah Wong

Research School of Chemistry
Australian National University
Canberra, A.C.T. 2601
Australia

ABSTRACT

An analysis is presented of the fragmentation of multiply-charged cations $AB^{(n+1)+}$ into $A^{n+} + B^+$ for cases in which the potential energy curve for the fragmentation can be satisfactorily described as arising from an avoided crossing between an attractive state corresponding to $A^{(n+1)+} + B$ and a repulsive state corresponding to $A^{n+} + B^+$. An important quantity is the Gill Δ parameter, given by the difference in ionization energies of A^{n+} and B. A large Δ corresponds to an early transition structure for the fragmentation reaction whereas a small Δ corresponds to a late transition structure. In the latter case, the value of Δ leads readily to estimates of the transition structure bond length and of the kinetic energy released during the fragmentation.

INTRODUCTION

There has been considerable recent theoretical and experimental interest in the gas-phase chemistry of dications.[1] Dications are usually thermodynamically unstable with respect to dissociation into two monocations but significant kinetic stability may result if sufficiently high barriers impede their fragmentation. In order to enable a reliable prediction of the stability of a dication to be made, an accurate theoretical description of this fragmentation process is very important. In this article, we use a recently introduced avoided-crossing model[2-5] to describe dicationic fragmentation and then generalize our treatment to fragmentation of certain more highly charged cations.

METHOD

The calculations referred to in this paper are of the standard <u>ab initio</u> molecular orbital type[6] with moderately large basis sets and with electron correlation incorporated using Møller-Plesset perturbation theory. Specific details are given in the original papers.[2-5]

219

DISCUSSION

We begin by examining some calculated transition structures for dicationic fragmentation.[2,3] In the case of symmetric fragmentations, we find that the internuclear separation in the transition structure is typically ~50% greater than that in the equilibrium structure. For example, for the fragmentation of He_2^{2+}, the bond length in the transition structure is 1.15 Å compared with 0.70 Å in the equilibrium structure. In asymmetric fragmentations, on the other hand, the bond length in the transition structure is often 2-3 times or more longer than the equilibrium structure. For example, for the deprotonation of AlH^{2+}, the relevant lengths are 3.59 and 1.65 Å. An even longer bond length (12.5 Å) is found for the transition structure for fragmentation of $MgH^{\cdot 2+}$.

Our preferred rationalization[2,3] for these unusually long transition structure bond lengths begins by considering the potential energy curve along the reaction coordinate for a dissociating AB^{2+} dication as arising from an avoided crossing between a repulsive state which correlates with $A^+ + B^+$ and an attractive state which correlates with $A^{2+} + B$. The asymptotic energy difference between the two curves (Δ_1) is equal to the difference in adiabatic ionization energies of A^+ and B:

$$\Delta_1 = IE_a(A^+) - IE_a(B) \tag{1}$$

When Δ_1 is large, the transition structure occurs early and with a shorter bond length (Figure 1). Conversely, we would predict that small Δ_1 values will be associated with late transition structures (Figure 2).

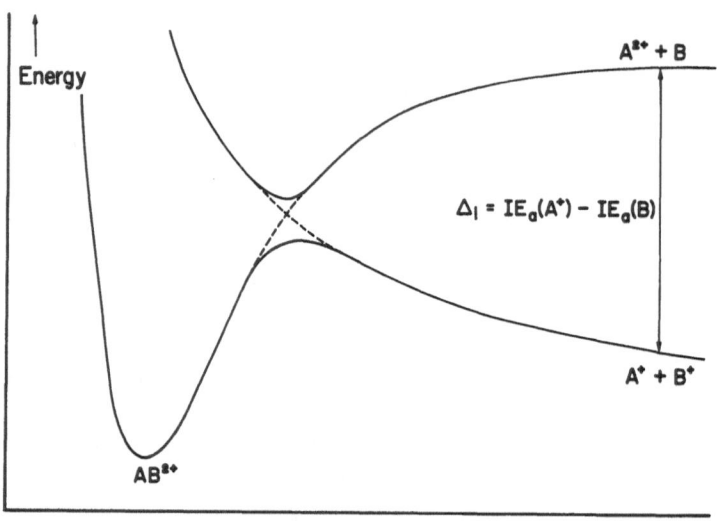

Fig. 1. Schematic potential energy curves describing fragmentation of a general AB^{2+} dication showing avoided crossing between the $A^{2+} + B$ and $A^+ + B^+$ diabatic potentials when Δ_1 is large.

This Gill Δ parameter is a very useful quantity. For example, it is easy to show that, for sufficiently late transition structures, the bond length in the transition structure is given by[2]

$$r_{TS} \simeq 1/\Delta_1 \tag{2}$$

This formula leads to a transition structure length of 12.40 Å for $MgH^{\cdot 2+}$, in close agreement with the directly calculated 12.45 Å. A striking feature is that this formula requires only knowledge of the ionization energies of the fragments formed.

The Δ parameter also provides an estimate of the kinetic energy released in dicationic fragmentations.[4] The kinetic energy release (T) in the case of a late transition structure is approximately equal to, and is bounded above by, Δ_1, i.e.

$$T \simeq \Delta_1 = IE_a(A^+) - IE_a(B) \tag{3}$$

provided that the coupling between the diabatic $A^{2+} + B$ and $A^+ + B^+$ curves is small and that the electron transfer occurs late, both of which will be the case if Δ_1 is small.

The prediction from equation (3) may be compared[4] with the observed[7] kinetic energy release in the case of the $SiH^{\cdot 2+}$ dication. In this case, Δ_1 is equal to the difference between the second ionization energy of Si (16.35 eV) and the (first) ionization energy of H (13.6 eV). This gives T as approximately equal to (and certainly no greater than) 2.75 eV, which is consistent with the experimental value[7] of 2.42 eV.

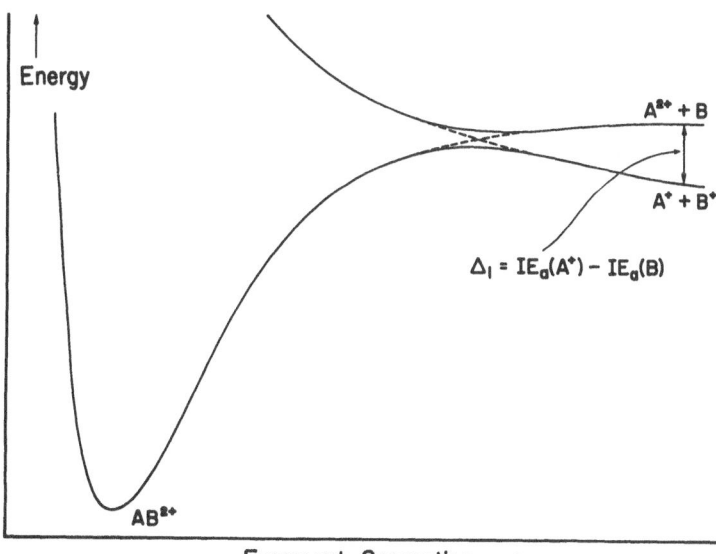

Fig. 2.   Schematic potential energy curves describing fragmentation of a general AB^{2+} dication showing avoided crossing between the $A^{2+} + B$ and $A^+ + B^+$ diabatic potentials when Δ_1 is small.

The avoided-crossing model yields detailed information concerning the mechanism of the fragmentation process. In the special case of deprotonation reactions, we find[4] that such reactions are less straightforward than is often assumed.

In cases where the transition structure for deprotonation occurs late on the reaction pathway, the deprotonation process for AH^{2+} involves (i) homolytic cleavage of the A–H bond to give $A^{\cdot 2+}...H^{\cdot}$; (ii) further stretching of $A^{\cdot 2+}...H^{\cdot}$, now dominated by a (weak) ion–induced-dipole potential with little change in energy; and (iii) a crossing to the $A^{+} + H^{+}$ diabatic potential curve yielding the dissociation products $A^{+} + H^{+}$. This unusual behaviour has important implications regarding the levels of theory required to describe adequately the deprotonation process, and this is discussed in detail elsewhere.[4]

The fragmentation process is further complicated in the case of polyatomic dications where the individual diabatic curves only <u>appear</u> to cross (because they are projections onto the energy–reaction-coordinate plane) while in full coordinate space they do not. The situation is exemplified by the case of deprotonation of $N_2H_6^{2+}$ (Figure 3).

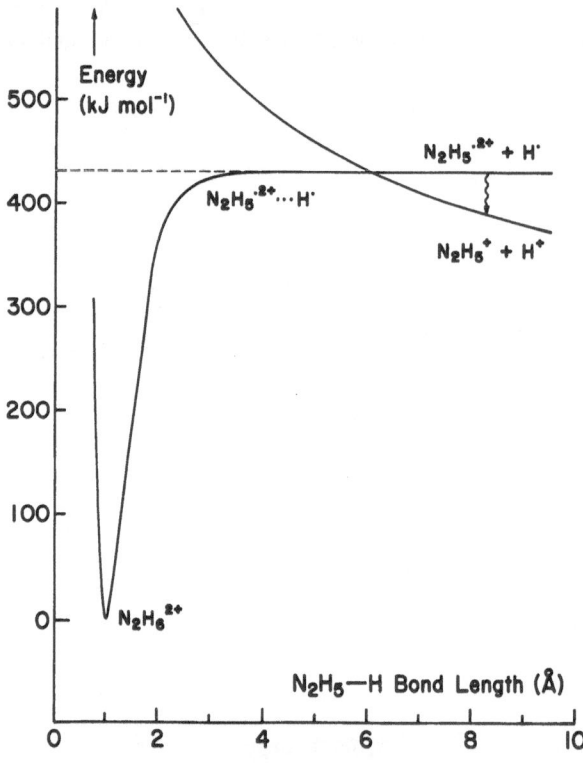

Fig. 3.   Potential energy curves describing deprotonation of $N_2H_6^{2+}$. The arrow shows the point at which spontaneous electron transfer can take place from the $N_2H_5^{\cdot 2+} + H^{\cdot}$ potential curve to the $N_2H_5^{+} + H^{+}$ potential curve.

Again, the fragmentation begins with an early steeply rising section during which the N_2H_5–H bond is homolytically cleaved. Then there is an almost flat plateau due to the very weak (r^{-4}) attraction between $N_2H_5^{\cdot 2+}$ and $H^{\cdot}$; and finally, it becomes energetically feasible for a spontaneous electron transfer to take place, thereby momentarily forming $N_2H_5^+ + H^+$ at the $N_2H_5^{\cdot 2+} + H^{\cdot}$ geometry. At this point, the dissociation becomes inevitable since the system will rapidly roll down onto the repulsive $N_2H_5^+ + H^+$ path and fragment.

It is easy to show[4] that the distance (r_{TS}) from the departing proton to the center of charge of A at the transition structure for deprotonation of AH^{2+} is given by

$$r_{TS} \approx 1/(\Delta_1 - \delta) \tag{4}$$

where δ (a positive number) is the difference between the vertical and adiabatic electron affinities of A^{2+}. This is the polyatomic analogue of equation (2).

Finally, we examine[5] more highly charged ions, specifically $SiHe^{\cdot 3+}$ and $SiHe^{4+}$.

The $SiHe^{\cdot 3+}$ trication has a moderately short bond length of 1.670 Å. Although fragmentation to $Si^{2+} + He^{\cdot +}$ is highly exothermic (by 724 kJ mol^{-1}), it is inhibited by an energy barrier of 100 kJ mol^{-1}. The $SiHe^{\cdot 3+}$ trication is therefore potentially observable in the gas phase. It could perhaps be generated through collision of the $Si^{\cdot 3+}$ trication with helium.

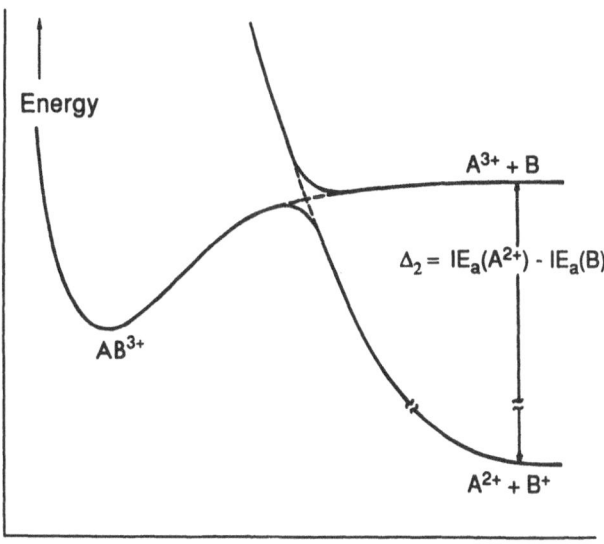

Fig. 4.   Schematic potential energy curves describing fragmentation of a general AB^{3+} trication showing avoided crossing between $A^{3+} + B$ and $A^{2+} + B^+$ diabatic potentials.

In cases where we can consider the potential energy curve along the reaction coordinate for a dissociating trication AB^{3+} as arising from an avoided crossing between a repulsive state which correlates with $A^{2+} + B^+$ and an attractive state which correlates with $A^{3+} + B$ (Figure 4), the transition structure bond length is given approximately by[5]

$$r_{TS} \simeq 2/\Delta_2 \qquad (5)$$

where Δ_2 is the difference between the adiabatic ionization energies of A^{2+} and B. For the case of $SiHe^{\cdot 3+}$, this equation predicts a transition structure bond length of 3.31 Å. in good agreement with the directly calculated value of 3.29 Å.

For the $SiHe^{4+}$ tetracation, the equilibrium bond length of 1.550 Å is comparable to that of a normal Si-H bond (e.g. 1.481 Å in SiH_4) and has been rationalized[5] in terms of favorable orbital interactions. Although the exothermicity of fragmentation to $Si^{\cdot 3+} + He^{\cdot +}$ is enormous (1615 kJ mol^{-1}), the barrier for such a process (76 kJ mol^{-1}) is sufficiently large that we believe that experimental observation will be feasible.

$SiHe^{4+}$, with just two valence electrons, is the smallest stable polyatomic tetracation yet reported.

For a dissociating AB^{4+} tetracation, in cases where the potential energy curve along the reaction coordinate can be considered as arising from an avoided crossing between a repulsive state which correlates with $A^{3+} + B^+$ and an attractive state which correlates with $A^{4+} + B$, the estimated transition structure bond length is given by[5]

$$r_{TS} \simeq 3/\Delta_3 \qquad (6)$$

where Δ_3 is the difference between the adiabatic ionization energies of A^{3+} and B. For $SiHe^{4+}$, the value of r_{TS} from equation (6) is 2.36 Å compared with the directly calculated value of 2.15 Å.

In general. if the potential energy curve for the fragmentation of a multiply-charged cation $AB^{(n+1)+}$ to give $A^{n+} + B^+$ can be satisfactorily described by an avoided crossing between diabatic curves corresponding to $A^{(n+1)+} + B$ and $A^{n+} + B^+$, the transition structure bond length may be estimated as

$$r_{TS} \simeq n/\Delta_n \qquad (7)$$

where Δ_n is the difference between the adiabatic ionization energies of A^{n+} and B.

CONCLUDING REMARKS

The important conclusion to emerge from this work is that the avoided-crossing model and the Gill Δ parameter are extremely useful in describing the fragmentation processes of multiply-charged cations.

REFERENCES

1. For a recent review, see W. Koch and H. Schwarz, in "Structure/Reactivity and Thermochemistry of Ions", P. Ausloos and S.G. Lias, eds., NATO ASI, D. Reidel Publishing Company, Dordrecht (1987).
2. P.M.W. Gill and L. Radom, Chem. Phys. Lett., 136, 294 (1987).
3. L. Radom, P.M.W. Gill, M.W. Wong and R.H. Nobes, Pure Appl. Chem., in press.
4. P.M.W. Gill and L. Radom, J. Am. Chem. Soc., submitted.
5. M.W. Wong and L. Radom, J. Am. Chem. Soc., in press.
6. W.J. Hehre, L. Radom, P.v.R. Schleyer and J.A. Pople, "Ab Initio Molecular Orbital Theory", Wiley, New York (1986).
7. W. Koch, G. Frenking, H. Schwarz, F. Maquin and D. Stahl, J. Chem. Soc., Perkin Trans. 2, 757 (1986).

THE NATURAL BOND ORBITAL LEWIS STRUCTURE CONCEPT

FOR MOLECULES, RADICALS, AND RADICAL IONS

Frank Weinhold and John E. Carpenter

Theoretical Chemistry Institute and Department of Chemistry,
University of Wisconsin, Madison, Wisconsin 53706

EMPIRICAL LEWIS STRUCTURE CONCEPTS

As the "Coulomb explosion"[1] and other techniques add to our knowledge of molecular geometry, it is appropriate to recall the debt of gratitude that many theoretical concepts owe to structural studies. Indeed, new structural data have often provided the principal stimulus for new chemical concepts. Even prior to the discovery of the electron in the last century, qualitative structural inferences based on stoichiometry, number of isomers, and other lines of indirect chemical evidence were giving rise to models of molecular connectivity and geometry (e.g., the tetrahedral carbon atom of van't Hoff and Le Bel[2]) that underlie current electronic theories of valence.

Pre-quantum-mechanical efforts to rationalize chemical phenomena culminated in the remarkable "Lewis structure" picture of electronic structure.[3] In this picture, the geometrical and chemical properties of stable, closed-shell molecules were linked to the patterns of dots and lines in the simple mnemonic diagrams that we teach to beginning chemistry students to this day. With the discovery of quantum theory and the development of X-ray crystallography, rotational spectroscopy, and other methods for precise determination of molecular structure, the Lewis structure concept was considerably extended. Pauling[4] and Slater[5] introduced the hybridization concept, allowing one to relate the angle θ_{12} between two bonds to the sp^λ composition of directed hybrid orbitals, based on formulas of the form

$$\cos \theta_{12} = -(\lambda_1 \lambda_2)^{-1/2} \qquad (1)$$

which express the orthogonality relationship between sp^{λ_1}, sp^{λ_2} hybrids (formed from the s and three p orbitals of the atomic valence shell). Pauling showed how the length and polarity of a bond in the Lewis structure could be related to empirical covalent radii and atomic electronegativities,[6] and Coulson subsequently employed bond-order–bond-length relationships to associate bond length variations with the formal number of multiple bonds of the Lewis structure.[7] Mulliken and Coulson further extended the localized Lewis structure concept by employing bond orbital functions,

$$\sigma_{AB} = c_A h_A + c_B h_B \qquad (2)$$

in which the "polarization" coefficients c_A, c_B of directed hybrids h_A, h_B of a heteropolar σ_{AB} bond are related to dipole moments or empirical electronegativity differences.[8] Using Eq. (1) and drawing upon a large body of structural data, Bent[9] obtained the important extension of the hybridization concept summarized in "Bent's Rule," showing how systematic changes in bond angles could be related to the hybridization differences in bonds to atoms of different electronegativity.

It is appropriate to observe that these extensions of the basic Lewis structure concept were largely stimulated by empirical structural data rather than by formal quantum theory. Pauling once remarked[10] to the effect that if chemists had invented orbitals, they would have invented tetrahedral orbitals first, and s and p orbitals would then be explained as hybrids of the tetrahedral orbitals! Even if this remark was offered whimsically, it reflects the strong emphasis on empirical structural data, and the relatively minor role played by rigorous *ab initio* quantum calculations, in the formative period when basic concepts of valence theory were forged.

NATURAL LEWIS STRUCTURES

As *ab initio* wavefunctions for polyatomic molecules began to achieve respectable chemical accuracy in the early 1970s,[11] it became possible to contemplate more direct methods of attaching definite theoretical meaning to the Lewis structure diagrams. A simple Lewis structure such as

$$\diagdown \ddot{\text{A}} - \text{B} \diagdown \tag{4a}$$

corresponding to the electron configuration $\dots (n_A)^2 (\sigma_{AB})^2 \dots$, can be associated with an approximate wavefunction ψ_L of the form

$$\psi_L = \mathcal{A}[\dots(n_A)^2(\sigma_{AB})^2\dots(\text{spin})] \tag{4b}$$

where $\mathcal{A}$ is the antisymmetrizer, and n_A, σ_{AB} represent localized one-center (lone pair) and two-center (bond) electron-pair functions built from directed atomic hybrids (h_A, h_A',...), (h_B, h_B',...) on atoms A, B, ... Given an accurate wavefunction ψ_{acc} for a molecule, one can ask which possible Lewis structure ψ_L, ψ_L', ψ_L'',... best corresponds to ψ_{acc}, and further, what are the *optimal* hybrids and polarization coefficients in (2), (4b) that give the *best possible* Lewis structure representation ψ_L?

A criterion of "best possible" might be based on expectation values of the Hamiltonian or other operators, but we have employed a criterion based on the electron density $\rho(\vec{r})$, which requires only the information contained in the first-order density matrix $\hat{\Gamma}$.[12] The general linear transformation from the basis set atomic orbitals to bond orbitals leads to orbitals such as (2) that are occupied in the formal Lewis structure ψ_L as well as to residual non-Lewis (NL) orbitals that make no contribution to ψ_L. Chief among the latter are the valence-shell *antibonds* σ^*_{AB}

$$\sigma^*_{AB} = c_B h_A - c_A h_B \tag{5}$$

as well as the extra-valence-shell "Rydberg" orbitals that complete the span of the input basis. The electron density $\rho_{acc}(\vec{r})$ associated with ψ_{acc} can be decomposed into its contribution from Lewis- and non-Lewis-type orbitals,

$$\rho_{acc}(\vec{r}) = \rho_L(\vec{r}) + \rho_{NL}(\vec{r}) \tag{6}$$

or, after integration over all space of the N-electron system,

$$N = N_L + N_{NL} \tag{7}$$

where N_L, N_{NL} are the number of electrons respectively associated with the Lewis and non-Lewis manifolds. The optimal choice of Lewis structure hybrids $\{h_i\}$ and polarization coefficients $\{c_i\}$ can be defined as that which *maximizes* the Lewis-type contribution ρ_L to the electron density,

$$\max_{\{h_i\},\{c_i\}} \int \rho_L(\vec{r})d^3\vec{r} = \max_{\{h_i\},\{c_i\}} N_L \tag{8}$$

subject to hybrid orthonormality requirements.

A maximum-occupancy criterion analogous to (8) is a distinguishing characteristic of "natural" orbitals in the classic Löwdin sense,[13] but the orbitals are here restricted to be of *localized* one-center or two-center type. The direct occupancy maximization (8) can be carried out numerically, but it is more efficient (and leads to practically equivalent results) to determine these bond orbitals by diagonalizing localized one-center and two-center blocks of the density matrix $\hat{\Gamma}$.[12] The numerical algorithm for determining these *natural bond orbitals* (NBOs) has been implemented in a program[14] that can be easily attached to a variety of electronic structure packages, and requires only the information contained in the first-order density matrix. Note that the NBO procedure makes no reference to the specific form of the input wavefunction ψ_{acc}, so that these orbitals can be obtained for any type of wavefunction calculation (correlated or uncorrelated) that leads to a density matrix.[15]

The set of optimal NBOs constitutes the *natural Lewis structure* of the molecule, and is found to converge to a unique limit as the variational accuracy of the wavefunction ψ_{acc} is improved.[16] The natural Lewis structure is generally in good agreement with the chemist's empirical Lewis structure model, and the optimal ψ_L generally provides an excellent approximation to the energy and electron density of the full ψ_{acc} (typically > 99%), as elementary valence theory would suggest. Table I illustrates the accuracy of the natural Lewis structure description (at the *ab initio* RHF/6-31G* level)[17] for a variety of simple molecules, showing the percentage of the total energy E(%Lewis) and electron density ρ(%Lewis) associated with ψ_L in each case. Table II exhibits the optimized sp^λ hybrids in C-H bonds of some representative hydrocarbon molecules, showing the close agreement with the idealized hybrid forms (sp^3, sp^2, etc.) predicted by simple valence theory. The slight departures from idealized hybrid forms, or the further departures that result when H atoms are replaced by atoms of differing electronegativity, are generally in good agreement with Bent's Rule. Furthermore, one can verify that these bond functions are highly transferable from one molecule to another,[18] in agreement with empirical Lewis concepts. Thus, the natural bond orbitals for stable, closed-shell molecules offer strong *ab initio* support for the empirical concepts embedded in the Lewis structure model, and they retain a close connection to molecular structure.

Table I *NBO decomposition of RHF/6-31G* wavefunctions (Pople-Gordon idealized geometry) for selected molecules, showing the percentage contribution of the Lewis-type ψ_L to the energy E and electron density ρ.*

molecule	E(%Lewis)	ρ(%Lewis)
BH_3	99.979	99.95
CH_4	99.981	99.97
NH_3	99.994	99.98
H_2O	99.996	99.98
HF	99.999	99.99
AlH_3	99.982	99.64
SiH_4	99.989	99.81
PH_3	99.994	99.90
H_2S	99.998	99.98
HCl	99.999	100.00
CF_3H	99.912	99.18
C_2H_6	99.927	99.74
C_2H_4	99.902	99.70
C_2H_2	99.884	99.75
H_2CO	99.911	99.42
C_6H_6	99.742	97.12

Table II *CH bond NBOs for selected organic molecules (RHF/6-31G* level, idealized Pople-Gordon geometry), showing polarization coefficients c_C, c_H and carbon hybridization sp^λ.*

molecule	polarization coefficients		carbon hybridization			
	c_C	c_H	h_C	%s	%p	%d
sp^3-like						
CH_4	0.7801	0.6257	$sp^{2.99}$	25.00	74.84	0.16
C_2H_6	0.7788	0.6273	$sp^{3.09}$	24.39	75.44	0.17
sp^2-like						
C_2H_4	0.7774	0.6290	$sp^{2.26}$	30.65	69.18	0.17
H_2CO	0.7606	0.6493	$sp^{1.90}$	34.35	65.43	0.22
C_6H_6	0.7842	0.6206	$sp^{2.34}$	29.88	69.97	0.15
sp-like						
C_2H_2	0.7881	0.6156	$sp^{1.08}$	47.91	51.96	0.13

EXTENSION OF LEWIS STRUCTURE CONCEPTS TO OPEN-SHELL SYSTEMS: DIFFERENT HYBRIDS FOR DIFFERENT SPINS

The principles that govern the geometries of simple closed-shell molecules appear in certain respects to break down in open-shell systems. Indeed, the very concept of a Lewis structure may be questioned in open-shell species, where the pairing of electrons is to some extent disrupted. Experimentally, it is found that open-shell species often exhibit "floppy," non-rigid structures and low barriers to interconversion among different isomeric forms, leading to complex vibrational dynamics.[19] Thus, the very notion of "molecular structure" appears in some measure to be less well specified in these systems, and the elementary Lewis structure concepts that are successful for closed-shell molecules seem to give much less guidance to the structure and dynamics of radicals.

Natural bond orbital analysis suggests how a simple modification of the Lewis structure concept can be applied to open-shell species.[16,20] In the open-shell case, the density matrices $\hat{\Gamma}^{(\alpha)}$, $\hat{\Gamma}^{(\beta)}$ of the α and β spin systems are generally *distinct*. In this case, application of the NBO procedure leads to *different* NBOs for the two spin sets (reminiscent of the "different orbitals for different spins" and related AMO and non-paired spin orbital methods)[21] representing two independent ways to form the hybrids and bonds of the molecule. In this context, one can recognize that the essential feature of a Lewis diagram is *the pattern of lines and dots* [two-center (bonded) and one-center (nonbonded) functions, respectively] that describe the electron distribution. In the closed-shell case, the "pairing" of α and β electrons may be considered to result from the special circumstance that these two Lewis patterns are coincident. However, in the open-shell case the α and β bond patterns can be entirely distinct, and the theoretical situation can then be described in terms of "different Lewis structures for different spins." Although this use of the term "Lewis structure" places primary emphasis on the *bonding pattern* of the Lewis diagram (rather than electron pairing *per se*), and thus differs somewhat from common usage, we believe it is the essential generalization needed to successfully extend Lewis structure concepts to open-shell radicals and radical ions.

In this spirit, we may write a Lewis-type wavefunction for an open-shell system symbolically in the form

$$\psi_L^{(open)} = A[\psi_L^{(\alpha)}\psi_L^{(\beta)}(\text{spin})] \qquad (9)$$

where $\psi_L^{(\alpha)}$ denotes the product of singly occupied spin NBOs of the α Lewis structure, and $\psi_L{}^\beta$ is the corresponding Lewis-type function for the β spin set. Numerically, the Lewis-type description (9) for an open-shell system is found to be generally of the same order of accuracy as that for corresponding closed-shell systems. For example, Table III compares the accuracy of the NBO Lewis description of ground state CO with that of various open-shell excited states of CO and CO^+ at the UHF/6-31G level, showing that the open-shell Lewis description $\psi_L^{(open)}$ is sometimes (e.g., $\tilde{A}\,^2\Pi$ CO^+) even of *higher* accuracy than that of corresponding closed-shell species. Although the accuracy of an open-shell Lewis-type description will vary from molecule to molecule (as is also the case for closed-shell molecules), the available numerical results suggest that the simple generalization (9) provides a surprisingly accurate description of radicals or radical ions in Lewis structure terms.

How can these results be put to qualitative conceptual use for predicting or rationalizing the structures of open-shell species? The Lewis structures $\psi_L^{(\alpha)}$, $\psi_L^{(\beta)}$ of the two spin sets may be considered to represent "competing" versions of the bonding in the molecule, each of which favors a particular arrangement of the nuclei. For example, a Lewis structure having bonds to four atoms (i.e., four singly-occupied two-center NBOs) from a central atom A would suggest sp^3 hybridization and tetrahedral geometry about that atom, whereas a Lewis structure with A bonded to three atoms would suggest sp^2 hybrids and trigonal planar geometry. If the Lewis structures $\psi_L^{(\alpha)}$, $\psi_L^{(\beta)}$ of the α and β spin sets differ in this way, the actual potential energy surface of the radical would be expected to resemble a compromise between the two closed-shell species having the corresponding Lewis structures. The classic Pauling-Wheland "resonance theory"[22] provides a useful analogy suggesting how properties of the radical could be inferred from the properties of the two distinct spin Lewis structures.

In the simple case of a radical cation A^+ derived from a closed-shell neutral species A by removal of an electron (say, from a doubly occupied orbital n_i), the α (majority) spin set retains the Lewis structure of the parent closed-shell A, whereas the ionized β system is expected to have a Lewis structure like that of the corresponding closed-shell dication A^{2+} (in which n_i is completely empty). Similarly, the properties of a neutral radical will appear to be a compromise between the "anion-like" α Lewis structure and "cation-like" β Lewis structure of the corresponding closed-shell ions.

To illustrate these concepts,[20] we consider the simple numerical example of the methyl radical CH_3 (UHF/6-31G* level). If we remove a hydrogen atom from neutral methane, holding all the remaining nuclei fixed in the tetrahedral methane geometry, we obtain the open-shell NBOs summarized in Table IV. This table shows the hybridization of the carbon atom in the two spin systems of CH_3, and compares these with the corresponding closed-shell ions, CH_3^-, CH_3^+, all in the fixed tetrahedral geometry of CH_4. The three carbon hybrids of the C-H bonds in CH_3^+ are nearly sp^2-like, while the vacant nonbonding orbital is almost pure p in character, suggesting the planar geometry to which this ion reverts when the nuclei are allowed to equilibrate. The corresponding hybrids of the CH_3^- anion are more nearly sp^3-like. Comparison of the open-shell CH_3 hybrids with the ion hybrids shows the clear similarity of the alpha system to the anion and the beta system to the cation. For example, the

Table III *Accuracy of the NBO Lewis structure description for ground and low-lying excited states of CO, CO^+ (HF/6-31G level), showing the percentage of the electron density ρ(%Lewis) associated with ψ_L in each state.*

species	state	NBO excitation	ρ(%Lewis)
CO	$\tilde{X}\,^1\Sigma$	–	99.922
	$\tilde{A}\,^1\Pi$	$n_C \rightarrow \pi^*$	99.603
	$\tilde{a}\,^3\Pi$	$\bar{n}_C \rightarrow \pi^*$	99.748
CO^+	$\tilde{X}\,^2\Sigma$	–	99.893
	$\tilde{A}\,^2\Pi$	$\bar{\pi} \rightarrow \bar{n}_C$	99.967
	$\tilde{D}\,^2\Pi$		99.224

Table IV *NBO hybrids in tetrahedral* CH_3, CH_3, *and* CH_3^- *at the UHF/6-31G* level (methane-like geometry), showing the hybrid* sp^λ *type and percentage p-character of carbon nonbonding* (n_C) *and C-H bond* (σ_{CH}) *orbital.*

molecule	n_C hybrid	σ_{CH} hybrid
$CH_3\ \alpha$	$sp^{3.9}$ (79.4)	$sp^{2.8}$ (73.4)
$CH_3\ \beta$	$sp^{16.}$ (93.3)	$sp^{2.2}$ (68.3)
CH_3^+	$sp^{41.}$ (96.6)	$sp^{2.1}$ (67.2)
CH_3^-	$sp^{4.6}$ (82.2)	$sp^{2.6}$ (72.4)

$\bar{n}_C$ radical hybrid *p*-character (93.3%) resembles that of the cation (96.6%), while the corresponding n_C radical hybrid (79.4%) resembles that of the anion (82.2%). The electronic properties of a methyl radical can thus be pictured in terms of a competition between the "cation-like" and "anion-like" spin systems. When the CH_3 geometry is allowed to relax, the radical adopts the planar geometry favored by the cation-like β spin system. However, the umbrella mode for deformation from planar to pyramidal geometry is of very low frequency in this radical (compared to that in the planar cation), reflecting the competing influence of the anion-like Lewis structure. Replacement of H by substituents that differentially *stabilize* one or the other of these Lewis structure will tend to displace the equilibrium geometry of the radical toward the corresponding idealized geometry. For example, neighboring electronegative atoms that act as strong sigma acceptors would be expected to favor pyramidalization (by Bent's Rule), as pointed out by Pauling.[23]

The DHDS (different hybrids for different spins) NBO method has been successfully employed to interpret the qualitative features of the bonding in a variety of open-shell neutral and radical ions, such as hydrated superoxide ions $O_2^-(H_2O)_n$,[24] ground and excited states of $(CO)_2^+$,[25] collision complexes of NO and HF,[26] and the hydroxymethyl radical.[20] These examples suggest that a relatively simple extension of Lewis structure concepts can provide a fruitful way to interpret the properties of a wide variety of reactive open-shell species.

NATURAL BOND ORBITAL ANALYSIS OF INORGANIC, METALLIC, AND ORGANOMETALLIC COMPOUNDS

Qualitative Lewis structure concepts were first developed to treat the chemistry of the common representative elements, particularly, H, C, N, O, and F. It is therefore not surprising that NBO analysis was first applied to simple organic and inorganic compounds involving these elements, and that natural Lewis structure concepts have proven particularly successful in treating such compounds. However, it has long been recognized that elements outside this narrow portion of the periodic table often appear to violate elementary Lewis structure principles, requiring the additional concepts of "3-center bonding," "expansion of the valence shell," "hypervalency," and so forth. Since the chemistry of metallic and semi-metallic compounds lies at the forefront of current research in catalysis, superconductivity, and materials technology, it is important to relate their behavior to that of simpler compounds which are well described in the Lewis framework. In this section we briefly describe some recent applications of NBO and natural population analysis to these more "difficult" compounds, seeking to understand the unique electronic and structural properties of species that violate elementary Lewis structure principles.

Second-Row Nonmetallic Elements and Hypervalency

The second-row representative elements P, S, Cl exhibit many departures from standard Lewis structure forms (particularly in bonding to oxygen or fluorine), as exemplified by PF_5, SF_6, ClF_2, and other "hypervalent" compounds. The bonding in these compounds has sometimes been interpreted in terms of expansion of the valence shell to include *d* orbitals, leading, for example, to octahedral sp^3d^2 hybridization in

SF$_6$. However, a comprehensive NBO and natural population analysis of the bonding in SF$_6$ has shown[27] that sp^3d^2 hybridization is unimportant in this case, and that the essential feature of hypervalency is the *ionic character of the bonding*. Thus, the covalent Lewis structure picture is better replaced in this case by an extreme ionic limit (using the "natural ionic hybrids" of isolated ions), with corrections for the partial back-transfer of charge from anionic ligands to vacant valence orbitals of the cationic central atom.

Even in ordinary Lewis structure compounds such as PH$_3$ or PF$_3$, the NBOs are often found to be more polar [i.e., larger c_B/c_A, where A is the second-row element] and the quality of the natural Lewis structure is found to be somewhat lower [i.e., larger percentage ρ_{NL} in Eq. (6)] than in corresponding first-row compounds. Table V compares the form of the optimal NBO for A-H bonds of simple first- and second-row hydrides at the RHF/6-31G* level, showing the gradual shift in the form of the NBOs as the electropositivity of the central atom increases. The differences in this table reflect the greater electronegativity differences and increased ionic character of the bonding involving electropositive second-row atoms. Indeed, as the ratio c_B/c_A of the optimal NBO coefficients in Eqs. (2), (5) increases, one sees a gradual change from covalency (with small ρ_{NL} and a "good" Lewis structure) toward the essentially ionic limit (with large ρ_{NL} and a "poor" Lewis structure) where the wavefunction ψ_L of Eq. (4b) is no longer a good starting point. NBO analysis permits quantification of this gradual transition between distinct bonding types.

A second characteristic of bonding involving second-row atoms is that *3s* and *3p* AOs are of significantly different radial diffuseness, so that sp^λ hybridization is less effective in promoting covalent bond formation.[28] As can be seen in Table V, the hybridization of the central atom in covalent second-row hydrides is generally closer to pure *s*-type (for lone pairs) or pure *p*-type (for bond pairs) than in analogous first-row compounds.[29] The hybridization shifts implied by Bent's Rule (related to electronegativity shifts) must therefore be added to the inherently higher *p*-character of bond hybrids in second-row atoms.

Reed and Schleyer[30] have analyzed many other interesting aspects of covalency and hypervalency in second-row compounds.

Boron Compounds

The NBO search for an optimal Lewis structure of one- and two-center bond functions typically leads to poor values of ρ_{NL} for boron hydrides and related group IIIA

Table V *NBOs for first- and second-row hydrides (RHF/6-31G* level, idealized Pople-Gordon geometry), showing polarization coefficients c_C, c_H and heavy atom hybridization sp^λ for σ_{AH} bonds.*

molecule	polarization coefficients		heavy atom hybridization			
	c_A	c_H	h_A	%s	%p	%d
BH$_3$	0.6598	0.7515	$sp^{2.00}$	33.26	66.40	0.34
CH$_4$	0.7801	0.6257	$sp^{2.99}$	25.00	74.84	0.16
NH$_3$	0.8296	0.5584	$sp^{2.96}$	27.00	72.77	0.20
H$_2$O	0.8616	0.5072	$sp^{3.01}$	24.90	74.86	0.24
HF	0.8825	0.4703	$sp^{3.93}$	20.20	79.50	0.26
AlH$_3$	0.5298	0.8481	$sp^{1.95}$	33.17	64.66	2.18
SiH$_4$	0.6263	0.7796	$sp^{2.93}$	25.00	73.25	1.75
PH$_3$	0.7036	0.7106	$sp^{3.40}$	22.38	76.21	1.41
H$_2$S	0.7578	0.6524	$sp^{4.28}$	18.71	80.13	1.16
HCl	0.8004	0.5995	$sp^{5.08}$	16.29	82.79	0.93

compounds, reflecting breakdown of the elementary Lewis structure picture. However, when the NBO search is extended to *three*-center functions τ_{ABC}, the resulting Lewis-type description is generally quite satisfactory. For example, in B_2H_6 the NBO Lewis structure includes both two-center $B-H_a$ and three-center $B-H_b-B$ bonds, in accord with the classic model,[31]

$$\psi_L = \ldots (\sigma_{BH_t})^2 (\sigma_{BH_t})^2 (\tau_{BH_bB})^2 (\tau_{BH_bB})^2 (\sigma_{BH_t})^2 (\sigma_{BH_t})^2 \ldots \qquad (10)$$

where H_t, H_b are terminal and bridging hydrogens, respectively. Table VI summarizes some details of the two- and three-center NBOs of diborane at the RHF/6-31G* level. In this case the value of $\rho(\%Lewis) = 99.64\%$ is comparable to that of other reasonable Lewis structures. NBO analysis thus provides direct numerical evidence of the importance of three-center τ_{BH_bB} bridge-bonding in boranes, and leads to quantitative Lewis structures (10) that are consistent with the generally accepted bonding picture for these compounds.

Lithium Compounds

The low electronegativity of lithium and other alkali metals (as compared, e.g., to hydrogen) leads typically to a bonding regime of large electronegativity differences and high ionic character in compounds involving bonding to non-metallic first-row elements. This ionic bonding regime gives rise to many surprising structures (as shown particularly by Schleyer and coworkers[32]) and many apparent violations of the octet rule and elementary Lewis structure principles. Even in the case of the apparently "normal" tetrahedral CLi_4 analog of methane, NBO analysis indicates that the bonding is essentially ionic in character, and thus that a simple Lewis structure picture based on analogous hydride molecules can be quite misleading. NBO analysis of CLi_6 and other "hyperlithiated" compounds of C, N, and O has shown[33] that the structures of these fascinating (theoretically predicted) species are better understood from the extreme ionic limit of bonding.

When lithium is bonded to atoms of similar electropositivity, quite different bonding types are encountered. A particularly astonishing structure has recently been discovered[34] for $SiLi_4$, which is predicted to have a "sawhorse" C_{2v} geometry, with all four Li atoms lying on one side of a plane passing through the Si atom. The contrast with the superficially analogous CH_4 structure is very striking, since either a purely covalent model (sp^3-hybridized) or purely ionic model $[Si^{-4}(Li^+)_4]$ of the bonding would have predicted a strong preference for tetrahedral geometry. NBO analysis indicates a surprisingly simple picture of the bonding: two of the three p orbitals of Si are used to form covalent (2-center, 2-electron) bonds (at approximately 90°) to two of the lithium atoms, and the third is used to form a 3-center, 2-electron bond (approximately linear) to the remaining two lithium atoms, the two remaining electrons occupying the s-like lone pair orbital on silicon,

$$\psi_L = \ldots (n_{Si})^2 (\sigma_{SiLi})^2 (\sigma_{SiLi})^2 (\tau_{LiSiLi})^2 \ldots \qquad (11)$$

Table VI *Natural bond orbitals of B_2H_6 (RHF/6-31G* level, experimental C_{2h} geometry), showing the polarization coefficients and boron hybrids for two-center and three-center bonds (of occupancy 1.994, 1.984, respectively).*

bond type	polarization coefficients		boron atom hybridization			
	c_B	c_H	h_B	%s	%p	%d
2c B-H	0.7018	0.7124	$sp^{2.17}$	31.45	68.32	0.24
3c B-H-B	0.5246	0.6701	$sp^{4.36}$	18.53	80.79	0.04

(Note that the valence electronic distribution shows only *four* electron pairs around silicon, despite the fancied resemblance to the VSEPR-like SF_4 structure.) This example illustrates the novel extensions of the Lewis structure concept that are needed to describe the bonding in molecules involving less common electronegativity combinations.

Titanium Compounds

Preliminary steps have been taken to apply the NBO method to transition metal systems.[35] We have carried out a series of *ab initio* SCF and MCSCF of the bonding in ground and low-lying states of some simple organotitanium compounds [$TiCH_2$, TiC_2H_4], using minimal basis and double zeta quality basis sets. The primary goal of this work was to investigate the validity and usefulness of Lewis structure concepts in describing the nature of the bonding in organometallic species of catalytic interest.

Geometry optimizations at both SCF and MCSCF levels using either minimal or double zeta (DZ) sets indicate that $TiCH_2$ has a planar C_{2v} structure in its ground ($^3A_2$) state, with the methylene geometry closely resembling that of the CH_2 group in ethylene. At the MCSCF/DZ level, the molecule is bound by almost 45 kcal/mol. The NBO occupancies at all levels confirm that an "ethylene-like" picture of the bonding between Ti and CH_2 aptly describes the electronic wavefunction. Ti is bonded to the CH_2 group by a double bond, where Ti contributes an sd hybrid to the σ bond and a pure d_{yz} orbital to the π bond. The two triplet-coupled electrons are found in sd and d_{xy} orbitals of the Ti atom, with the remaining two d orbitals unoccupied. Two of the three lowest triplet states of the other symmetry types are formed by exciting the triplet electrons into the low-lying unoccupied d orbitals, with little effect on the $Ti=CH_2$ covalent bonding. Likewise, the lowest singlet states are formed by simply recoupling the two unpaired nonbonding electrons of the titanium atom, and so are only slightly raised in energy from the corresponding triplet states. This picture suggests that the "ethylene-like" $Ti=CH_2$ double bond is an extremely robust feature of the electronic structure, and that $TiCH_2$ would exhibit a series of low-lying (within ~20 kcal/mol) excited states having rather similar potential energy curves. The striking similarities between the NBO Lewis structures for $TiCH_2$ and ethylene suggest that the chemical bonding characteristics of a Ti (or TiH_2) are analogous to those of a CH_2 group.

The striking chemical similarities of a Ti and a CH_2 group are confirmed in calculations on Ti + C_2H_4. In this calculation, titanium approaches the ethylene double bond in a T-shaped (C_{2v}) geometry, the geometry being reoptimized at each distance of separation between the Ti and the CC midpoint. The titanium atom is found to add directly to ethylene, breaking the double bond, and forming TiC_2H_4, cyclotitanoethane. The NBO structure of TiC_2H_4 bears a remarkable resemblance to that of cyclopropane [except that ring strain ("bond bending") at the Ti atom is significantly reduced by d hybridization]. We conjecture that low-lying singlet and triplet excited states of this species

(12)

would also exhibit the robust cyclopropane-like structure, with various spin pairings and occupancies of the two unshared Ti electrons among the available metal d orbitals.

The surprising success of a simple NBO Lewis picture such as (12) in aptly representing the essential features of the MCSCF/DZ wavefunction for a complex organometallic system is evidence that the Lewis structure concept is alive and well, and will continue to play a useful role in molecular structure studies.

REFERENCES

1. See, e.g., I. Plesser, Z. Vager, and R. Naaman, *Phys. Rev. Lett.* **56**, 1559 (1986).
2. J. H. van't Hoff, *Archiv. ne'erland* **9**, 445 (1874); J. A. Le Bel, *Bull Soc. Chim.* **22**, 337 (1874).
3. G. N. Lewis, *J. Am. Chem. Soc.* **38**, 762 (1916); G. N. Lewis, *Valence and the Structure of Atoms and Molecules* (The Chemical Catalog Co., New York, 1923).
4. L. Pauling, *J. Am. Chem. Soc.* **53**, 1367 (1931); cf. also Coulson, Ref. 7, pp. 203-205.
5. J. C. Slater, *Phys. Rev.* **37**, 481 (1931).
6. L. Pauling, *The Nature of the Chemical Bond* (Cornell U. Press, Ithaca, N.Y., 1960).
7. C. A. Coulson, *Valence*, 2nd ed. (Oxford Univ. Press, New York, 1961), p. 270.
8. R. S. Mulliken, *J. Chem. Phys.* **3**, 573 (1935); Ref. 7, pp. 107-110.
9. H. A. Bent, *Chem. Rev.* **61**, 275 (1961).
10. L. Pauling, in, *Correspondence Between Concepts in Chemistry and Quantum Chemistry* (Technical Note No. 16, Quantum Chemistry Group, Uppsala University, Uppsala, Sweden, 1958), Part II, p. 73.
11. H. F. Schaefer III, *The Electronic Structure of Atoms and Molecules: A Survey of Rigorous Quantum Mechanical Results* (Addison-Wesley, Reading, MA, 1972).
12. J. P. Foster and F. Weinhold, *J. Am. Chem. Soc.* **102**, 7211 (1980); A. E. Reed and F. Weinhold, *J. Chem. Phys.* **78** 4066 (1983); A. E. Reed, R. B. Weinstock, and F. Weinhold, *J. Chem. Phys.* **83**, 735 (1985).
13. P.-O. Löwdin, *Phys. Rev.* **97**, 4066 (1983).
14. A. E. Reed and F. Weinhold, *QCPE Bull.* **5**, 141 (1985).
15. For a review of the NBO formalism, see A. E. Reed, L. A. Curtiss, and F. Weinhold, University of Wisconsin Theoretical Chemistry Institute Report WIS-TCI-727 (1987); *Chem. Rev.*, (to be published).
16. J. E. Carpenter and F. Weinhold, University of Wisconsin Theoretical Chemistry Institute Report WIS-TCI-689 (1985), unpublished.
17. For the standard *ab initio* computational methods and basis set designations referred to herein, see W. J. Hehre, L. Radom, P. v. R. Schleyer, and J. A. Pople, *Ab Initio Molecular Orbital Theory* (John Wiley, New York, 1986).
18. J. E. Carpenter and F. Weinhold, *J. Am. Chem. Soc.* **110**, 368 (1988).
19. See, e.g., J. K. Kochi, *Adv. Free-Radical Chem.* **5**, 189 (1975).
20. J. E. Carpenter and F. Weinhold, *J. Mol. Struct. (THEOCHEM)* **165**, 189 (1988).
21. See, e.g., R. Pauncz, *The Alternant Molecular Orbital Method* (W. B. Saunders Co, Philadelphia, 1967); J. W. Linnett, *The Electronic Structure of Molecules. A New Approach* (Methuen, London, 1964).
22. G. W. Wheland, *Resonance in Organic Chemistry* (John Wiley, New York, 1955).
23. L. Pauling, *J. Chem. Phys.* **51**, 2767 (1961).
24. L. A. Curtiss, C. A. Melendres, A. E. Reed, and F. Weinhold, *J. Comp. Chem.* **1**, 294 (1986).
25. J. T. Blair, J. C. Weisshaar, J. E. Carpenter, and F. Weinhold, *J. Chem. Phys.* **87**, 392 (1987); J. T. Blair, J. C. Weisshaar, and F. Weinhold, *J. Chem. Phys.* **88**, 1467 (1988).
26. K. J. Rensberger, J. T. Blair, F. Weinhold, and F. F. Crim (in preparation).
27. A. E. Reed and F. Weinhold *J. Am. Chem. Soc.* **108**, 3586 (1986).
28. W. Kutzelnigg, *Angew. Chem., Int. Ed. Engl.* **23**, 272 (1984).
29. The molecules SiH_4 and AlH_3 are apparent exceptions, but these more ionic species exhibit rather large departures from the idealized Lewis form (cf. ρ_L, Table I) that also reflect the "strain" in using the geometrically optimal sp^3 or sp^2 hybrids.
30. A. E. Reed and P. v. R. Schleyer, *J. Am. Chem. Soc.* **109**, 7362 (1987).
31. See, e.g., F. A. Cotton and G. Wilkinson, *Advanced Inorganic Chemistry* (John Wiley Interscience, New York, 1962), p. 198ff.
32. P. v. R. Schleyer, in, P.-O. Löwdin and B. Pullman (eds.), *New Horizons in Quantum Chemistry* (D. Reidel Publishing Co., New York, 1983), pp. 95-109.
33. A. E. Reed and F. Weinhold, *J. Am, Chem. Soc.* **107**, 1919 (1985).
34. P. v. R. Schleyer, talk presented to the WATOC (World Association of Theoretical Organic Chemists) Congress, Budapest, Hungary, August, 1987; P. v. R. Schleyer and A. E. Reed, *J. Am. Chem. Soc.* **110**, 4453 (1988).
35. J. E. Carpenter, *Ph.D. Thesis* (University of Wisconsin, Madison, 1987).

STUDIES IN THE PAIRED ORBITAL METHOD III:

THE STRUCTURE OF THE PO WAVEFUNCTION

Ruben Pauncz

Department of Chemistry,Technion
Israel Institute of Technology
Haifa, 32000 Israel

1. INTRODUCTION

In the theoretical treatment of atoms and molecules the self-consistent field method is an excellent starting point. The wavefunction is given in the form of a single determinant in which $\underline{n}$ orbitals are doubly occupied. For the sake of simplicity we shall restict our treatment to the case where the number of electron is even (N=2n). The orbitals are determined from the minimization of the total energy of this wavefunction. In most cases the orbitals are given as linear combinations of given basic orbitals, let us denote the number of basic orbitals as $\underline{M}$. The corresponding variational equations for the best coefficients have been derived by Roothaan[1], Hall[2]. After solving the equations one obtains n orbitals which are doubly occupied and in addition $\underline{M-n}$ orbitals which do not have immediate physical signifi- cance. The latter are called virtual orbitals.

The SCF solution gives good results for the total energy, bond lengths and some other properties of the molecules. The small error in the total energy (0.5 %) is still too large when we would like to calculate transition energies, dissociation energies, and so on. It is necessary to go beyond the SCF method. The difference between the SCF energy and the best energy obtained in the given basis using the non-relativistic Hamiltonian is called the correlation energy. Several methods have been suggested to treat the electronic correlation problem. One should remember that the single deter- minantal description already takes into account to some extent the corre- lation between electrons with parallel spins because the wavefunction is antisymmetric and therefore the probability of finding two electrons with the same spin in the neigborhood of the same point is zero. The single determinant does not describe properly the correlation between electrons with antiparallel spins.

The best possible result within the given basis set can be obtained if one forms all the possible determinants (or certain combinations of them, called configuration wavefunctions) and one looks for the best combination of these functions. The method is called the full configuration interaction method (full CI). A practical limitation of this method is the fact that the number of configurations rises very sharply with the number of elec- trons. Another drawback of the configuration interaction treatment is the fact that is very difficult to visualize a wavefunction which is a linear combination of several thousands (or even a million) configurations.

Löwdin[3] suggested a simple method for the improvement of the one determinantal representation. One should relax the restriction that each of the n orbitals occurs twice in the wavefunction. One can assign different sets of orbitals to be associated with alpha and beta spins. The method is called different orbitals for different spins (DODS). The single determinant constructed in this way is not a pure spin eigenfunction, but a definite spin state is obtained using the projection operator method of Löwdin[4].

A simple variant of the DODS method was the alternant molecular orbital method (AMO) suggested by Löwdin[5]. The basis of the method and its early development are given in the book by Pauncz[6]. Recent advances of the DODS approach have been reviewed by Mayer[7]. In a quite recent paper Karadakov[8] has derived equations for obtaining the best DODS orbitals. The AMO method was quite successful for alternant systems.

The paired orbital method[9] is a variant of the DODS method. The wavefunction is formally similar to the one used in the AMO method and the corresponding energy expression is identical with the one derived by Pauncz, et al[10], and de Heer and Pauncz[11]. The difference between the two approaches is in the selection of the orbital pairs. In the case of AMO this selection was dictated by the alternancy symmetry[12]. Pauncz, Kirtman and Palke[13] have given an algorithm for the determination of the orbital pairs in the general case and an illustrative calculation was given for the case of water molecule. Harrison and Handy[14] have performed a full CI and so one can compare the results to the best possible treatment. The PO method using five non-linear parameters recovered about 20% of the correlation energy obtained with the full CI treatment (256473 configurations). Pauncz[15] has derived the expressions for the derivatives of the energy with respect to the non-linear parameters and he proved that the SCF energy is a maximum with respect to the non-linear parameters of the PO method. References [13] and [15] will be referred as I and II, respectively.

The aim of the present paper is to analyse of structure of the MO wavefunction. By expanding the spatial part and looking at the structure of the projected spin function one can obtain the wavefunction as a linear combination of certain configuration functions. The coefficients of these configurations are determined by the values of the non-linear parameters. This analysis will yield a comparison between the PO method and the configuration interaction treatment.

2. THE PAIRED ORBITAL WAVE FUNCTION

The wavefunction used in the PO method is of the following form:

$$\Psi = N \hat{A} \; \Phi \; \hat{O}_s \; \alpha(1)\ldots\alpha(n) \; \beta(n+1)\ldots\beta(2n), \qquad (1)$$

where $\hat{A}$ is the antisymmetrizer, N is a normalization constant and $\hat{O}_s$ is the spin projection operator. We shall consider the singlet state only (S=0). Φ is a spatial (freeon) wavefunction which is a product of one-electron orbitals:

$$\Phi = u_1(1)\ldots u_n(n) \; v_1(n+1)\ldots v_n(2n). \qquad (2)$$

In the PO (and in the AMO) method the u_i's and v_i's are formed from a set of orthogonal one-electron orbitals in the following way:

$$u_i = a_i \; \psi_i + b_i \; \psi_{i'} \quad \text{and} \quad v_i = a_i \; \psi_i - b_i \; \psi_{i'}, \qquad (3)$$

where $a_i = \cos\theta_i$ and $b_i = \sin\theta_i$, ψ_i (i=1,...,n) is a doubly occupied orbital in the single determinantal SCF wavefunction and $\psi_{i'}$ is a virtual orbital whith which it is paired. For the AMO method the pairing was obtained from the alternancy symmetry. In the PO method the pairs are generated by maximizing the sum of electron repulsion integrals:

$$C = \Sigma \ (ii|i'i') \tag{4}$$

The algorithm for obtaining the paired orbitals is described in I. The ψ_i's are chosen as localized orbitals and the ψ_i's occupy the same region of space as their partners.

As an example we shall choose the water molecule (N=10). The projected singlet spin eigenfunction is given as follows:

$$\hat{O}_s \ \theta = (10 \ [\alpha^5] \ [\beta^5] - 2 \ [\alpha^4\beta] \ [\alpha\beta^4] + [\alpha^3\beta^2] \ [\alpha^2\beta^3] - [\alpha^2\beta^3] \ [\alpha^3\beta^2]$$

$$+2 \ [\alpha\beta^4] \ [\alpha^4\beta] -10 \ [\beta^5] \ [\alpha^5])/60 \tag{5}$$

Here $[\alpha^i\beta^j]$ (i+j=n) is the sum of all primitive spinfunctions which have $\underline{i}$ α's and j β's. The first bracket refers to the first $\underline{n}$ electrons, and the second bracket to the last $\underline{n}$ electrons.

3. DECOMPOSITION OF THE WAVEFUNCTION INTO CONFIGURATIONS

Using eqs. (2) and (3) we can write the spatial part of the wavefunction in the following form:

$$\Phi = \prod_{i=1}^{n} \ (a_i\psi_i(i)+b_i\psi_{i'}(i))(a_i\psi_i(n+i)-b_i\psi_{i'}(n+i)) \tag{6}$$

The function Φ can be decomposed into functions according to the total number of $\underline{b}$'s occurring in them. Φ_r is the sum of all the functions obtained from eq. (6) which contain $\underline{r}$ $\underline{b}$'s.

$$\Phi = \Sigma \ \Phi_r \tag{7}$$

Pauncz[10] has shown that in the case of even number of electrons (N=2n) and S=0 only the even $\underline{r}$'s will contribute to the final (antisymmetrized and projected) wavefunction. This means that we shall have no contribution from singly, triply, etc. excited configurations.

Φ_0 is essentially a single determinant with n doubly occupied orbitals, it includes the factor $(a_1 a_2 \ldots a_n)^2$. Similarly Φ_n is a single determinant with $\underline{n}$ doubly occupied orbitals, all these orbitals belong to the virtual set and the factor is $(b_1 \ldots b_n)^2$.

For the remaining Φ_r's let us use the following notation: (i/j) is the sum of all functions (including the proper factors) in the expansion in which $\underline{i}$ functions among the first $\underline{n}$'s and $\underline{j}$ functions among the last $\underline{n}$'s are replaced by their antibonding partners (i+j=r). The sign factor of (i/j) is $(-1)^j$. As an example for the case of water molecule (N=10, n=5) Φ_4 can be written as follows:

$$\Phi_4= (4/0)-(3/1)+(2/2)-(1/3)+(0/4) \tag{8}$$

The number of terms for a given Φ_r (r < n) is obtained as:

$$\sum_{i=0}^{r} \ \binom{n}{i}\binom{n}{r-i} = \binom{2n}{r} \tag{9}$$

The total number of terms in the expansion of the spatial part is given as $(2^{2n}/2)$. We have 2n factors each having two terms and total number of even terms is equal to the number of terms coming from the odd Φ_r's.

Another way of analysing the ϕ_r's is to decompose them according to the number of singly occupied orbitals (s.o.o.). As $\underline{r}$ is even, this number can be 0, 4, 8,... etc. We can obtain the number of terms with a given value of s.o.o's using simple combinatorial arguments. Let us choose again the example of N=10 and ϕ_4. Its decomposition according to the number of s.o.o.'s is given as follows:

# of s.o.o.	0	4	8
# of terms	$\binom{n}{2}=10$	$\binom{n-2}{1}\binom{n}{2}=10*3=30$	$\binom{n-4}{0}\binom{n}{4}=5$

We shall outline the arguments for obtaining these values:

a) <u># of s.o.o.=0</u>. In ϕ_4 we have 4 electrons in orbitals from the upper set (virtual orbitals). Each of the orbitals is doubly occupied, so there will be 2 (doubly occupied) orbitals from the upper set and these can be chosen in $\binom{n}{2}$ =10 different ways.

b) <u># of s.o.o.=4</u>. Orbitals i,j,i',j' are singly occupied. These indices can be chosen in $\binom{n}{2}$ different ways. From the remaining (n-2) we have to choose 1 more orbital in the upper set which will be doubly occupied ($\binom{n}{1}$). The total number will be the product of these two factors.

c) <u># of s.o.o.=8</u>. Orbitals i,j,k,l,i',j',k',l' are singly occupied. The 4 indices can be chosen in ($\binom{n}{4}$) different ways.

Table 1. Decomposition of the spatial function for N=10

# s.o.o.	0	4	8	
Φ_0	1			
Φ_2	5	10		
Φ_4	10	30	5	
Φ_6	10	30	5	
Φ_8	5	10		
Φ_{10}	1			
#	32	80	10	122
#	32	320	160	512

Table 2. Decomposition of the spatial function for N=12

s.o.o.	0	4	8	12	
Φ_0	1				
Φ_2	6	15			
Φ_4	15	60	15		
Φ_6	20	90	30	1	
Φ_8	15	60	15		
Φ_{10}	6	15			
Φ_{12}	1				
#	64	240	60	1	365
#	64	960	960	64	2048

Tables 1 and 2 show the decomposition of the spatial functions for $N=10$ and $N=12$, respectively. The numbers appearing in Tables 1 and 2 give the number of terms which yield configurations. The actual number of terms for $\underline{2k}$ s.o.o. is obtained by multiplying the number in the table by 2^k. The reason is that in the actual expansion we shall have all the terms which are obtained from a given selection of the indices $i_1,\ldots,i_k,i_{1'},\ldots,i_{k'}$ by applying the elements of the set of permutations $\underline{C}$:

$$C=(e,(i_1,n+i_1))\ldots(e,(i_k,n+i_k)) \tag{10}$$

Here $\underline{e}$ is the identity permutation. The last two rows in the tables give the total number of terms for each column, the first one gives the configurations, and the second one the total number of terms. One can easily see that these numbers are given by the formulas:

$$\text{no of terms for } \underline{2k} \text{ s.o.o.} = 2^{n-k} \binom{n}{k} \tag{11}$$

$$\text{total no of terms} \quad " \quad = 2^n \binom{n}{k} \tag{12}$$

Finally the total number of orbital configurations is obtained by adding these numbers. This leads to the final formulas:

$$\text{total no of configurations} = (3^n+1)/2 \tag{13}$$

$$\text{total no of terms} = 2^{2n-1} \tag{14}$$

4. INCLUSION OF THE PROJECTED SPIN FUNCTION

The total wavefunction is obtained after multiplying the spatial part by the projected spin function and antisymmetrizing the product. One of the advantages of the use of the projected function is the fact that even we started with the projection operator corresponding to $N=2n$ electrons, in the case of doubly occupied orbitals the projection operator (after antisymmetrization) yields the same result as obtained using the one which operates on the singly occupied orbitals only. In the case of Φ_0 we are left after projection and antisymmetrization still with a single determinant with doubly occupied orbitals.

For Φ_2 we are left with the sum of 4 projected functions for a given selection of the indices i,j,i',j'. It is quite interesting that their sum yields again a projection of a single function. Table 3 illustrates the result for $i=1,j=2,i'=3,j'=4$. The new projection can be obtained from the starting one (apart from the factor -2) by applying the transposition $(2,3)$. When we look at the final wavefunction we could apply this permutation to the spatial part and write the projected spinfunction in the starting form:

$$\Psi = \hat{A} \ \psi_1\psi_1,\psi_2\psi_2, \ \hat{O}_s\alpha\alpha\beta\beta \tag{15}$$

Table 3. The sum of 4 projected functions for $N=4$

$$\theta_1=(2\alpha\alpha\beta\beta-\alpha\beta\alpha\beta-\alpha\beta\beta\alpha-\beta\alpha\alpha\beta-\beta\alpha\beta\alpha+2\beta\beta\alpha\alpha)/6$$

$$(1,3)\theta_1=(2\beta\alpha\alpha\beta-\alpha\beta\alpha\beta-\beta\beta\alpha\alpha-\alpha\alpha\beta\beta-\beta\alpha\beta\alpha+2\alpha\beta\beta\alpha)/6$$

$$(2,4)\theta_1=(2\alpha\beta\beta\alpha-\alpha\beta\alpha\beta-\alpha\alpha\beta\beta-\beta\beta\alpha\alpha-\beta\alpha\beta\alpha+2\beta\alpha\alpha\beta)/6$$

$$(1,3)(2,4)\theta_1=(2\beta\beta\alpha\alpha-\alpha\beta\alpha\beta-\beta\alpha\alpha\beta-\alpha\beta\beta\alpha-\beta\alpha\beta\alpha+2\alpha\alpha\beta\beta)/6$$

---

sum $\quad (2\alpha\alpha\beta\beta-4\alpha\beta\alpha\beta+2\alpha\beta\beta\alpha+2\beta\alpha\alpha\beta-4\beta\alpha\beta\alpha+2\beta\beta\alpha\alpha)/6=$

$\quad -2(2\alpha\beta\alpha\beta-\alpha\alpha\beta\beta-\alpha\beta\beta\alpha-\beta\alpha\alpha\beta-\beta\beta\alpha\alpha+2\beta\alpha\beta\alpha)/6= \ -2(2,3)\theta_1$

One has to observe the order of the spatial functions in the new wave-function. The projected antisymmetrized function has a very simple physical meaning: One combines the spin of the first two orbitals to a resultant triplet and the same applies to the last two orbitals. Finally the two triplets are combined into a singlet wavefunction.

For 8, 12,.etc s.o.o. we no longer have this simple relation. The wavefunction will be a combination of several projected wavefunction, e. g. for 8 s.o.o. we shall have contribution from all the 14 projected functions.

5. SUMMARY

The PO wavefunction can be decomposed into the sum of doubly, four times, etc excited configurations. Only those configurations will appear in which we transfer electrons from orbitals i_1, ...into $i_{1'}$,.... We have obtained a closed formula for the number of configurations for the decomposition of the spatial function. The final configurations are obtained by usng the appropriate projected spin functions (or sum of projections) and by antisymmetrizing the whole product.

ACKNOWLEDGMENT

Part of this research was performed during the stay of the author at the Department of Applied Mathematics, University of Waterloo, Waterloo. The author would like to express his gratitude to Profs. J.Cizek, J. Paldus, E. Vrscay and Ms. F. Vinette for their kind hospitality and for helpful discussions. The author is grateful for the support of this research by the Fund for Advance of Research at the Technion.

REFERENCES

1   C. C. J. Roothaan, Revs. Mod. Phys. 23,69 (1951)
2   G. G. Hall, Proc. Roy. Soc. A 202, 336 (1951)
3   P. O. Löwdin, Phys. Rev. 97, 1474,1490,1509 (1955)
4   P. O. Löwdin, Revs. Mod. Phys. 34, 520 (1962)
5   P. O. Löwdin, Symposium on Molecular Physics at Nikko, Japan, Maruzen Co
          Tokyo, (1954), p. 13
6   R. Pauncz, The Alternant Molecular Orbital Method, W. Saunders, Co.
          Philadelphia (1967)
7   I. Mayer, Adv. Quantum Chem. 12, 189 (1980)
8   P. Karadakov, Int. J. Quantum Chem. 30, 239 (1986)
9   R. Pauncz, M. B. Chen and R. G. Parr, Proc. Natl. Ac. Sci. USA 79,705
          (1986)
10 R. Pauncz, J. de Heer and P. O. Löwdin, J. Chem. Phys. 36, 2247 (1961)
11 J. de Heer and R. Pauncz, J. Chem. Phys. 39, 2314 (1963)
12 C. A. Coulson and G. S. Rushbroke, Proc. Camb. Phil. Soc. 36, 193
          (1940)., J. Koutecky, J. Paldus and J. Cizek, J. Chem. Phys. 83, 1722
          (1985)
13 R. Pauncz, B. Kirtman and W. E. Palke, Int. J. Quantum Chem. 21S, 533
          (1987)
14 R. J. Harrison and N. Handy, Chem. Phys. Letters, 95, 386 (1983)
15 R. Pauncz, Theor. Chim. Acta (in press)
16 R. Pauncz, ref. 6 ,p. 81

PARTICIPATION OF A HYDROGEN-BRIDGED INTERMEDIATE ION

[O=C(H)-H .. CO]$^{+\cdot}$ IN THE LOSS OF CO FROM IONIZED GLYOXAL ?

Paul J. A. Ruttink

Theor. Chem. Group
Dept.of Chemistry, State University of Utrecht
Paddualaan 8
De Uithof, Utrecht
The Netherlands

1. INTRODUCTION

In a recent paper [1] concerned with the identification of dihydroxyacetylene, HO-CC-OH, as a stable neutral molecule in the gas phase by neutralizing the corresponding cation [HO-CC-OH]$^{+\cdot}$ (1) in a NRMS experiment [2], three other isomeric $C_2H_2O_2^{+\cdot}$ ions were briefly investigated, *i.e.* ionized glyoxal, [CHO-CHO]$^{+\cdot}$(2), ionized hydroxyketene, [HOC(H)=C=O]$^{+\cdot}$ (3) and [CH$_2$-O-CO]$^{+\cdot}$ (4). It was further suggested that ionized glyoxal, (2), isomerizes into a stable hydrogen bridged intermediate with a -C..H..C- bridge, *viz.* [O=C(H)-H ..CO]$^{+\cdot}$ (2a) prior to its unimolecular dissociation into $CH_2O^{+\cdot}$ + CO.

Experiments (Collisional Activation and Metastable Ion spectra) rule out the participation of ions (1), (3) and (4) in the mechanism for the metastable loss of CO from ionized glyoxal, (2) which was further observed to proceed at the thermochemical threshold and to be associated with an unusually small Kinetic Energy Release (KER), $T_{0.5} = 0.3$ meV [3].

Although it becomes increasingly apparent that hydrogen bridged radical cations play an important role as stable intermediates in the gas phase ion chemistry of molecular ions of initially conventional structure [4,5], so far only -O..H..O- bridged species (like [CH$_2$=C(H)-O..H..OH$_2$]$^{+\cdot}$[6]) and -C..H..O- bridged species (like CH$_2$=C(H)..H..OH$_2$]$^{+\cdot}$ and the closely related [C$_2$H$_4$ / H$_2$O]$^{+\cdot}$ ion-dipole complex [7]) have been closely examined. This was done by combining the results of mass spectrometric experiments with those of *ab ibitio* Molecular Orbital theory (MO) calculations.

In this contribution the participation of the -C..H..C- bridged ion (2a) on the glyoxal radical cation potential energy surface will be explored using *ab initio* MO theory and next a rationale for the very small amount of kinetic energy released in the dissociation of [CHO-CHO]$^{+\cdot}$ will be provided in terms of the Adiabatic Channel Theory [8] potential energy curves for the rotational states of the dissociating ion.

2. THEORY

In a previous study [9] the average KER <T> for a MI fragmentation process was calculated for the distonic ion $CH_2CH_2OH_2^{+\cdot}$. However, this model cannot be used directly in our case for the following reasons.

i) The $(CHO)_2^{+\cdot}$ ion has to rearrange before dissociating into $H_2CO^{+\cdot}$ and CO.

ii) The CO molecule has only a small dipole moment. Therefore the ion-dipole attraction will not dominate the long range interaction between the dissociating fragments as in the case of $C_2H_4^{+\cdot}$ / H_2O.

Therefore we will discuss the extensions to this model needed to describe the dissociation of $(CHO)_2^{+\cdot}$ in some detail.

The calculation of $<T>$ involves two steps. First, the potential energy surface has to be explored in order to determine the critical points. This is done by *ab initio* MO theory, including electronic correlation effects Second, statistical theory (RRKM / QET) [10] is used in order to obtain the microcanonical rate constant for the dissociation $k_{diss}(E)$ as a function of the internal energy E. The number of open channels is obtained by using the Adiabatic Channel Theory [8], where the effective potential energy curves are obtained by diagonalizing the nuclear Hamiltonian for a number of R-values. Each rovibrational state is assumed to yield a contribution to the KER which equals the height of the centrifugal barrier due to the end-over-end rotation of the complex. Finally these KER contributions are averaged over all internal energies which are compatible with the experimental time frame for MI fragmentations, *viz* 10^{-6} s. In the following these two steps will be discussed separately.

2.1 THE POTENTIAL ENERGY SURFACE

In order to determine the critical points *ab initio* MO calculations were performed using the program system GAMESS [11] with the standard 4-31G basis in the Restricted Hartree-Fock (RHF) method. Since the surface appeared to be rather flat, it turned out to be difficult to characterize the reaction path leading from $(CHO)_2^{+\cdot}$ to $H_2CO^{+\cdot}$ + CO uniquely. Moreover, 6-31G* / CI calculations performed at the 4-31G / RHF optimized geometries show large electronic correlation effects on the relative energies of the critical points. This effect may be traced to the inability of the RHF method to account for the localization of the charge in $(CHO)_2^{+\cdot}$ even if the geometry of this ion is allowed to distort from the C_{2h}-symmetry of the neutral molecule to C_s - symmetry as found for the ion. Thus if in the HF - configuration the charge is more or less localized on one O - atom, this implies that there is a low-lying single excitatiion carrying the charge on the other O - atom. Therefore MRDCI calculations [12] were performed using these two configurations as references. By applying size-consistency corrections in order to remedy the lack of additivity of CI calculations for the separated fragments, satisfactory results may then be obtained for $(CHO)_2^{+\cdot}$ in comparison to the dissociation limits $H_2CO^{+\cdot}$ + CO, $HCOH^{+\cdot}$ + CO and $HCO^{\cdot}$ + HCO^{+}.

For the size consistency corrections two possible generalizations of the method due to Pople were used [13]. These methods differ in the choice of the reference function used for calculating the CI correlation energy ΔE_{CI} and the angle θ, as defined by

$$\Delta E_{CI} = E_{CI} - < \Psi_0 | H | \Psi_0 > / < \Psi_0 | \Psi_0 > \tag{1}$$

$$\theta = \arccos < \Psi_0 | \Psi_{CI} > \tag{2}$$

where Ψ_0 is the rteference function

$$\Psi_0 = \Sigma_R \, c_R \, \Phi_R \tag{3}$$

The first choice corresponds to determining the c_R by diagonalizing the reference part of the H-matrix. In this case E_0 is the lowest eigenvalue of the reference Hamiltonian. Since in our case the second reference configuration is singly excited with respect to the first (RHF) configuration Φ_{HF}, this choice leads to $\Psi_0 = \Phi_{HF}$ and $\Delta E_{CI} = E_{CI} - E_{HF}$. The second choice corresponds to projecting the CI function to the reference configuration space. The c_R are then identical to the CI-coefficients of the reference configurations. If the MRDCI coefficient of the second reference configuration appears to be large, the two methods will yield significantly different results.

The computational results obtained so far, including 4-31G / RHF Zero Point Vibrational Energy (ZPVE) corrections with a scaling factor of 0.9 [14], are given in Table 1 and the proposed reaction scheme is shown in Fig. 1. From these results we draw the following tentative conclusions.

i) Apart from $(CHO)_2^{+\cdot}$ the two size-consistency correction methods yield virtually the same results. For $(CHO)_2^{+\cdot}$ the first method seems to be preferable.

ii) Isomers III and V (the -C..H..C- bridged complexes) are stable with respect to dissociation.

iii) The overall isomerization I → V may proceed at internal energies close to the thermochemical threshold for $H_2CO^{+\cdot}$ + CO, which is calculated to lie 10-12 Kcal/mol below the other dissociation limits (the experimental values are 6-7 Kcal/mol [15,16]).

iv) The dissociation dynamics are primarily determined by the last step V → VI, which has no reverse activation energy. However, the potential energy well for $[O=C(H)-H ..CO]^{+\cdot}$ (isomer V) is very shallow and therefore it communicates very easily with the ion dipole complex $[HCO^{\cdot}..HCO^+]$ (isomer IV).

v) An interesting aspect of the route shown in fig. 1 is the result that the isomerization barrier for $HCOH^{+\cdot} \rightarrow H_2CO^{+\cdot}$ is substantially lowered by the presence of the CO molecule. We find virtually no barrier for

$$HCOH^{+\cdot}..CO\ (III) \rightarrow TS2 \rightarrow HCO^{\cdot}..HCO^+\ (IV) \rightarrow TS3 \rightarrow OC(H)-H^{+\cdot}..CO\ (V)$$

whereas a barrier of ca. 45 Kcal/mol has been calculated for the isomerization $HCOH^{+\cdot} \rightarrow H_2CO^{+\cdot}$ [16,17]. This is due to the high proton affinity of CO which causes a temporary transfer of a proton to CO during the isomerization III → IV → V

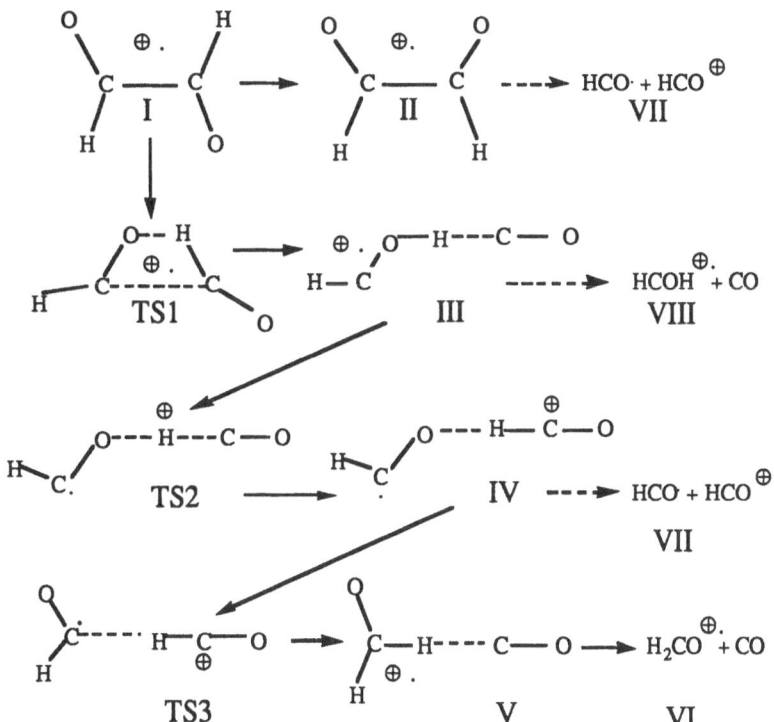

Figure 1, Reaction scheme for the MI dissociation of the glyoxal radical cation.

Table 1 Energies (hartree) and relative energies (Kcal/mol) w.r.t. $H_2CO^{+\cdot} + CO$

	4-31G / RHF	6-31G* / corr 1[a]	6-31G* / corr 2	ZPVE	$E_{rel}(1)$[b]	$E_{rel}(2)$
I	-225.87620	-226.84674	-226.82577	22	-7	6
II	-225.87273	-226.83734	-226.81938	22	-1	10
TS1	-225.86168	-226.83109	-226.83054	20	0	1
III	-225.90191	-226.83648	-226.83633	20	-3	-3
TS2	-225.86317	-226.83139	-226.83120	18	-2	-2
IV	-225.89801	-226.83965	-226.83772	18	-7	-6
TS3	-225.89781	-226.83662	-226.83662	17	-6	-6
V	-225.90527	-226.83891	-226.83803	20	-4	-4

	4-31G / RHF	6-31G* / corr	ZPVE	E_{rel}
VI	-225.89448	-226.83035	19	0
VII	-225.84667	-226.80731	17	12
VIII	-225.88005	-226.81471	19	10

a) 6-31G* / corr n : MRDCI with two reference configurations including size-consistency corrections according to method n (n=1,2), as described in the text For the dissociation limits HFSDCI calculations were performed. The 6-31G* / corr results include standard Pople size consistency corrections [13].

b) $E_{rel}(n)$: relative energies in Kcal/mol, including ZPVE corrections. n=1,2 corresponds to the size consistency correction method chosen (see a)).

2.2 STATISTICAL THEORY

By analogy to the method used in [9], the interaction potential used in the nuclear Hamiltonian [18] is given by [19]

$$V_{int}(R) = -\mu \cos\theta / R^2 - Q(3\cos^2\theta - 1) / 2 R^3 - \alpha / 2 R^4 \qquad (4)$$

Here α, μ and Q are the (isotropic) polarizability, the dipole moment and the quadrupole moment of CO respectively and θ is the angle determining the orientation of the CO molecule in the complex, where the ion is simply represented by a point charge. By diagonalizing the nuclear Hamiltonian for several values of R we find the effective potential energy cureves $V_{eff}^q(R)$ to be used for the calculation of the number of open channels in the RRKM canonical rate constant expression

$$k_{diss}(E) = G_t^*(E - E_o) / h N(E) \qquad (5)$$

Since these calculations are analogous to tthose for $CH_2CH_2OH_2^{+\cdot}$ discussed earlier no details will be given here. The results show that the contribution of the ground state of the hindered rotor to the number of open channels is much smaller than for the $CH_2CH_2OH_2^{+\cdot}$ system due to the smallness of the dipole moment of CO. The contribution of the rotational excitations of $H_2CO^{+\cdot}$ is about the same as for $C_2H_4^{+\cdot}$ in $CH_2CH_2OH_2^{+\cdot}$ because the rotational constants of $H_2CO^{+\cdot}$ and $C_2H_4^{+\cdot}$ are roughly the same. However, the rotational constant of CO is much smaller than those of H_2O. As a consequence, several rotational excitations of CO also contribute to G_t^*. The result is that the total number of open channels for the two systems are of comparable magnitude in the internal energy range of interest.

The level density N(E) for the OC(H)–H$^{+\cdot}$..CO complex was only calculated using harmonic frequencies for all nuclear degrees of freedom. Together with the relatively small value found for the dissociation energy of this complex this leads to a much smaller level density for $OC(H)-H^{+}\cdot..CO$ than for $CH_2CH_2OH_2^{+}\cdot$. We thus find significantly <u>lower</u> values for E_{max} and for $<T>$ than in the case of $CH_2CH_2OH_2^{+}\cdot$. This result is unsatisfactory since the experimental $T_{0.5}$ - values yield $<T> = 0.6$ meV and $<T> = 0.4$ meV [20] for $(CHO)_2^{+}\cdot$ and $[CH_2CH_2OH_2]^{+}\cdot$ respectively.

The discrepancy between the experimental and the calculated $<T>$ values for $(CHO)_2^{+}\cdot$ is probably due to the inaccuracy in the evaluation of the level density for $OC(H)-H^{+}\cdot..CO$. Some factors affecting this level density are the following.

i) The calculated dissociation energy of the complex may be too low. This may lead to a severe decrease in N(E). Unfortunately no experimental value for the heat of formation of the complex is available.

ii) The force constant evaluation yields several low frequencies, corresponding to the bending modes of the -C..H..C- bridge. Better results will therefore probably be obtained by treating these modes as (hindered) rotations instead of harmonic vibrations.

iii) It is not quite clear from our results whether the dissociation step is really the rate-determining step in the overall reaction leading from $(CHO)_2^{+}\cdot$ to $H_2CO^{+}\cdot + CO$. Since the transition states TS1 - TS3 have roughly the same energy as the dissociation limit, the correspondin isomerisations may compete with the dissociation reaction, thereby lowering the dissociation rate constant for a fixed value of E. Moreover, there are several other possibilities for isomerisation of the H-bridged complex, leading to e.g. $OC(H)-H^{+}\cdot..OC$ or to a C-C bonded complex between $H_2CO^{+}\cdot$ and CO. Since 6-31G* / CI calculation show that these isomers are in the same energy range as $OC(H)-H^{+}\cdot..CO$ itself, their effect should also be investigated by calculating the appropriate transition state energies. The dissociation rate constant may then be calculated using the Transition State Switching model [21], which predicts an effective increase in the level density of the dissociating ion and thus also in $<T>$, if there are other isomers with which the ion may interconvert easily.

3. CONCLUSIONS

In the present stage of this investigation no detailed results can be given for the average KER. However, some general conclusions may still be drawn from the results obtained so far.

i) The average KER for MI fragmentations proceeding without reverse activation energy may be directly related to the centrifugal barriers originating from the end-over-end rotation of the dissociating complex.

ii) As in the $CH_2CH_2OH_2^{+}\cdot$ system there is no reverse activation energy for the last step in the dissociation reaction, because the H-bridged intermediate is primarily electrostatically bound.

iii) The small average KER found for the $(CHO)_2^{+}\cdot$ dissociation may be explained in part by the small rotational constant of the CO molecule. In contrast, the small KER for ion-dipole complexes results primarily from the strong long range attraction between the ion and the dipole.

iv) Since according to our calculations the ions $HCO^{+}..HCO\cdot$ (IV) and $O=C(H)-H^{+}\cdot..CO$ (V) may interconvert easily and since these ions are equally stable, the MI dissociation rate constant may be substantially affected by the isomerizatuion IV $\leftrightarrow$ V. This will lead to an increase in the average internal energy needed for the MI dissociation to take place and consequently to an increase in the KER.

Finally, some remarks concerning the computation of the effective potential energy curves are in order.

i) Since the number of open channels is sensitive to the form of the effective potential energy curves, it is important to use a quantum mechanical model for their calculation.

ii) In our nuclear Hamiltonian the Coriolis coupling terms [18,22] have been neglected. However, for the excited states of the hindered rotor curve crossings will occur between states with the same azimuthal quantum number Ω [18]. This may be seen directly if we construct a correlation diagram relating the free rotor levels of the molecule and the corresponding harmonic oscillator levels of the complex before dissociation. By including the Coriolis coupling terms many of these crossings will be avoided. Whether this will have a substantial effect on the average KER is a matter for further investigation.

ACKNOWLEDGEMENT

The author is very grateful to Dr. J.K.Terlouw for his assistance in setting up this investigation and for providing him with the experimental data prior to publication.

REFERENCES

[1] Terlouw, J.K., Burgers, P.C., van Baar, B.L.M., Weiske, T., Schwarz, H., Chimia **1986**, *40*, 357

[2] a). Wesdemiots, C., McLafferty, F.W., submitted to Chem. Rev.
b). Terlouw, J.K., Schwar z, H., submitted to Angew. Chem. Int. Ed. Engl.

[3] Terlouw, J.K., private communication

[4] Heinrich, N., Schmidt, J., Schwar z, H., Apeloig, Y., J. Am. Chem. Soc. **1987**, *109*, 1217

[5] Burgers, P.C., Holmes, J.L., Hop, C.E.C.A., Postma, R., Ruttink, P.J.A., Terlouw, J.K., J. Am. Chem. Soc. **1987**, *109*, in press

[6] a) Postma, R., Ruttink, P.J.A., van Duijneveldt, F.B., Terlouw, J.K., Can. J. Chem. **1985**, *63*, 2758
b) Postma, R., van Helden, S.P., van Lenthe, J.H., Ruttink, P.J.A., Terlouw, J.K., Holmes, J.H., submitted to Org. Mass Spectrom.

[7] a) Postma, R., Ruttink, P.J.A., van Baar, B.L.M., Terlouw, J.K., Holmes, J.H., Burgers, P.C., Chem. Phys. Letters **1986**, *123*, 409
b) Postma, R., Ruttink, P.J.A., Terlouw, J.K., Holmes, J.L. J. Chem. Soc. Commun. **1986**, 683

[8] Quack, M., Troe, J. Ber. Bunsenges. Phys. Chem. **1974**, *78*, 240, ibid. **1975**, *79*, 469

[9] Ruttink, P.J.A. J. Phys. Chem. **1987**, *91*, 703

[10] Robinson, P.J., Holbrook, K.A. Unimolecular Reactions, 1st ed.; Wiley-Interscience, New York, 1972
Forst, W. Theory of Unimolecular Reactions, 1st ed.; Academic Press, New York, 1973

[11] Dupuis, M., Spangler, D., Wendolowski, J. NRCC Software Catalog vol I, program no. QG01 (GAMESS), 1980
Guest, M.F., Kendrick, J. GAMESS User Manual, An Introductory Guide, CCP/86/1 Daresbury Laboratory, 1986

[12] Saunders, V.R., van Lenthe, J.H. Mol. Phys. **1983**, *48*, 923

[13] Pople, J.A., Seeger, R., Krishnan, R. Int. J. Quantum Chem. Symp. **1977**, *11*, 149

[14] Pople, J.A., Schlegel, H.B., Krishnan, R., DeFrees, D.J., Binkley, J.B., Frisch M.J.,Whiteside, R.A., Hout jr, R.F., Hehre, W.J. Int. J. Quantum Chem. Symp. **1981**, *15*, 269

[15] Rosenstock, H.M., Draxl, K., Steiner, B.W., Herron, J.T., J. Phys. Chem. Ref. Data **1977**, 6, Suppl. 1

[16] Bouma, W.J., Burgers, P.C., Holmes, J.L., Radom, L., J. Am. Chem. Soc. **1986**, *108*, 1767

[17] Osamura, Y. Goddard, J.D. Schaefer III, H.F. Kim, K.S. J. Chem. Phys. **1981**, *74*, 617

[18] Brocks, G., van der Avoird, A., Sutcliffe, B.T., Tennyson, Mol. Phys. **1983**, *50*, 1025

[19] McLean, A.D., Yoshimine, M. J. Chem. Phys. **1967**, *47*, 1927
van der Avoird, A., Wormer, P.E.S., Mulder, F., Berns, R.M. Top. Curr. Chem. **1980**, *93*, 1

[20] Terlouw, J.K., Heerma, W., Dijkstra, G.
Org. Mass Spectrom. **1981**, *16*, 326
[21] Bowers, M.T., Jarrold, M.F., Wagner-Redeker, W., Kemper, P.R.,
Bass, L.M. Faraday Discuss. Chem. Soc. **1983**, *75*, 57
[22] Holmgren, S.L., Waldman, M., Klemperer, W.J.
J. Chem. Phys. **1977**, *67*, 4414
Tennyson, J., Sutcliffe, B.T. J. Chem. Phys. **1982**, *77*, 4061

STUDIES OF ION CLUSTER STRUCTURES BY MS/MS METHODS

Chava Lifshitz and Muhammad Iraqi

Department of Physical Chemistry and The Fritz Haber Research Center for Molecular Dynamics, The Hebrew University of Jerusalem, Jerusalem 91904, Israel

ABSTRACT

A tandem mass spectrometer has been employed to study cluster ions by collisionally activated dissociations (CAD), unimolecular fragmentations of metastable ions (MIKES), collision induced dissociative ionization (CIDI) of neutral fragments and collision induced charge stripping (CS). These MS/MS techniques have been applied to clusters incorporating several solvent molecules about a proton, $H^+(A)_l(W)_m$ which are of importance in the lower ionosphere and to small carbon cluster ions, C_n^+. Information on the structures and fragmentations of these cluster ions obtained by these techniques will be discussed.

INTRODUCTION

Cluster ions have aroused great interest in recent years.[1] We have concentrated recently on studies of two types of ion clusters – i) clusters incorporating several solvent molecules about a proton,[2] $H^+(A)_l(W)_m$, where $A = CH_3CN$, CH_3OH, CH_3COCH_3 or NH_3 and $W = H_2O$; ii) carbon ion clusters, C_n^+.[3] Information on the former category can give insight into the transition from gas phase to solution phase chemistry.[4,5] Such clusters are also important for the understanding of the chemistry of the lower ionosphere.[4,6-8] Small carbon cluster ions are interesting because of their importance in catalysis, combustion, and astrochemical environments.[9,10] The thermochemical stability of mixed proton bound clusters has been studied extensively by high pressure ion source mass spectrometry, particularly by Kebarle and coworkers,[11] and Meot-Ner and coworkers.[4,5] *Ab initio* calculations[4,12] have been carried out on $H^+(CH_3CN)_l(W)_m$ and $H^+(NH_3)_l(W)_m$ cluster ions, respectively, to determine their structures and bond dissociation energies. *Ab initio* calculations have also been performed on small carbon cluster ions (C_n^+, n = 2-10).[13]

We have applied MS/MS methods to study structures and fragmentations of the two types of ion clusters.[2,3] We find this to be of importance particularly since, on the one hand, for the mixed proton bound clusters calculations show that several isomeric structures may be of rather closely similar stability.[12] On the other hand, linear and/or cyclic structures have been suggested[9,10,13] for small C_n^+ ions. Considerable progress in many areas of gas phase ion chemistry has come through the experimental use of tandem (MS/MS) mass spectrometers.[14] These experiments allow mass selection of the reactant ion and thus provide a unique identification of the cluster ion studied. We have employed several methods:[2,3] a) Unimolecular fragmentations of the cluster ions were studied by MIKES (Mass analyzed Ion Kinetic Energy Spectroscopy) and by CIDI (Collision Induced Dissociative Ionization) of the neutral fragments formed. b) Collisional processes of high energy (keV) cluster ions were studied. These include CS (Charge Stripping) and CAD (Collisionally Activated Dissociation). The latter method is known[14] to yield very important information on structures of isomeric ions. More recently, we have incorporated into our tandem mass spectrometer a high-pressure, temperature-variable ion source.[15] This ion source has allowed formation of fairly large proton bound clusters and most importantly, as will be shown here, it has allowed us to regulate the relative contributions of isomeric cluster ions, as judged by the temperature variation of CA (Collisional Activation) spectra.

EXPERIMENTAL

Measurements were performed on a VG ZAB-2F double focussing mass spectrometer of reversed geometry.[16] For MIKES, the magnetic field was set to select the ions of desired m/z value under investigation; ionic products of their decompositions in the second field-free region, between the magnetic and electrostatic analyzers, were detected by scanning the electric sector potential under conditions of good energy resolution with the β-slit partially closed. Collisional activation (CA) spectra[17] were obtained by using He as the collision gas on 6-8 keV ions. The desired operating conditions have been described in the literature for metastable ion (MI), CA and CS spectra.[18-20] The pressure in the collision cell was adjusted for CA spectra such that the primary ion beam intensity fell by not more than 30% of its original intensity. The value of Q_{min} = 15.7 eV for the process $C_7H_8^+ \rightarrow C_7H_8^{2+}$ in toluene was used as a standard to calibrate the energy scale in charge-stripping measurements.[20].

Several different ion sources were employed. The early experiments on proton bound acetonitrile/water mixed cluster ions[2] employed the chemical ionization (CI)[21] cell which comes with the instrument. Its lowest workable temperature is ~100°C. The more recent experiments on $H^+(A)_l(W)_m$, A = CH_3OH, CH_3COCH_3 or NH_3, employed an improved high-pressure, temperature-variable ion source with coaxial electron entrance and ion exit apertures. Its construction details have been described by van Koppen et al.[15] The available temperature range for the mixed water containing clusters is +5°C – +90°C. When neat clusters $H^+(A)_l$ were studied, the temperature could be reduced further without freezing out the reagents. The pressure range employed was 0.1-1 torr. An extraction voltage of 10 volts was applied to the ion exit plate. Ions move through the source at constant velocity due to the presence of a constant drift potential gradient. An electron energy of 150 eV was employed to ionize the reactant gas mixture. Water rich mixed clusters were best prepared by introducing the base A into the ion source and pumping on it, before introducing pure water into the ion source containing a background of A. Alternatively, premixed water rich solutions (for example of ammonia) were employed.

The carbon cluster ions were prepared by dissociative electron ionization (EI) of overcrowded perchlorinated hydrocarbons, for example: C_{10}^+ and C_7^+ were made from $C_{10}Cl_8$ isomers, while C_{14}^+ and C_{11}^+ were made from $C_{14}Cl_{10}$. The original EI source of the ZAB-2F instrument was employed in these experiments.

The nature of the neutral product of the unimolecular fragmentation of protonated acetonitrile was determined[2] by CIDI.[22] The mass-selected protonated acetonitrile and its ionic fragmentation products were prevented from entering the collision cell situated in the second field-free region of the ZAB-2F by applying to the collision cell a positive voltage V_C greater than the ion acceleration voltage V_A. The neutral species resulting from the unimolecular dissociations of the protonated acetonitrile suffered collisionally induced dissociative ionization (CIDI), using helium as the collision gas, and could be identified from the positive ions which were generated.

RESULTS AND DISCUSSION

a) Mixed Proton Bound Cluster Ions, $H^+(A)_l(W)_m$

We have obtained systematic data on these cluster ions, where A=CH₃OH, CH_3COCH_3 and NH_3.[23] Data concerning mixed acetonitrile-water proton bound clusters have already been published.[2] We will present here only some of the saliant features of the published and unpublished results concerning these ions. CA results for ions $H^+(A)_3(W)_1$ are compiled in Table 1.

Table 1. Spectra of $H^+(A)_3(W)_1$ Ions

Ion	$H^+(CH_3CN)_3(H_2O)$		$H^+(NH_3)_3(H_2O)$		$H^+(CH_3OH)_3(H_2O)$		$H^+(CH_3COCH_3)_3(H_2O)$	
	m/z	abundance	m/z	abundance	m/z	abundance	m/z	abundance
$H^+(A)_3$	124	0	52	100	97	100	175	100
$H^+(A)_2(W)$	101	100	53	20	83	16	135	30
$H^+(A)_2$	83	50	35	11	65	10	117	79
$H^+(A)(W)$	60	25	36	3	51	1	77	2
$H^+(A)$	42	25	18	-	33	1	59	5

All of these cluster ions demonstrate a strong water loss peak except for $H^+(CH_3CN)_3(H_2O)$. The fact that the acetonitrile cluster ion does not lose water has been noticed earlier[6] for low-energy CA spectra. It has been ascribed to an ion structure in which H_3O^+ forms the core ion and the acetonitrile molecules are symmetrically attached to the three hydrogen atoms sharing the positive charge. This structure (Fig. 1)

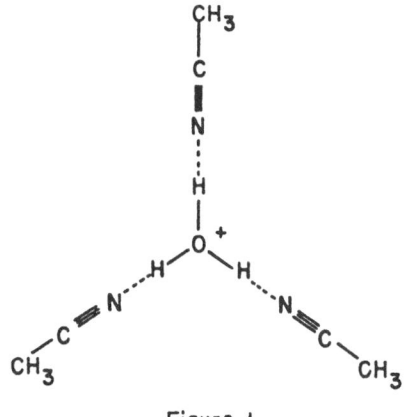

Figure 1

has since been calculated[4] by *ab initio* methods and has been found to be the most stable tetramer. With respect to the stability of acetonitrile containing clusters, two factors appear to be important: the high proton affinity of CH_3CN (189.2 kcal/mol) compared to H_2O (166.5 kcal/mol) and the hydrogen bonding of water.[4] Despite the large difference in the proton affinities of H_2O and CH_3CN the proton is located on the water to allow for a network of strong hydrogen bonds, pushing the CH_3CN molecules to the periphery. In the CA spectra these peripheral CH_3CN molecules come off first. Only after the first CH_3CN has come off to form $H^+(CH_3CN)_2(H_2O)$ is a water molecule able to come off by being exposed to the He target. In contrast to the behavior of mixed acetonitrile-water clusters, it has been suggested by Meot-Ner[5] that in methanol-water clusters the methanol molecules are near the charged center ion. As a result, the methanol containing structure which is analogous to the one containing acetonitrile (Fig. 1) is probably of comparable stability to the isomeric structure of

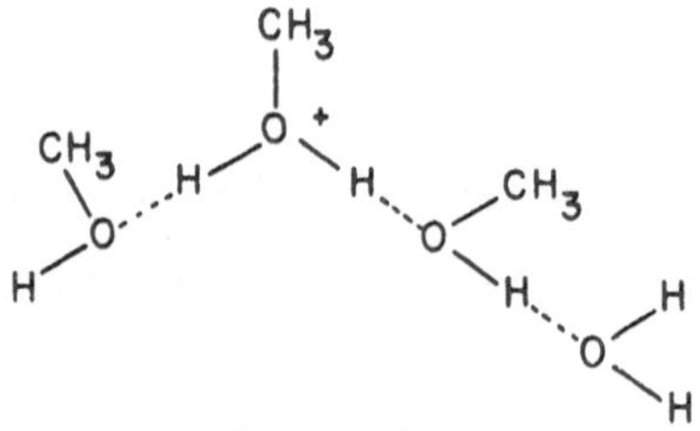

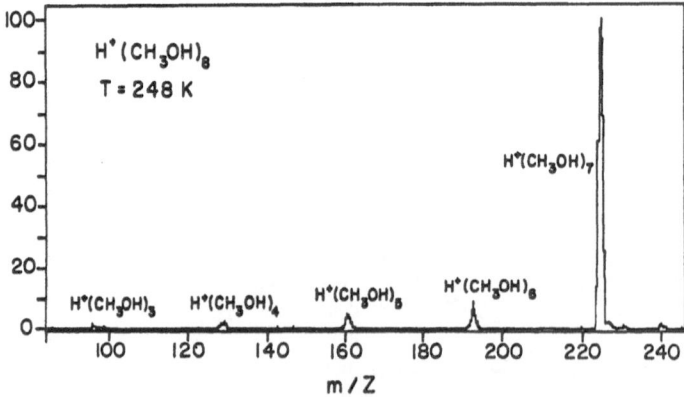

Figure 2

Figure 2, in which water is in the periphery. This suggestion of Meot-Ner[5] is corroborated by our CA spectra (Table 1) which for $H^+(CH_3OH)_3(H_2O)$ give as the largest peak the one due to water elimination, $H^+(CH_3OH)_3$. It is interesting to notice that $H^+(CH_3COCH_3)_3(H_2O)$ also demonstrates a strong water loss peak, in spite of the fact that acetone cannot easily form a network of hydrogen bonds.

We have next tested the temperature variable high pressure ion source by running CA spectra on neat proton bound methanol and ammonia clusters. These spectra do not demonstrate so-called "magic" numbers but rather a series of successive losses of the ligand molecules at continuously reduced intensities, with the highest mass peak – the cluster ion minus one ligand molecule – being the most intense. Examples are shown for $H^+(CH_3OH)_8$ and $H^+(NH_3)_6$ in Figures 3 and 4, respectively.

Figure 3

254

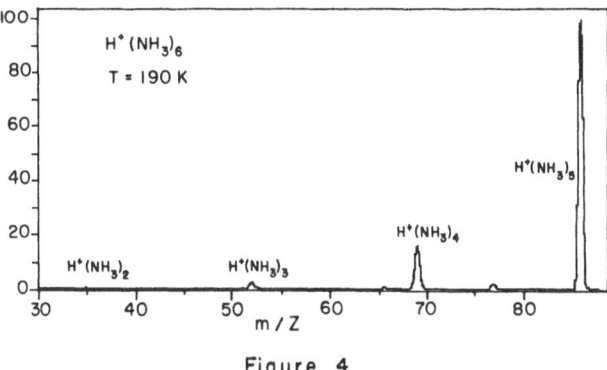

Figure 4

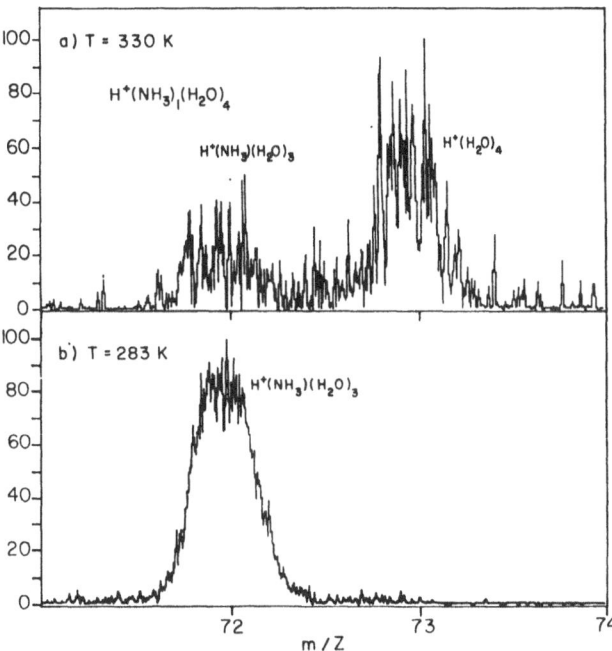

Figure 5. CA Spectra of $H^+(NH_3)_1(H_2O)_4$ as a function of Ion-Source Temperature: a) 330K; b) 283 K.

Structures and stabilities of mixed ammonia-water proton bound clusters, $H^+(NH_3)(H_2O)_m$ (with one ammonia molecule) have been calculated by *ab initio* methods.[12] There are several isomeric structures of comparable stability for a given number of water ligands, m. We have studied the temperature dependence of CA spectra for m=4 and results are shown in Figure 5. At low ion source temperatures (~10°C) the spectrum is dominated by the water loss peak (at m/z = 72, Fig 5(a)) while at high temperatures (50-60°C) the ammonia loss peak becomes of comparable intensity or even dominant (Fig. 5(b)). One way of explaining these data is that at lower temperatures the isomeric clusters present at equilibrium are dominated by the structure (Fig. 6) in which NH_4^+ is the core ion, while at higher temperatures less stable isomers,

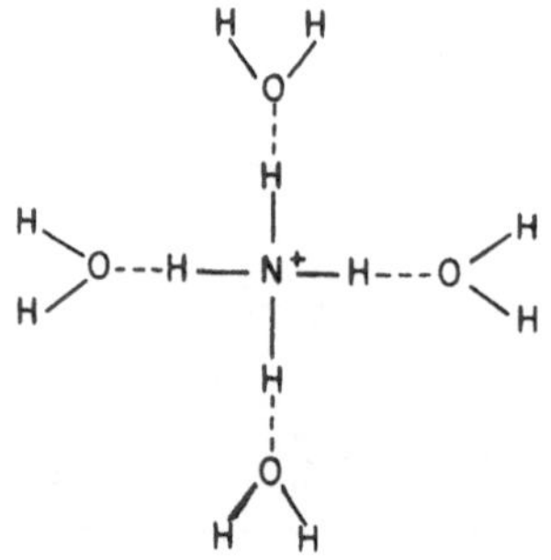

Figure 6

in which H_3O^+ is the core ion, for example those depicted by Figures 7 or 8, contribute. The possible contribution to CA spectra of $H^+(NH_3)(H_2O)_m$ ions, of an ammonia loss channel, is of great importance for the determination of ammonia vapour abundances in the atmosphere by so-called PACIMS (passive chemical ionization mass spectrometry) and ACIMS (active chemical ionization mass spectrometry).[24,25]

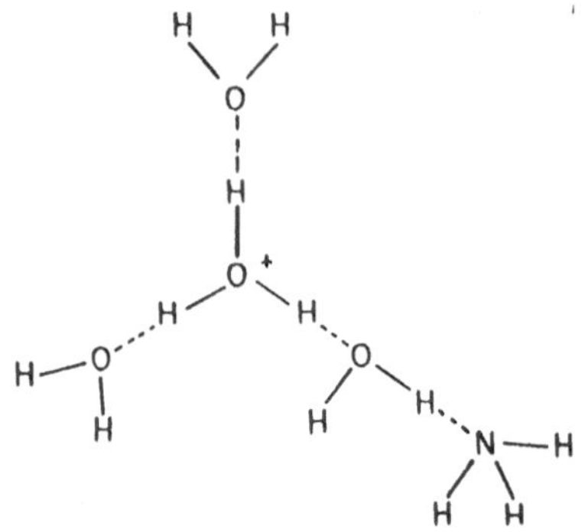

Figure 7

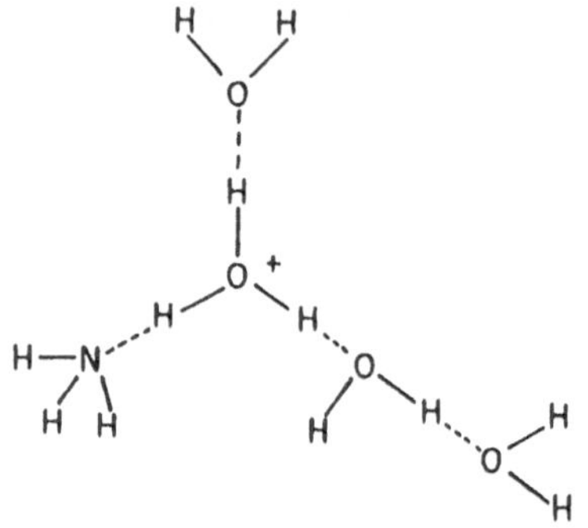

Figure 8

Finally, just a few words on alternative MS/MS methods. We are in the process of determining kinetic energy releases (KERs) for unimolecular dissociations of these ions using MIKES. These clusters belong to the category of ion-dipole complexes which demonstrate very unusual dissociation dynamics.[26,27] CIDI can also be very useful as we have demonstrated for the dissociation of protonated acetonitrile.[2] Formation of CH_3^+ from CH_3CNH^+ has been of crucial importance[7] for the identification of acetonitrile in mixed clusters in the stratosphere. By reionization of the neutral fragment formed with CH_3^+ we were able[2] to identify it as HNC rather than HCN, i.e., the reaction is: $CH_3CNH^+ \rightarrow CH_3^+ + HNC$.

Ionized carbon clusters have been studied previously by laser vaporization of a graphite rod to generate an ion plasma.[28] Our experiments employ an entirely different route for studying C_n^+ species – exhaustive chlorine elimination from perchloro conjugated hydrocarbons.[3] $C_{10}Cl_8$ isomers – octachloropentafulvalene and octachloronaphthalene – were found[3] to be good sources for C_{10}^+ and C_7^+, while $C_{14}Cl_{10}$ – perchlorophenanthrene was found[29] to be a good source for C_{14}^+ and C_{11}^+. This alternative way of forming small carbon cluster ions allowed us to find out whether their fragmentation behavior was diffferent or similar to that of ions generated by the laser vaporization techniques. Furthermore, previous experimental studies of these cluster ions concentrated on their photofragmentation,[9] their ion-molecule reactions and low energy collision induced dissociations.[10] It has been demonstrated however,[30] that high-energy CA spectra are very informative in cluster ion structure determinations. We have applied the techniques of MIKES, CA spectoscopy and CS to these small carbon cluster ions. The MIKES method has been applied in the meantime[31] also to laser-generated carbon cluster ions. Our unimolecular (MIKES) fragmentation patterns and CA spectra are summarized in Table 2.

Table 2. MIKE and CA Spectra of C_n^+ Ions

Size of neutral lost	Size of initial cluster ion (n)							
	MIKES				CA			
	7	10	11	14	7	10	11	14
1	-	-	0.85	-	0.1	0.01	0.41	-
2	0.2	-	-	-	0.19	0.01	-	-
3	0.8	1.00	0.15	1.00	0.55	0.59	0.39	0.7
4					0.14	0.05	0.1	0.07
5					0.02	0.11	0.03	0.02
6						0.12	0.05	0.09
7						0.1	0.01	0.07
8						0.006	0.004	0.02
9								0.02
10								0.007
11								0.001

Previous experiments on laser generated C_n^+ ions[9,10] and *ab initio* calculations have indicated that two factors are important and contribute to the fragmentation behavior of carbon cluster ions. The first factor is the extra stability of the neutral cluster C_3 and the second factor is the fact that larger clusters are easier to ionize than smaller ones. The "magic" fragmentation pattern of small laser generated carbon cluster ions is thus characterized by the preponderance of C_3 loss. This is evident in the photofragmentation patterns,[9,32] in low-energy collision induced dissociations[10] and in unimolecular metastables (MIKE spectra).[31] Our data on dissociative electroionization generated C_n^+ ions are no different in this respect. This leads to the occurrence, in the mass spectra of C_nCl_m compounds, of fairly abundant C_n^+ and C_{n-3}^+ ions. The abundant loss of C_3 is also apparent in our MIKE and CA spectra (Table 2).

A structural change from linear to monocyclic rings in the cluster ions between n=9 and 10 has been suggested on the basis of changes in the reactivity,[10] and in the photodissociation cross section[9] and by *ab initio* calculations. We have determined[3] the minimum amount of translational energy lost (Q_{min}) for $C_{10}^+ \rightarrow C_{10}^{2+}$, 15.5 ± 0.2 eV, by charge stripping (CS) to be very nearly the same[20] as that for C_9^+. On the other hand, the unique loss of one carbon atom from C_{11}^+ (Table 2), which has been noticed earlier[9,10,31] may reflect the special stability of the C_{10}^+ ring structure formed.[13] MIKE spectra reflect the lowest energy dissociation route and very often involve dissociative isomerizations. On the other hand, CA spectra are good indicators of ion structures, since simple bond cleavages, even if energetically less favorable, become more pronounced.[14] It is thus quite significant that the CA spectrum of C_{11}^+ demonstrates an increased abundance of C_3 loss relative to C_1 loss while the opposite is true for the MIKE spectrum. C_8^+ is most probably linear,[13] and C_8^+ formation could take place by simple bond cleavage while C_{10}^+ formation would involve a (lower energy) dissociative isomerization. Ion cluster fragmentations under high energy (keV) CA conditions have been rationalized[30] using an "instantaneous" dissociation model. Further studies on C_nCl_m isomers are underway[29] to clarify some of these points.

ACKNOWLEDGEMENT

The research on ionospheric cluster ions was supported by a grant from the National Council for Research and Development, Israel and the GSF München, Germany. The construction of the high-pressure temperature variable ion-source was partially supported by a grant from the United States-Israel Binational Science Foundation (BSF), Jerusalem. We thank Dr. F. Arnold and Prof. M.T. Bowers for helpful discussions. Special thanks are due to Dr. Petra A.M. van Koppen for her patient guidance in the construction of the high pressure ion source.

REFERENCES

1. T.D. Mårk, *Cluster ions: production detection and stability.* Int. J. Mass Spectrom. Ion Processes 79 : 1 (1987).

2. M. Iraqi and C. Lifshitz, *Some stratosphere-related aspects of the gas-phase ion chemistry of acetonitrile-water mixtures,* Int. J. Mass Spectrom. Ion Procces 71 : 245 (1986).

3. C. Lifshitz, T. Peres, S. Kababia and I. Agranat, C_{10}^+ *and* C_7^+ *carbon cluster ions from overcrowded octachloropentafulvalene and octachloronaphthalene,* Int. J. Mass Spectrom. Ion Processes (1987), in press.

4. C.A. Deakyne, M. Meot-Ner (Mautner), C.L. Campbell, M.G. Hughes and S.P. Murphy, *Multicomponent cluster ions. 1. The proton solvated by* CH_3CN/H_2O, J. Chem. Phys. 84 : 4958 (1986).

5. M. Meot-Ner (Mautner), *Comparative stabilities of cationic and anionic hydrogen-bonded networks. Mixed clusters of water-methanol,* J. Am. Chem. Soc. 108 : 6189 (1986).

6. H. Böhringer and F. Arnold, *Acetonitrile in the stratosphere – implications from Laboratory Studies.* Nature 290 : 321 (1981).

7. H. Schlager and F. Arnold, *Balloon-borne fragment ion mass spectrometry studies of Stratospheric positive ions: unambiguous detection of* $H^+(CH_3CN)_l(H_2O)_m$ *clusters,* Planet. Space Sci. 33 : 1363 (1985).

8. R.G. Keesee and A.W. Castleman, Jr., *Ions and cluster ions: Experimental studies and atmospheric observations.* J. Geophys. Res. 90 : 5885 (1985).

9. M.E. Geusic, M.F. Jarrold, T.J. McIlrath, R.R. Freeman and W.L. Brown,

Photodissociation of carbon cluster ions, J. Chem. Phys. 86 : 3862 (1987).

10. S.W. McElvany, B.I. Dunlap and A. O'Keefe, *Ion molecule reactions of carbon cluster ions with D_2 and O_2,* J. Chem. Phys. 86 : 715 (1987).

11. J.D. Payzant, A.J. Cunningham and P. Kebarle, *Gas phase solvation of the ammonium ion by NH_3 and H_2O and stabilities of mixed clusters $NH_4^+(NH_3)_n(H_2O)_w$,* Can. J. Chem. 51 : 3242 (1973).

12. C.A. Deakyne, *Filling of solvent shells about ions. 2. Isomeric clusters of $H_2O)_n(NH_3)H^+$,* J. Phys. Chem. 90 : 6625 (1986).

13. K. Raghavachari and J.S. Binkley, *Structure, stability and fragmentation of small carbon clusters,* J. Chem. Phys. 87 : 2191 (1987).

14. C. Lifshitz, *Recent progress in gas-phase ion chemistry,* Int. Rev. in Phys. Chem. 6 : 35 (1987).

15. P.A.M. van Koppen, P.R. Kemper, A.J. Illies and M.T. Bowers, *An improved high-pressure, temperature-variable ion source with coaxial electron beam/ion exit slit,* Int. J. Mass Spectrom. Ion Processes 54 : 263 (1983).

16. R.P. Morgan, J.H. Beynon, R.H. Bateman and B.N. Green, *The MM-ZAB-2F double-focussing mass spectrometer and MIKE spectrometer,* Int. J. Mass Spectrom. Ion Phys. 28 : 171 (1978).

17. F.W. McLafferty, *"Interpretation of Mass Spectra",* Benjamin, New York (1973).

18. J.K. Terlouw, P.C. Burgers and H. Hommes, *Structure and formation of $[C_4H_5O]^+$ ions.* Org. Mass Spectrom. 14 : 387 (1979).

19. P.C. Burgers, J.L. Holmes, J.E. Szulejko, A.A. Mommers and J.K. Terlouw, *The gas phase ion chemistry of acetyl cation and isomeric $[C_2H_3O]^+$ ions,* Org. Mass Spectrom 18 : 254 (1983).

20. M. Rabrenović, C.J. Proctor, T. Ast, C.G. Herbert, A.G. Brenton and J.H. Beynon, *Charge stripping of hydrocarbon positive ions.* J. Phys. Chem. 87 : 3305 (1983).

21. F.H. Field and J.L. Franklin (Ed.), *"Ion Molecule Reactions",* Plenum Press, New York, 1 : 261 (1972).

22. P.C. Burgers, J.L. Holmes, A.A. Mommers and J.K. Terlouw, *Neutral products of ion fragmentations: HCN and HNC identified by collisionally induced dissociative ionization,* Chem. Phys. Lett. 102 : 1 (1983).

23. C. Lifshitz and M. Iraqi, to be published.

24. H. Ziereis and F. Arnold, *Gaseous ammonia and ammonium ions in the free troposphere,* Nature 321 : 503 (1986).

25. H. Ziereis and F. Arnold, *Combined measurements of ammonia and nitric acid vapours in the free trophosphere using aircraft-borne ACIMS,* J. Geophys. Res. (1988), to be submitted.

26. P.J.A. Ruttink, *Kinetic energy release in ion-dipole fragmentations. The $C_2H_4^+/H_2O$ system,* J. Phys. Chem. 91 : 703 (1987).

27. J.-D. Shao, T. Baer, J.C. Morrow and M.L. Fraser-Monteiro, *The dissociation dynamics of energy selected ion dipole complexes: I. The cyclopropane ion water complex $[c-C_3H_6^+-OH_2]$,* J. Chem. Phys. 87 : 5242 (1987).

28. S.W. McElvany, W.R. Creasy and A.O'Keefe, *Ion-molecule reaction studies of mass selected carbon cluster ions formed by laser vaporization,* J. Chem. Phys. 85 : 632 (1986).

29. C. Lifshitz and I. Agranat, to be published.

30. R.B. Freas, B.I. Dunlap, B.A. Waite and J.E. Campana, *The role of cluster ion structure in reactivity and collision-induced dissociation. Application to cobalt/oxygen cluster ions in the gas phase.* J. Chem. Phys. 86 : 1276 (1987).

31 P.P. Radi, T.L. Bunn, P.R. Kemper, M.E. Molchan and M.T. Bowers, *A new*

method for studying carbon clusters in the gas phase: Observation of size specific neutral fragment loss from metastable reactions of mass selected C_n^+, $n \le 60$, J. Chem. Phys. (in press).

32. W.L. Brown, R.R. Freeman, K. Raghawachari and M. Schlüter, *Covalent group IV atomic clusters*, Science 235 : 860 (1987).

NONADIABATIC EFFECTS IN UNIMOLECULAR REACTIONS

OF IONIZED MOLECULES

J.C. Lorquet, B. Leyh-Nihant* and F. Remacle*

Département de Chimie, Université de Liège
Sart-Tilman, B-4000 Liège 1, Belgium

We have studied several unimolecular reactions of ionized molecules which exhibit remarkable effects. According to ab initio calculations, the mechanism of these reactions is controlled by a nonadiabatic interaction, i.e., a crossing between two potential energy surfaces leading in most cases to an electronic predissociation. Our aim is to derive criteria which could help us to distinguish between adiabatic and nonadiabatic reactions. Particularly interesting are the remarkably strong isotope effects which sometimes affect the three observable quantities which characterize a reaction.

(i) The breakdown graph, i.e., the yield of fragmentation as a function of the internal energy of the molecular ion. A particularly striking example is offered by the methylnitrite ion which dissociates as follows:

$$CX_3 ONO^+ \longrightarrow [CX_3 O]^+ + NO \qquad (X = H \text{ or } D) \qquad (1)$$

The normal compound produces fragments having the global structure CH_3O^+ with a maximum yield of 60% [1,2]. By contrast, the maximum yield with which the corresponding fragments are produced from the perdeuterated species is at most equal to 5% according to some authors [1] (12% according to others [2]).

(ii) The rate constant. (The experimental measurements are limited to the so-called "metastable range", i.e., around 10^6 s^{-1}). The rate constant of reaction (1) for the same perdeuterated ion $CD_3 ONO^+$ is found [1] to remain constant over a remarkably large range of internal energies. Another example is offered by the perdeuterated formaldehyde ion. The reaction

$$CX_2 O^+ \longrightarrow XCO^+ + X \qquad (2)$$

takes place with a rate constant of about 5×10^4 s^{-1}, whereas the normal compound dissociates at least more than one order of magnitude faster [3].

(iii)   The translational energy released on the fragments.   The reaction

$$XOCO^+ \longrightarrow XOC^+ + O \qquad\qquad\qquad (3)$$

gives rise to a metastable signal for both isotopic species. However, the kinetic energy release is much larger for the deuterated than for the normal compound [4].

Reactions (1), (2) and (3) are all thought to involve a transition from one potential energy surface to another, i.e., to involve an electronic predissociation. However, this does not preclude the use of simple statistical methods. A statistical theory valid for nonadiabatic reactions has been developed [5]. The rate constants are expressed in terms of energy level densities, just as in the usual RRKM theory. Depending on the coupling scheme, a variety of equations can be generated.

For instance, in the simplest possible case, i.e., that of a molecule made up of a reaction coordinate R accompanied by a bath of u oscillators, and exhibiting a surface crossing at an energy E_c above its zero-point energy, one has:

$$k^{pred}(E) = [2/h\ N(E)] \int_0^{E-E_c} dE_u\ N_u^*(E_u)\quad p(E-E_c-E_u)\quad (4)$$

where N(E) represents the density of states at energy E of the set $\{R, u\}$, N_u^* represents the density of states of the set u alone, evaluated with frequencies appropriate to the crossing point, and p is the probability of undergoing a nonadiabatic transition between the two crossing potential energy surfaces. The latter quantity can be obtained either from the Landau-Zener equation or from the more accurate weak-coupling formula. Other formulas are available to describe more complicated situations.

Equation (4) accounts very well [6] for the isotope effect which affects reaction (3). The heavier compound is found at any energy to dissociate more slowly than the lighter one. Conversely, in order to bring about dissociation at a given rate (e.g., 10^6 s^{-1}), the internal energy has to be higher for the deuterated species. As a result, the released translational energy is also higher.

The interpretation of the other examples is less straightforward and requires additional complications in the coupling scheme and resulting equations. Two such effects are particularly important.

a)   When the two interacting electronic states admit different equilibrium positions along a particular degree of freedom y, the most favourable reaction path often involves tunneling. If now the "displaced" coordinate y involves the motion of a hydrogen atom, then this tunneling process becomes sensitive to isotopic substitution. This effect, together with the normal (RRKM-like) isotope effect on the denominator of eq. (4), is thought [7] to account for the very strong isotope effect described in (i).

b)　When the asymptotic expression of the potential energy curves along the reaction coordinate R are very different for both electronic states (i.e., because one of them describes a short-range whereas the other describes a long-range interaction), then the position of the crossing point R_C varies with the internal energy. This has immediate consequences on the value of the off-diagonal matrix element $V_{12}(R_C)$ and thus on the transition probability p. This effect is thought [7] to account for the weak dependence of the predissociation rate constant k^{pred} with respect to the internal energy E. The additional channels which open up as E increases are characterized by a smaller transition probability than (and hence are less efficient than) the low-energy ones.

ACKNOWLEDGEMENTS

B.L.N. and F.R. are indebted to the F.N.R.S. (Belgium) for the award of a fellowship. This work has been supported by the Belgian Government (Action de Recherche Concertée) and by the Fonds de la Recherche Fondamentale Collective.

REFERENCES

[*]　Aspirant F.N.R.S., Belgium
[1]　J.P. Gilman, T. Hsieh and G.G. Meisels, J. Chem. Phys., **78**, 3767 (1983).
[2]　T. Baer and O. Dutuit, private communication.
[3]　R. Bombach, J. Dannacher, J.P. Stadelmann and J. Vogt, Chem. Phys. Lett., **77**, 399 (1981).
[4]　P.C. Burgers, J.L. Holmes and A.A. Mommers, Int. J. Mass Spectr. Ion Proc., **54**, 283 (1983).
[5]　J.C. Lorquet and B. Leyh-Nihant, submitted to J. Phys. Chem.
[6]　F. Remacle, D. Dehareng and J.C. Lorquet, submitted to J. Phys. Chem..
[7]　B. Leyh-Nihant and J.C. Lorquet, submitted to J. Chem. Phys.

POTENTIAL ENERGY SURFACES AND DISSOCIATION MECHANISMS

OF THE $C_2 H_4^+$ ION

M. Oblinger, A.J. Lorquet and J.C. Lorquet

Département de Chimie, Université de Liège
Sart-Tilman
B-4000 Liège 1, Belgium

For several years, the potential energy surfaces and reaction mechanisms of the ethylene cation have been investigated in our laboratory by ab initio calculations at the configuration interaction (CI) level. A preliminary paper has already been published [1]. However, a more accurate investigation is under way on which we report briefly here, although our current calculations and conclusions are by no means final.

The $C_2H_4^+$ ion is known [2] to dissociate via two competing channels:

$$C_2 H_4^+ \longrightarrow C_2 H_2^+ + H_2 \quad (A.E. = 13.14 \text{ eV}) \quad (1)$$
$$C_2 H_4^+ \longrightarrow [C_2 H_3^+] + H \quad (A.E. = 13.22 \text{ eV}) \quad (2)$$

While the structure of the $C_2 H_2^+$ fragment is certainly that of acetylene, that of $[C_2 H_3^+]$ is unknown. The vinyl ion can exist in two forms of nearly equal energy: the classical H_2CCH^+ and the bridged HC^HCH^+ structures. Both forms are known to interconvert easily, with a very low barrier to rearrangement [3]. Coulomb explosion experiments indicate that the nonclassical bridged structure predominates [4].

The onsets of both reactions fall in the Franck-Condon range of the first excited electronic state $\tilde{A}\ ^2B_{3g}$. The potential energy surface of this state dissociates smoothly and without an energy barrier via a simple CH bond cleavage, i.e., generates classical H_2CCH^+ ions [1]. However, we feel that this process, although rapid, cannot compete with a still faster internal conversion to the ground electronic state via a conical intersection which takes place at short CC distances [5]. The presence of this conical intersection is confirmed by unpublished ab initio calculations in our laboratory [6].

Once in its ground $\tilde{X}$ $^2B_3$ electronic state, the ion can isomerize [1] from the (nonplanar) $H_2CCH_2^+$ geometry to the ethylidene structure H_3CCH^+. The transition state has a bridged structure $H_2C^HCH^+$, and its energy is lower than that of either dissociation asymptote. Thus, the dissociation mechanisms of each structure have to be studied individually.

1) Reaction (1) raises few problems. The only possibility is a 1,1 elimination from a hydrogen molecule from the H_3CCH^+ isomer. Our CI calculations predict a barrier for this process. However, the height of this barrier is observed to decrease steadily as the quality of the calculation increases. Thus, by extrapolation, we feel that a highly accurate calculation would predict either a steadily increasing potential energy curve with no reverse activation energy barrier or, possibly, a very small barrier, as suggested by Powis [7].

2) Reaction (2) raises more difficult problems since several mechanisms seem to compete.

a) Stretching one of the CH bonds of the lowest energy structure $H_2CCH_2^+$ of the electronic ground state ($\tilde{X}$ $^2B_3$) generates fragments having the classical structure H_2CCH^+ via a curve crossing with the potential energy curve of the first excited state [1]. This mechanism thus involves a nonadiabatic interaction. This curve crossing is converted into a conical intersection when the ion becomes nonplanar. We are as yet unable to estimate the magnitude of the rate constant in such a case.

b) The classical H_2CCH^+ fragment can also be generated by hydrogen atom loss from the ethylidene H_3CCH^+ structure. The CI calculations predict a barrier which, this time, remains fairly stable (≈ 0.3 eV) when the quality of the calculation is improved. Hence, this mechanism might contribute, but with a threshold higher than the thermochemical requirement.

c) Finally, a third, unconventional, mechanism is being currently studied. Calculations (which, as yet, are in a primitive stage) reveal that it is possible to break one of the CH bonds of the bridged transition state $H_2C^HCH^+$ to generate nonclassical bridged HC^HCH^+ fragments. In contradistinction to mechanism (2b), this process involves apparently no activation energy.

ACKNOWLEDGEMENTS
This work has been supported by the Belgian Government (Action de Recherche Concertée) and by the Fonds de la Recherche Fondamentale Collective.

REFERENCES

[1] C.Sannen, G.Raşeev, C.Galloy, G.Fauville and J.C. Lorquet, J.Chem. Phys.,**74**, 2402 (1981); J.C. Lorquet, C.Sannen, G. Raşeev, J. Am. Chem. Soc.,**102**, 7976 (1980).
[2] R. Stockbauer and M. G. Inghram, J. Chem. Phys.,**62**, 4862 (1975).

[3]   J. Weber, M. Yoshimine and A.D. McLean, J. Chem. Phys.,**64**, 4159 (1976); G.P. Raine and H.F. Schaefer III, J. Chem. Phys., **81**, 4034 (1984).

[4]   E.P. Kanter, Z. Vager, G. Both, and D. Zajfman, J. Chem. Phys., **85**, 7487 (1987).

[5]   H. Köppel, Chem. Phys., **77**, 359 (1983).

[6]   C. Galloy, unpublished work from this laboratory.

[7]   P.D. Lightfoot, C.J. Danby and I. Powis, Chem. Phys. Lett. **96**, 232 (1983).

THE UNIMOLECULAR CHEMISTRY OF [1,2-PROPANEDIOL]$^{+\cdot}$:

A RATIONALE IN TERMS OF HYDROGEN BRIDGED RADICAL CATIONS

Johan K. Terlouw

Analytical Chemistry Laboratory, Utrecht University
Croesestraat 77a, 3522 AD Utrecht, The Netherlands

The low-energy dissociations of radical cations in the gas phase often only occur from isomeric ions generated via extensive rearrangement processes. This is especially true for, but by no means limited to, those ions which have half-lives of $> 10^{-5}$ s, i.e., the metastable ions which decompose in the drift regions of a mass spectrometer [1]. The quasi-equilibrium theory provides a rationale in that it stipulates that dissociation of metastable ions is strongly dependent on the overall activation energy rather than on the mechanistic complexity of the reaction. Among the many well-documented cases, ionized methyl isobutyrate, $(CH_3)_2CHCOOCH_3$, provides a classical example [2]. The loss of $CH_3\cdot$ from these ions occurs only by C-C cleavage in the isomeric radical cation $[CH_3CH_2CH=C(OH)(OCH_3)]^{+\cdot}$, whose formation involves inter alia a 1,2 shift of the protonated ester moiety.

In the context of rearrangement processes hydrogen bridged radical cations are increasingly being proposed as stable intermediates in the dissociation pathways of a wide range of radical cations [3], along the lines developed by Morton [4]. Using this concept we have rationalised the loss of HCO· from ionized glycol according to the following Scheme [3]:

$$[CH_2OHCH_2OH]^{+\cdot} \rightarrow [CH_2=O\cdot\cdot H\cdot\cdot O(H)-CH_2]^{+\cdot} \rightarrow$$
$$[H-C=O\cdot\cdot H\cdot\cdot O(H)-CH_3]^{+\cdot} \rightarrow CH_3OH_2^+ + HCO\cdot$$

More importantly, it was shown from a combination of mass spectrometry based experiments and high level molecular orbital theory calculations that the hydrogen bridged species proposed to be involved, as well as the isomeric ions $[CH_3-O\cdot\cdot H\cdot\cdot O=CH_2]^{+}\cdot$ and $[CH_2=CH(H)-O\cdot\cdot H\cdot\cdot OH_2]^{+\cdot}$, exist as thermodynamically stable species in deep potential wells [3,5]. For the latter ion it could further be shown that well below the dissociation level of lowest energy requirement, i.e. the formation of $[CH_2=CHOH]^{+\cdot} + H_2O$, it can freely interconvert with distonic ions of the type $[CH_2=CH(OH)(OH_2)]^{+\cdot}$ and ion-dipole complexes between ionized vinyl alcohol and a water molecule [5].

Inspired by these results the possible role of other stable hydrogen bridged radical cations in the six unimolecular dissociation reactions of ionized 1,2-propanediol was investigated.This molecule is the next homologue of glycol and so may be expected to behave similarly. Moreover its behaviour closely resembles that of $[CH_2=C(H)O\cdot\cdot H\cdot\cdot O(H)CH_3]^{+\cdot}$ which is already known to exist as a stable species in the gasphase [6] and whose isomerization behaviour is expected to be related to that of the [vinylalcohol/water]$^{+\cdot}$ system [5].

For metastable 1,2-propanediol molecular ions the following reactions have been established:

$$CH_3C(=O)OH_2^+ \quad + CH_3\cdot \qquad (1)$$
$$CH_3C(OH)=CH_2^{+\cdot} + H_2O \qquad (2)$$
$$CH_2=CHOH^{+\cdot} \quad + CH_3OH \qquad (3)$$
[1,2-propanediol]$^{+\cdot}$
$$CH_3\text{-}C^+=O \qquad + H_2O + CH_3\cdot \qquad (4)$$
$$CH_2=C=O^{+\cdot} \qquad + H_2O + CH_4 \qquad (5)$$
$$CH_3OH_2^+ \qquad + CH_3CO\cdot \qquad (6)$$

The calculated appearance energies for the above reactions agree quite well with the experimental values, with the exception of reaction (2), the formation of m/z 58 ions; however, the ionization energy of 1,2-propanediol is 9,9 eV (ΔH_f^0 1,2-propanediol$^{+\cdot}$= 534 kJ/mol) and so this reaction is exothermic. All other processes have activation energies of only ~0,2 eV. By combining these results with those from other mass spectrometric techniques (metastable ion (MI), collisional activation (CA), collision induced dissociative ionization (CIDI) [2b], neutralization reionization (NR) spectrometry [2]) and the classical method of isotopic labelling a unified mechanism could be derived [8] for the complex unimolecular chemistry of ionized 1,2-propanediol in terms of the participation of stable hydrogen-bridged radical cations interrelated via 1,4- ,1,5- and 1,6-hydrogen shifts. This is shown in Figure 1.

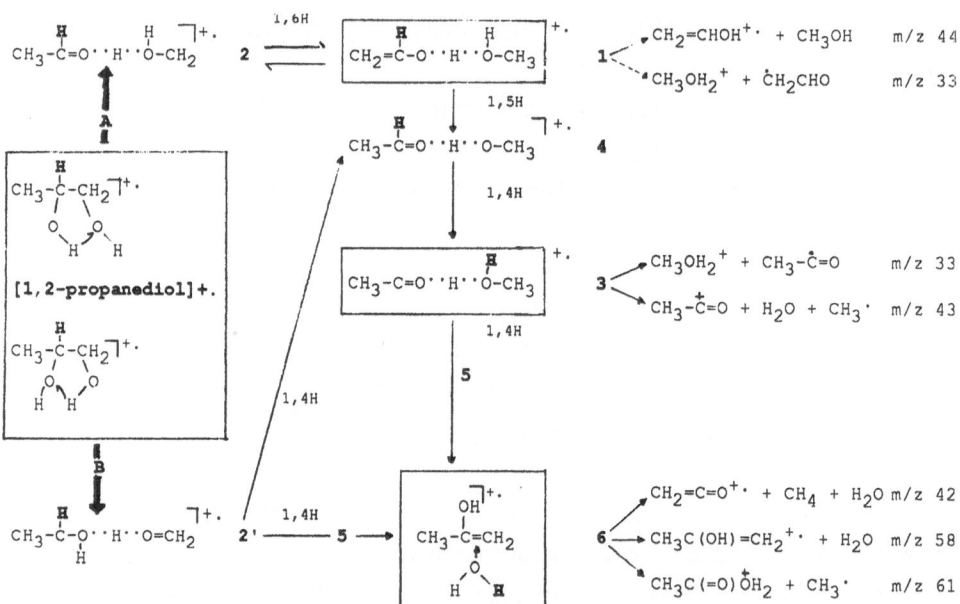

ΔH_f^0 [1,2-propanediol]$^{+\cdot}$ = 534 kJ/mol, ΔH_f^0 2,2',1,4,3 and 6 were estimated as 471, 528, 431, 512, 478 and 375 kJ/mol.

Figure 1. The unimolecular chemistry of [1,2-propanediol]$^{+\cdot}$

Key intermediates involved are the stable hydrogen-bridged ions $[CH_2=C(H)-O\cdots H\cdots O(H)CH_3]^{+\cdot}$, **1**, which was generated independently from [4-methoxy,1-butanol]$^{+\cdot}$ (loss of C_2H_4), $[CH_3-C=O\cdots H\cdots O(H)CH_3]^{+\cdot}$,**3**, and the related ion-dipole complex $[CH_2=C(OH)CH_3/H_2O]^+$,**6**.

The latter species serves as the precursor for the loss of $CH_3\cdot$ and in this reaction the same non- ergodic behaviour is observed [8] as in the loss of $CH_3\cdot$ from the ionized enol of acetone [9].

State of the art *ab initio* calculations on hydrogen bridged radical cations and ion-dipole complexes on similar but less complex systems [3,10] have shown that such species are surprisingly stable, they thus form an important justification for Fig.1. The proposal is amenable to further refinements but these best come from high level *ab initio* computations which for a system of this size are currently very expensive and laborious, especially where the calculation of transition state energies for the various hydrogen shifts is concerned.

REFERENCES

1. for a recent review see: J.L. Holmes, Org. Mass Spectrom. **20**, 169 (1985).

2. E. Göksu, T. Weiske, H. Halim and H. Schwarz, J. Am. Chem. Soc. **106**, 1167 (1984).

3. P.C. Burgers, J.L. Holmes, C.E.C.A. Hop, R. Postma, P.J.A. Ruttink and J.K. Terlouw, J. Am. Chem. Soc. **109**, 7315 (1987) and references cited therein.

4. T.H. Morton, Tetrahedron **38**, 3195 (1982).

5. R. Postma, S. van Helden, J.H. van Lenthe, P.J.A. Ruttink, J.K. Terlouw and J.L. Holmes, Org. Mass Spectrom. **23**, in press (1988).

6. a) J.K. Terlouw, W. Heerma, P.C. Burgers and J.L. Holmes, Can. J. Chem. **62**, 289 (1984).

    b) R.Postma, P.J.A. Ruttink, F.B. van Duijneveldt, J.K. Terlouw and J.L. Holmes, Can J. Chem. **63**, 2798 (1985).

7. for recent reviews see: a) C. Wesdemiotis and F.W. McLafferty, Chem. Rev. **87**, 405 (1987);

    b) J.K. Terlouw and H. Schwarz, Angew. Chem. Int. Ed. Engl. **26**, 805 (1987).

8. B.L.M. van Baar, P.C. Burgers, J.L. Holmes and J.K. Terlouw, Org. Mass Spectrom.,submitted for publication.

9. C. Lifshitz, Int. J. Mass Spectrom Ion Phys. **43**, 179 (1982) and references cited therein.

10. for recent references see: a) R. Postma, P.J.A. Ruttink, B. van Baar, J.K. Terlouw, J.L. Holmes and P.C. Burgers, Chem. Phys. Lett. **123**, 409 (1986); b) R. Postma, P.J.A. Ruttink, J.K. Terlouw and J.L. Holmes, J. Chem. Soc. Chem. Commun. **1986**, 683; c) N. Heinrich and H. Schwarz, Int. J. Mass Spectrom. Ion Proc. **79**, 295 (1987); d) see also ref. 5 for a theoretical assessment of the relationship between distonic ions, hydrogen bridged species and ion-dipole complexes in the $[CH_2=CHOH/H_2O]^{+\cdot}$ system.

THE ENERGETIC AND STRUCTURAL INTERPRETATION OF ION

PHOTODISSOCIATION SPECTRA

S.P. Goss and J.D. Morrison

Department of Chemistry
La Trobe University
Bundoora, Victoria, 3083, Australia

Introduction

Ion Photodissociation (IPD) experiments permit optical spectroscopy to be performed on gas phase ions. Unlike classical absorption spectroscopy it is the effect of the photon, rather than the change in photon flux through the sample, that is measured. Charged particles are easier to count than photons, and the signal to noise aspects of distinguishing a small current from zero are much better than for a small change in a very large signal.

IPD spectroscopy measures the energetic differences between ionic states, and the relative transition probabilities. The data can be interpreted to yield dissociation thresholds, information about the relative positions of the potential surfaces involved and in favourable cases the resolution of vibrational, rotational and fine structure allows the calculation of force constants, rotational constants, and the lifetimes of predissociating states. The experiment can be extended to give branching ratios for the competing processes of fluorescence and predissociation. The instrumentation and some of the results of IPD have been reviewed elsewhere.(1)

The most closely related experimental techniques to IPD spectroscopy are photoionization (PI) mass spectrometry and photoelectron spectroscopy (PES). In both of these, ionization is produced directly by photon absorption, and transitions take place from the neutral molecule in its electronic ground state to the various states of the ion accessible in the Franck-Condon region. In the IPD technique, various states of the ion are formed similarly in the ion source by vertical excitation, but using electron impact which is not subject to optical selection rules. They then undergo mass analysis, where some relaxation of excited states may take place, thus modifying the initial population of vibronic levels. The mass analysed ion beam, populating a range of states and levels is the precursor to the photodissociation process.

This is illustrated in Figure 1. Two Franck Condon regions are involved, the first for the initial electron impact ionization, the second for the photo absorption.

The Relationship Between PE and IPD Spectral Structure.

The first step in interpreting any IPD spectrum is to identify the contributing electronic transition(s). PES is invaluable in that most species of interest have been previously studied by this technique, and the molecular orbitals contributing to each band assigned. Koopman's theorem(2) allows the equation of molecular orbital eigenvalues to vertical ionization potentials (IPs) and assignment of the corresponding ion states can be made. The close relationship between IPD band structure and PES was recognised by Dunbar in his early low resolution studies(3).

To observe an IPD spectrum, there must be a dissociation limit either in the dissociating upper state, ($\tilde{B}$ in Figure 1.) or accessible to it by a predissociation mechanism ($\tilde{A}$).

The PE spectrum is unfortunately not an infallible guide to the IPD spectrum and in several instances reported in the literature(4-11) IPD spectra exhibiting coincident features have been interpreted in substantially different ways. There has been no general discussion in the literature to date of the interpretation of high resolution IPD spectra. It becomes very clear that seeking point to point correspondence with the PE spectrum, offset by the adiabatic ionization potential, is an over-simplification. It fortuitously met the case of CH_3I^+, one of the first reported high resolution spectra.

The relationship between PE spectral structure and IPD band shape is illustrated in Figure 2. Two cases are shown for (a) direct dissociation, and for (b) predissociation. The ionic dissociation limits are indicated by D in each case. In case (a) the dissociation limit occurs in the middle of the product state band, in case (b) the dissociation limit is found in a third state which interacts electronically with the product state.

In case (a_i), the precursor state has little vibrational structure populated, in this case, the IPD spectrum traces out the profile of the product state continuum above its dissociation limit, and appears with the energy offset by the adiabatic ionization potential. The same applies to case (b_i) but in this case the IPD spectrum reproduces the vibrational and, if the lifetime of the state is long enough, the PQR rotational structure. This is clearly what is found for CH_3I^+.

In case (a_{ii}), the precursor state comprises a skewed vibrational distribution, but the dissociating product state has a very shallow potential minimum. In this case, the IPD spectrum traces out the reverse of the vibrational envelope of the precursor state, and lies on the energy scale at the value corresponding to the separation of the highest populated band of the precursor band from the product state dissociation limit. It is believed that this represents the situation in O_3^+.

Case (b_{ii}) is the more complex instance of all levels of the precursor band undergoing transition to all levels of the product band. A broad band corresponding to the convolution of the precursor band with the product band results, giving a most complex rovibronic spectrum. This is observed for O_2^+ and for SO_2^+. If the band profile for either the precursor or product ion states is known, it can be used to deconvolute the IPD spectrum to obtain the other band profile.

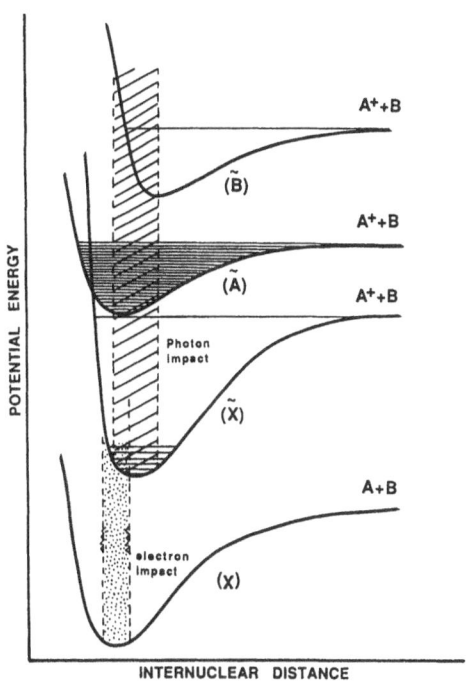

Figure 1.   Potential energy diagram illustrating direct dissociation and predissociation for a hypothetical ion AB⁺.

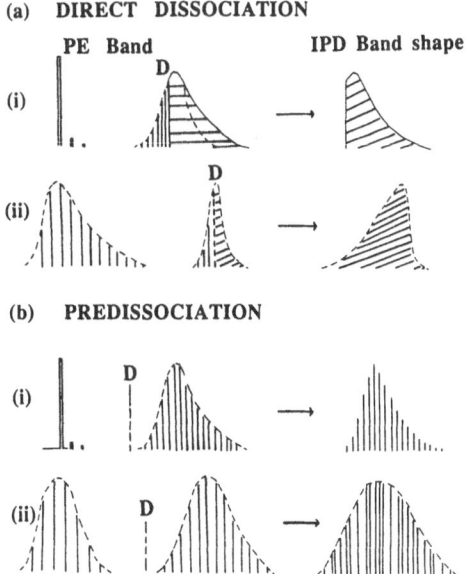

Figure 2.   Four cases of the relationship between PE and IPD spectral structure.

We have assumed in the above that the Franck-Condon factors for the photodissociating transitions are the same as those from the neutral ground state. This is generally not the case, although a reasonable assumption for states which undergo little geometry change on ionization. When several ionic states contribute to a PE band, optical selection rules may preclude some transitions and result in an altered band shape. Similarly, photodissociation may occur from some excited states, accessible by electron impact though optically forbidden, which will not appear in the PE spectrum. This is thought to be the case in the photodissociation of the halogen molecular ions, Cl_2^+ Br_2^+ and I_2^+.(12)

Onset Energies and Thermochemistry

The above discussion shows that care must be taken in interpreting IPD spectroscopic data to give dissociation thresholds. Appearance potentials for fragment ions from PI mass spectra, give upper bounds to dissociation limits. In IPD spectrometry, particularly in case (a_{ii}), the product fragment ion onset may set a lower bound, certainly not an upper one, and the spectral feature with thermochemical significance is more likely to be the high energy tailout of the IPD band. We believe that it is the failure to recognize this which has led to the confusing thermochemistry reported for the bond dissociation energy $D_0(O_2+O^+)$ for O_3^+.

The bond dissociation energy of neutral ozone, $D_0(O_2+O)$, as determined thermodynamically(13) is generally accepted to be 1.05 ± 0.02 eV. A lower value with an upper limit of 0.761 eV is proposed by Vestal and co-workers in the interpretation of their IPD studies upon O_3^+(4,6) and O_3^-(14).

The experimental results of Vestal et.al. for O_3^+ are in good agreement with Goss and Morrison for O_3^+ and their total photodissociation cross sections for O_3^- with Cosby et.al. (15,17) and Lee and Smith (18).

If the IPD onsets are interpreted as lower bounds, rather than upper bounds the apparent contradictions are resolved, and the accepted value confirmed.

The difficulties of interpreting IPD spectra of class (b(ii))

SO_2^+ produces a most complex IPD spectrum, consisting of many thousands of overlapping bands extending from ~2.3eV to at least 3.3 eV. This is a clear indication that the dissociation takes place by a predissociation mechanism. Since the first dissociation limit of the ground state of this ion lies at 15.93 eV, the PE spectrum shows that the precursor(s) leading to the IPD structure in this energy range can be only the Ã and/or B̃ states of the ion, and that the product metastable states can be only the C̃, D̃ and possibly Ẽ which then predissociate into the continuum of the X̃ state.

Examination of the bands at high resolution shows that they are all triangular in shape, being shaded to the red end of the spectrum, thus indicating that the bond lengths in the upper state have increased.

Figure 3 shows the IPD spectrum for SO_2^+. Curve (a) is linearized to an energy spacing of 0.8 meV. With the exception of the region above 2.97 eV, where the scatter is due to poor S/N ratio, all the fine structure is real. Curve (b) is an expanded scan of the region from 2.6238 - 2.6688 eV

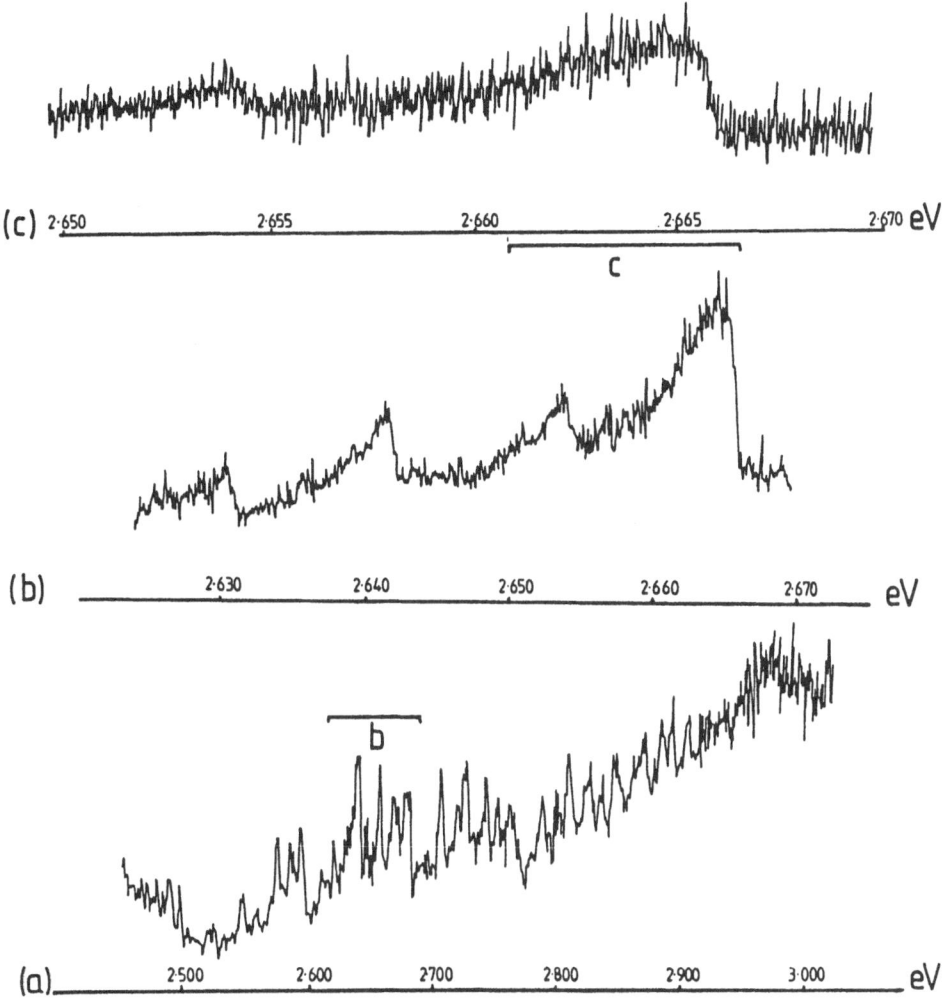

Figure 3. The IPD spectrum of $SO_2^+ \rightarrow SO^+ + O$
(a) over the range 2.54–3.11 eV linearized to 0.8 meV,
(b) finer detail scan over the range 2.6238–2.6698 eV after
16 iterations of summed pair smoothing, and
(c) the same region sampled at 16.7 μeV with a bandwidth of
37 μeV.

and shows unresolved rotational structure for several vibrational bands. Curve (c) is a further expansion of this region, sampled at 16μeV, with a bandwidth of 37μeV. There is such a wealth of detail in this spectrum that it at first proved impossible to relate it to the photoelectron spectrum available at this time, recorded(19,20) at an energy resolution of ~35meV. Heavy smoothing of the IPD data was necessary to achieve any sort of comparison with the PES. Assignment of the bands in this IPD spectrum was difficult, but it seemed that a best fit was achieved by assuming that the observed transitions were predominantly $\tilde{C} \leftarrow \tilde{B}$. Paulson et al, on the other hand, interpreted their results for SO$_2$$^+$ in terms of $\tilde{C} \leftarrow \tilde{A}$ transitions citing evidence suggesting that the $\tilde{B}$ state was short lived, and would have reverted to the $\tilde{X}$ state by fluorescence before mass analysis.(11)

Recently there appeared a higher resolution (13 meV) PE spectroscopic study of SO$_2$,(21) which allowed us to calculate the IPD spectrum by convoluting the profile of the precursor bands with those of the resulting states. The results of this are shown in Figure 4, firstly (a) using both $\tilde{A}$ and $\tilde{B}$ state vibrational levels, then (b) the $\tilde{A}$ levels only and (c) the $\tilde{B}$

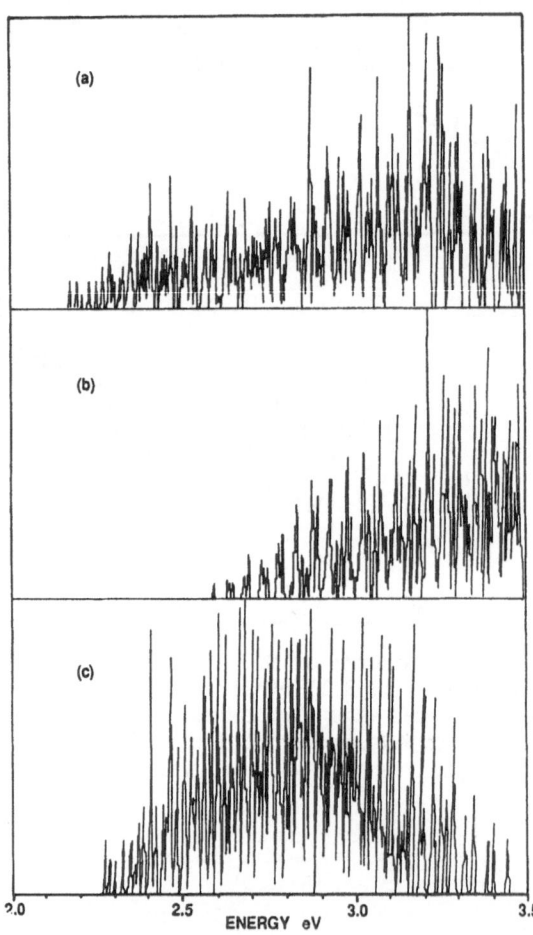

Figure 4. The IPD spectra calculated from SO$_2$ PES assuming no change in Frank–Condon factors
(a) best fit obtained with $\tilde{C} \leftarrow \tilde{A}$, $\tilde{D} \leftarrow \tilde{B}$. (b) $\tilde{C}$, $\tilde{D}$ and $\tilde{E} \leftarrow \tilde{A}$ only, (c) $\tilde{C}$, $\tilde{D}$ and $\tilde{E} \leftarrow \tilde{B}$ only.

levels only. Comparison of these with the experimental spectrum suggests that significant relaxation has taken place before the photodissociation process, but favours case (a) where photodissociation takes place from both the Ã and B̃ ion states.

We have shown previously, in the case of O_2^+(22) and of the halogen molecular ions, Cl_2^+, Br_2^+ and I_2^+(12), that the population of the precursor ions can be modified by varying the energy of the electrons used to produce the initial ionization, and that this was sufficient to establish the precursor state with certainty. Unfortunately, because of the form of the probability laws for direct electron impact ionization, the modification in the population is not great, as can be seen in Figure 5. The populations in this figure are calculated, assuming that $P(I) \propto (E-E_n)$. In practice, the electron energy spread in the ionizing beam will smear out these differences illustrated in Figure 6 even more in the close vicinity of threshold.

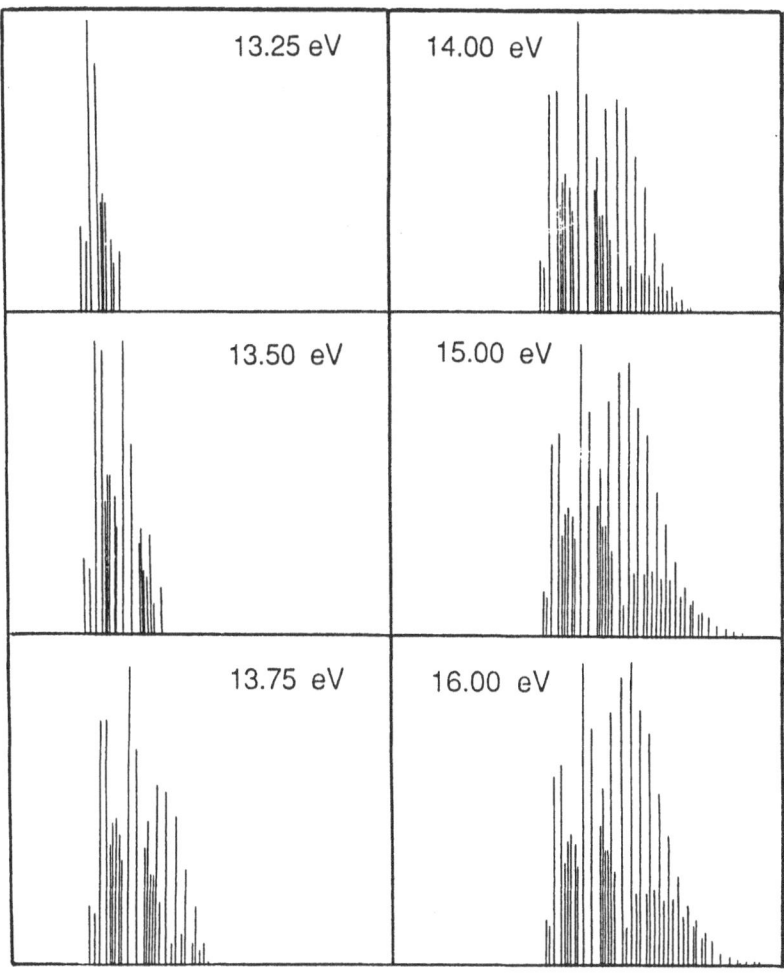

Figure 5. The population of precursor state of SO_2^+ as a function of ionizing electron energy, calculated using a linear ionization probability law.

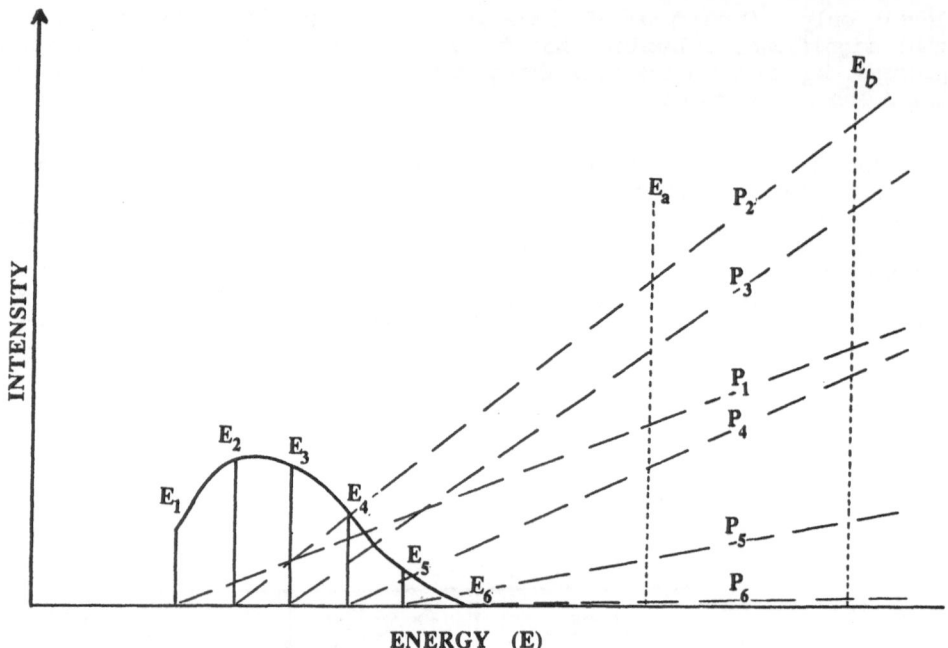

Figure 6.   Diagram illustrating the method of calculating the vibronic populations in figure 5.

Photon impact ionization would be advantageous in this regard since its threshold law approximates a step function of the energy, but the ion currents attainable in PI work are 1000 fold less than those in electron impact.   Multiphoton laser ionization offers the most interesting possibilities for selected state ionization(23).

Figure 7 shows a section of the IPD spectrum of SO_2^+ sampled with a range of ionizing electron energies under otherwise identical conditions. An immediately obvious effect is a reduction in ion current.   When the spectra are normalized with respect to ion current, some peaks are persistent, some enhanced and others suppressed.   The persistent peaks are interpreted as being due to a strong transition from low lying vibrational levels accessible with photons of the energy used.   The suppressed peaks are interpreted as due to transitions from higher vibronic excitations of the precursor state to higher levels of the predissociating state.   The enhanced peaks are attributed to the low lying relaxed states increasingly populated with a change to the initial energy distribution.   This method does not grossly simplify the IPD spectrum, but does reduce the number of potential candidate levels of the initial state(s) contributing to an observed spectral feature.

We have also explored the use of resonant charge transfer, as a means of ionization, in order to populate selectively specific precursor vibrational levels.   Xe and CO_2 have ionization energies in the region of the $\tilde{A}$ and $\tilde{B}$ bands of SO_2^+.   The PE spectra of these are superimposed in Figure 8.   The $^2P_4$ band of Xe$^+$ almost coincides with the $\tilde{B}(020)$ band, while the lowest lying band of CO_2^+ is close to $\tilde{B}(080)$.   (This numbering is taken from Wang et.al (21) and is one vibrational quantum less than in a previous publication(10).   If Xe gas is added to the SO_2 being admitted to the electron impact ion source, the first and most obvious effect is a dramatic reduction in the number of SO$^+$ photodissociated fragment ions at a given photon energy.   We attribute this to the collisional deactivation of the $\tilde{A}/\tilde{B}$ state populations.

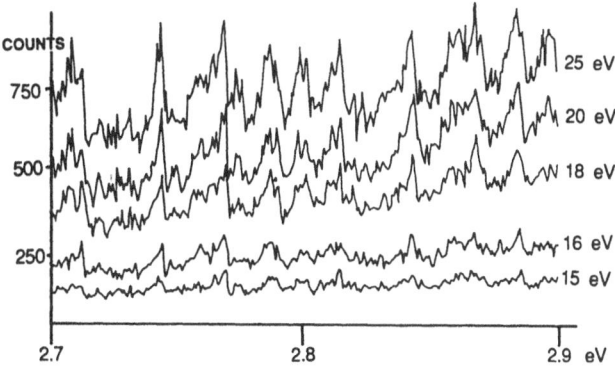

Figure 7. IPD spectra recorded for SO_2^+ at different ionizing electron energies.

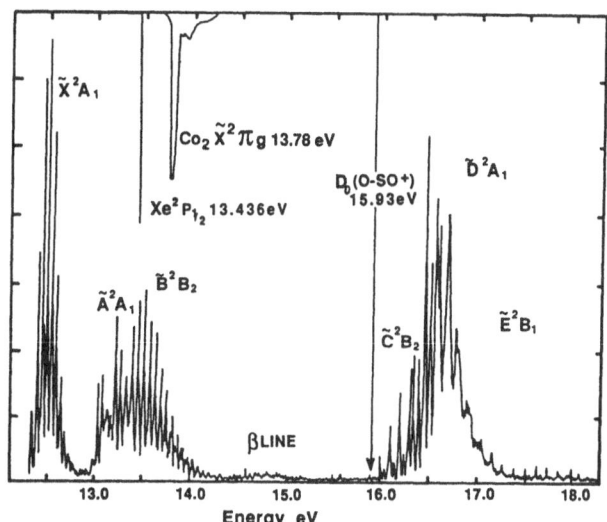

Figure 8. PE spectrum for SO_2 redrawn from ref. 21 with reversed energy scale. Superposed on the figure are the photoelectron spectra of Xe and CO_2, showing states of the ions available for charge transfer to the $\tilde{A}$ and $\tilde{B}$ states of SO_2^+. The width of the CO_2^+ $\tilde{X}$ band is an over-estimate but will include rotational broadening. The Xe^+ $^2P_{1/2}$ level is sharply defined.

Using both charge transfer gases at a ratio of 10/1 to the SO_2, a change is observed in the IPD spectra. (See Figure 9) By a process of spectrum stripping, it is possible to reduce the contribution to the IPD spectrum for SO_2^+ ions formed directly by electron impact, and leave that due to charge transfer. This results in a much simplified spectrum. As for the spectra obtained by varying the electron energy of ionization, some peaks are enhanced, some are unaltered, and some suppressed. In the case of Xe/SO_2 we believe that the contribution of highly excited vibrational

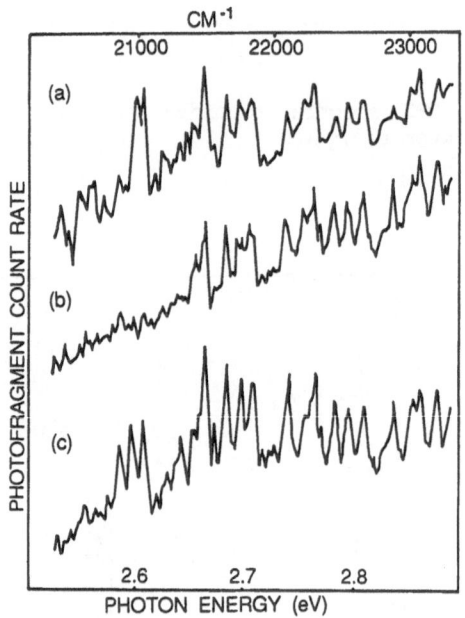

Figure 9. IPD spectra for SO_2^+ formed by charge exchange with (a) Xe^+ (b) CO_2^+, and by direct electron impact (c).

levels of the $\tilde{B}$ state has been removed. The remaining transitions producing the spectrum have their origin in low vibrational levels of the $\tilde{B}$ state and higher levels of the $\tilde{A}$ state. The process of charge exchange is either resonant or endothermic and it is unlikely that levels of the precursor higher than 13.436 eV will be populated. Any contribution at higher energies will be due to electron impact. In the case of CO_2/SO_2, we believe that the wiping out of the structure below 2.66 eV can be explained only by assuming that the transitions observed are due to $\tilde{D} \leftarrow \tilde{B}$, and that the $\tilde{A}$ state levels are being selectively deactivated.

These experiments were carried out using the tandem triple quadrupole mass spectrometer. It is hoped that the problems due to collisional deactivation will be reduced, and a much wider range of charge transfer ions made possible, by a new quinque quadrupole mass spectrometer, nearly complete, which will allow the various regions of primary ionization, charge transfer ionization, and photodissociation to be separated.

References and Notes

1.(a) Morrison, J.D., Chap. in "Gas Phase Ion Chemistry and Mass Spectrometry", ed. J. Futrell. Wiley 1986.
  (b) Moseley, J.T., in Photodissociation and Photoionization, ed. K.P. Lawley, Wiley 1985.
  (c) Dunbar, R.C., in Kinetics of Ion Molecule Reactions, ed. P. Ausloos, Plenum Press, NY. 1979.
      and references therein.

2. Koopman, T., *Physica,* $\underline{1}$, 104 (1943).

3. Dunbar, R.C., *J.A.C.S.* $\underline{95}$, 6191 (1973).

4. Vestal, M.L. and Mauclaire, G.H., *J. Chem. Phys.,* $\underline{67}$, 3767 (1977).

5. Goss, S.P. and Morrison, J.D., *J. Chem. Phys.,* $\underline{76}$, 5175 (1982).

6. Hiller, J.F. and Vestal, M.L., *J. Chem. Phys.,* $\underline{77}$, 1218 (1982).

7. Moseley, J.T., Ozenne, J.B. and Cosby, P.C., *J. Chem. Phys.,* $\underline{74}$, 3767 (1981).

8. Goss, S.P. and Morrison, J.D., *Int. J. Mass Spec. and Ion Proc.,* $\underline{64}$, 213 (1985).

9. Thomas, T.F., Dale, F. and Paulson, J.F., *J. Chem. Phys.,* $\underline{79}$, 4079 (1983).

10. Goss, S.P. and Morrison, J.D., *J. Chem. Phys.,* $\underline{84}$, 2423 (1986).

11. Thomas, T.F., Dale, F. and Paulson, J.F., *J. Chem. Phys.,* $\underline{84}$, 1215 (1986).

12. McLoughlin, R.G., Morrison, J.D. and Smith, D.L., *Int. J. Mass Spec. Ion Proc.,* $\underline{58}$, 201 (1984).

13. Gole, J.L. and Zare, R.N., *J. Chem. Phys.,* $\underline{57}$, 5331 (1972). $D_0(O_2-O)$ is calculated and the original calorimetric determinations are referenced.

14. Hiller, J.F. and Vestal, M.L., *J. Chem. Phys.,* $\underline{74}$, 6096 (1981).

15. Cosby, P.C., Bennett, R.A., Petersen, J.R. and Moseley, J.T., *J. Chem. Phys.,* $\underline{63}$, 1612 (1975).

16. Cosby, P.C., Ling, J.H., Petersen, J.R. and Moseley, J.T., *J. Chem. Phys.,* $\underline{65}$, 5267 (1976).

17. Cosby, P.C., Moseley, J.T., Petersen, J.R. and Ling, J.H., *J. Chem. Phys.,* $\underline{69}$, 2771 (1978).

18. Smith, G.P. and Lee, L.C., *J. Chem. Phys.,* $\underline{71}$, 2323 (1979).

19. Eland, J.H.D. and Danby, C.J., *Int. J. Mass. Spec. and Ion Phys.,* $\underline{1}$, 111 (1968).

20. Lloyd, D.R. and Robert, P.J., *Mol. Phys.,* $\underline{26}$, 225 (1973).

21. Wang, L., Lee, Y.T. and Shirley, D.A., *J. Chem. Phys.,* $\underline{87}$, 2489 (1987).

22. McGilvery, D.C., Morrison, J.D. and Smith, D.L., *J. Chem. Phys.,* $\underline{68}$, 4759 (1978).

23. Weinkauf, R., Walter, K., Boesl, V. and Schbag, E.W., *Chem. Phys. Letters,* $\underline{141}$, 267 (1987).

THE SEARCH FOR NOVEL SMALL MOLECULES AND RADICALS

John L. Holmes

Chemistry Department
University of Ottawa
Ottawa, Ontario, Canada K1N 6N5

Advances in the study of ion structures in the gas phase have shown that small organic cations of unusual structure can readily be prepared by simple mass spectrometric experiments. Such species are well exemplified by the wide range of ylid and distonic ions which have been fully characterized by both theory and experiment; e.g. the homologous ions $CH_2XH^{+\cdot}$, $CH_2CH_2XH^{+\cdot}$ (X = OH, halogen etc.).

The possibility that the neutral counterparts of these ions could be generated in the gas phase by electron transfer became apparent from the work of Porter's group and of McLafferty and his coworkers at Cornell; the experimental technique, named neutralization-reionization mass spectrometry, [NRMS], has recently been reviewed[1-3].

In these experiments, fast (kV energies), beams of mass selected ions are allowed to interact in a collision cell with a stationary target gas which is to act as the electron transfer agent. Alkali metals, mercury, zinc and xenon have successfully been employed. After surviving ions have been electrostatically deflected away, the beam of neutral species enter a second collision cell where they are ionized (and fragmented) by collision with a target gas, typically He, O_2 or NO_2. The resultant beam of ions is mass analyzed by, in most cases, an (energy selector) electric sector and detected thereafter by an electron multiplier or similar device. The experimental arrangement is illustrated in the Figure below and is a typical example of apparatus employing a suitably modified, commercially available mass spectrometer[4]. The apparatus shown in the Figure is based on a VG Analytical ZAB-2F mass spectrometer[5], a reversed geometry (B/E), double focusing instrument. Also included is a field ionization grid[6], used for the detection and analysis of neutral species in high Rydberg states, very close to their ionization energy.

A good example of the potential of this new technique was the first demonstration that the carbonic acid molecule, H_2CO_3, is stable as an isolated species in the gas phase[7]. In these experiments, ammonium bicarbonate, NH_4HCO_3, was sublimed into a mass spectrometer ion source. Ions of m/z 62, having the formula $[H_2CO_3]^{+\cdot}$ and dissociation characteristics appropriate to the structure $[O=C(OH)_2]^{+\cdot}$, were mass selected and passed into a collision cell containing Xe gas. Ions were displaced from the continuing beam path and the collision induced reionization of the neutral

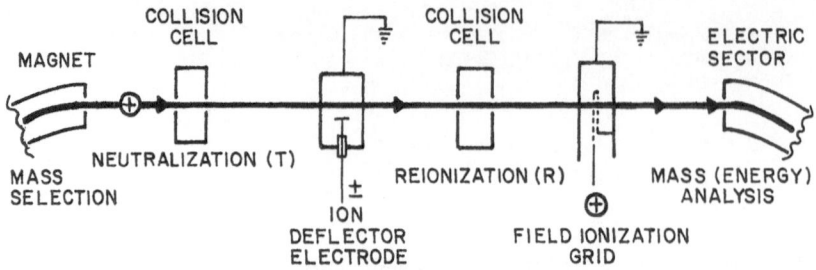

Fig. 1   Accessories in the second field free region of a double focusing
         (B/E) mass spectrometer for NRMS experiments.

species showed unambiguously, the regeneration of $[O=C(OH)_2]^{+\cdot}$.

   However, it must be emphasized that at this early stage in the technique's development, the interpretation of NR mass spectra is not a light
undertaking and healthy contraversies have appeared in the literature.

   Let us suppose that a selected ion, m/z = $M_1^{+\cdot}$, has interacted with
the collision gases (T = electron transfer target, R = reionization gas)
as in equations (1) and (2), i.e. a recovery signal, m/z = $M_1^{+\cdot}$, has been
observed.

$$M_1^{+\cdot} + T \longrightarrow M_1 + T^{+\cdot} \qquad\qquad (1)$$

$$M_1 + R \longrightarrow M_1^{+\cdot} + R + e \qquad\qquad (2)$$

For the experiment to be considered a success, the following conditions
must have been met:

   The input ion flux must consist solely of ions having (a) the
correct m/z value, (b) the correct molecular formula and (c) the desired ion structure.

(a)   Except in extreme cases, the unique selection of the correct m/z
      value should not be problematic provided that the entry to cell 1
      (or a beam defining aperture between the magnet and cell 1) does
      not permit the entry of ions having adjacent m/z values (i.e.
      $M_1^{+\cdot} \pm 1$).

(b)   The common types of interference vis-à-vis molecular formula are
      (i) overlap ions containing naturally occurring isotopes e.g.
      $^{13}CC_2H_5^+$ accompanying $C_3H_6^{+\cdot}$; $C_2H_4^{18}O^{+\cdot}$ accompanying $C_2H_6O^{+\cdot}$; and
      (ii) interference from parallel fragmentation routes e.g. when the
      selected ion, $M_1^{+\cdot}$, is a fragment ion from a larger molecular ion,
      then in some cases the desired ion is seriously interfered with by
      another species of the same nominal mass, but having a different
      molecular formula; e.g. the m/z 41 ion derived from oxazole, C_3H_3NO
      is a mixture of C_2HO^+ and $C_2H_3N^{+\cdot}$ (losses of $HCN, H^\cdot$ and CO respec-
      tively). This second interference should, in principle, be a matter

for routine caution but the first problem, of isotopic overlap, should never be underestimated. Generally, whenever a peak at $[M_1-1]^+$ is present in the mass spectrum from which $M_1^{+\bullet}$ is selected, then the total NRMS cross section for the $[M_1-1]^+$ ion should be measured, relative to that for $M_1^{+\bullet}$. Such NRMS cross sections can differ by factors of 10^2 and so small $[M_1-1]^+$ ion contents can never-theless provide a significant ^{13}C overlap signal with $M_1^{+\bullet}$.

(c) Assuming that the above problems do not apply (or have been fully identified and characterized), it remains to be demonstrated that the selected ion, $M_1^{+\bullet}$, also has <u>solely</u> the desired structure. This is no easy task and its solution depends upon there being adequate mass spectrometric information for the unequivocal identification of the desired ion and its <u>stable isomeric forms</u>. Moreover, a detailed knowledge of the mechanism of the fragmentation which generates the ion $M_1^{+\bullet}$ may also be necessary for a full evaluation of the problem.

A good example of these difficulties is encountered in the continuing story of the attempt to identify the neutral ylide CH_2ClH by NRMS. The situation can be summarized as follows: The ylid <u>ion</u> is readily generated by the fragmentation[8]

$$ClCH_2COOH^{+\bullet} \longrightarrow \overset{\bullet}{C}H_2\overset{+}{C}lH + CO_2$$

The NRMS experiments of Danis et al.[9] (with T=Zn and R=He), showed a surviving $[C,H_3,^{37}Cl]^{+\bullet}$ ion of m/z = 52, but they were undecided as to its structure, $CH_3Cl^{+\bullet}$ or $\overset{\bullet}{C}H_2\overset{+}{C}lH$. Terlouw et al[10] (using T=Xe, R=He) proposed that the neutral ylide had not been generated because the reionization mass spectrum did not contain the doubly charged ion $[C,H_3,Cl]^{++}$, a characteristic of the ylid ion and not of ionized methyl chloride. Further experiments, $[T=Hg, R=He$ or $O_2]$, from the Cornell group[11], which included observations on deuterium labelled ions, led to the conclusion that CH_2ClH molecules had been generated. Most recently[12], from a very detailed study of the C,H_3,Cl system, with emphasis on items (b) and (c) above, it was found that <u>both</u> difficulties contributed significantly to the observations. Moreover, when the surviving ions from a (T=Xe, R=O_2) experiment were transmitted onwards to a <u>third</u> collision cell after the electric sector, collisionally dissociated and the fragments mass analyzed, only fragment ions characteristic of $[CH_3Cl]^{+\bullet}$ could be identified. At the same time, a closely similar experiment (T=Hg, R=O_2) by Wesdemiotis et al.[13] provided equally convincing evidence for the existence of the ylide as a stable intermediate! A possible resolution of this impasse may well come from a better understanding of the role played by the electron transfer target.

It should be noted that if the neutralization step is indeed a vertical process, then according to the high level ab initio molecular orbital theory calculations of Yates et al.[14], stable CH_2ClH ylide molecules (which lie in a well only 15 kJ mol^{-1} deep with respect to decomposition into $CH_2 + HCl$) cannot be generated. These difficulties notwithstanding, it will be as well to keep an open mind with regard to this area of research until more definitive results are obtained.

The final topic of this short report will be a review of the status of the methonium radical, $CH_5^{\bullet}$, a species also predicted not to be stable[15], but which has been observed as an excited (high Rydberg) species in NRMS experiments.

In 1980 Williams and Porter[16] and in 1984 Gellene and Porter[17] investigated the methonium ($CH_5^{\bullet}$) radical by neutralized ion beam techniques. In the earlier report[16] the angular scattering of the neutral products from charge exchange between $[CH_5]^+$ and $[CD_5]^+$ ions having 5 keV translational energy and alkali metal vapors (Na and K) was measured. In the second report the neutral species from charge exchange between 6 keV $[CD_5]^+$ ions and K atoms were reionized by collisions with NO_2 target gas, and the resulting fast ions were mass analyzed by observing their deflection in a uniform electric field. These experiments showed, in keeping with the theoretical predictions[15], that $CH_5^{\cdot}$ radicals formed by the vertical neutralization of the ground state $[CH_5]^+$ ions are not stable. They dissociated mainly into $CH_4 + H^{\cdot}$ products[18] with 2.65 ± 0.15 eV as the maximum kinetic energy release[16]. The <u>vertical</u> electron affinity of $[CH_5]^+$ was estimated to be 5.3 ± 0.15 eV.

In conflict with these results Griffiths et al.[19] have recently reported the recovery of parent ions in NR experiments with 6 keV $[CH_5]^+$ and $[CD_5]^+$ ions. They used a modified double focussing mass spectrometer having B/E geometry (VG Analytical ZAB-2F). The modifications consisted of a collision gas cell preceded by ion beam deflector electrodes mounted in the second field free region of the instrument. They admitted a gas of high ionization energy, N_2 (IE=15.6 eV), Kr (IE=14.0 eV) or Xe (IE=12.1 eV), into the collision cell. Gas diffused therefrom into the region between the magnet and the ion beam deflection plates and acted as the neutralizing agent. The fast neutrals which were produced were then ionized by collisions with the gas in the collision cell. The resulting ions were mass analyzed in the usual way via their kinetic energy spectrum, obtained by scanning the electric sector of the mass spectrometer. The observation of a recovery signal for the parent ion led them to conclude that neutral $CH_5^{\cdot}$ species, having a lifetime of at least 390 ns, were formed. From the net kinetic energy loss undergone by the $[CH_5]^+$ ions in the neutralization-reionization process, they estimated the ionization energy for the metastable $CH_5^{\cdot}$ species to be 8.0 ± 0.5 eV. Only the kinetic energy spectrum of the $[CH_5]^+$ recovery peak was shown and no description was given of the fragment ion peaks $[CH_4]^+$, $[CH_3]^+$, etc., or those from $[CD_5]^+$ ions.

Motivated by these contradictory results Selgren and Gellene[20] have more recently repeated the experiments reported in references 16 and 18. With Na atoms as neutralization target (IE=5.1 eV) they reproduced the results of refs. 16 and 18. However, with Zn atoms (IE=9.3 eV), as neutralizing agent they obtained completely different observations; the neutral beam profile showed a much larger kinetic energy release in the dissociation of $CH_5^{\cdot}$ than when Na was the target. The Zn/NO_2 NR mass spectrum of $[CD_5]^+$ contained a recovery peak, which was the <u>most intense</u> peak in the spectrum. Moreover, the shape of the $[CD_4]^+$ fragment peak reflected a very large kinetic energy release in the dissociation $CD_5^+ \rightarrow CD_4 + D^{\cdot}$. The poor energy resolution of their apparatus did not allow a precise determination of the released kinetic energy, but the results were compatible with 8 eV as the most probable value. It was considered that this is the largest possible kinetic energy release for a fragmenting $CH_5^{\cdot}$ neutral having slightly less energy than the difference between the ground state of $[CH_5]^+$ and the neutral dissociation products $CH_4 + H^{\cdot}$ (or $CH_3^{\cdot} + H_2$) ($\Delta H_f[CH_5]^+$ = 912.5 kJ. mol^{-1}; $\Delta H_f[CH_4] + \Delta H_f[H]$ = 143.1 kJ. mol^{-1}). Their explanation was that high Rydberg $CD_5^{\cdot}$ radicals ($CD_5^{\cdot}(HR)$) were formed in $[CD_5]^+$/ Zn collisions. Such a species can be metastable in the μs time scale and can predissociate into non-Rydberg fragments, $CH_4 + D^{\cdot}$, giving up to 8 eV kinetic energy release. They argued that in the case of Na and other targets having a low ionization energy, resonant charge exchange leading to the unstable ground Rydberg level of $CH_5^{\cdot}$ is the most important process. In the case of targets with higher ionization energy (IE > 6-7 eV) the formation of metastable $CH_5^{\cdot}(HR)$ becomes the most important process. They thus attributed the recovery of the parent ion in the neutralization-reionization experiments of Griffiths et al.[19] to the formation of metastable $CH_5^\cdot$ (HR) radicals.

By exploiting specific features of our ZAB 2F mass spectrometer which has been modified for NRMS experiments[4] we have investigated and augmented[21] the results of Selgren and Gellene[20] and Griffiths et al.[19]. The resolution of our apparatus permits the determination of the kinetic energy released in the $CH_5^\cdot \rightarrow CH_4 + H^\cdot$ dissociation with ± 0.1 eV precision. With the field ionization device[6] fast, high-Rydberg species can selectively be ionized and detected. It can therefore be unequivocally determined whether metastable $CH_5^\cdot$ (HR), ($CD_5^\cdot$ (HR)) species are produced in collisions between $[CH_5]^+$ ($[CD_5]^+$) ions of keV translational energy and high ionization energy targets, such as Xe, Kr, NO, etc.

The salient results can be summarized as shown in the Table and in point form below.

CH_5^+ ions (and deuterated analogues) were generated in all cases by high pressure chemical ionization in the mass spectrometer ion source.

Present results[21] showed that any NRMS signal at m/z 17 arises solely from $^{13}CH_4^{+\cdot}$ ions, co-transmitted with CH_5^+.

The kinetic energy released in the fragmentation of the unstable $CH_5^\cdot$ species is independent of the target gas T (see the NRMS of CH_5^+ illustrated below).

The only report of stable $CH_5^\cdot$ radicals is that of Selgren and Gellene[20], T = Zn; these $CH_5^\cdot$ radicals were high Rydberg species. In the present work no high Rydberg $CH_5^\cdot$ could be detected.

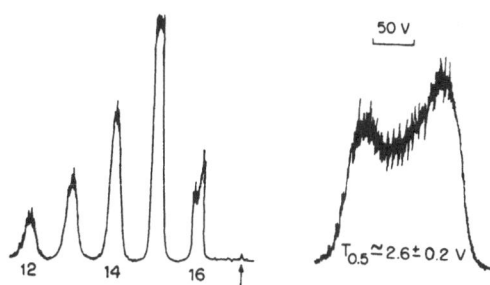

Fig. 2   NRMS (T=Xe, R=O_2) of CH_5^+ ions[21]. The m/z 16 peak, enlarged, is shown to the right.

Future prospects

It appears from work reported to date, that the investigation of novel neutral species by the NR mass spectrometric technique may result in their being observed not only in their ground state but possibly in excited states (as well). Important questions to be answered are (a) how the neutralization process depends upon the energetics of the electron transfer and upon the initial quantum state of the transferred electron; (b) how the neutralization process relates to the energy content of the projectile ion (ro-vibrational and electronic states) and whether this affects the "verticality" of the neutralization step.

Table 1   Neutralization-Reionization Experiments with CH_5^+ Ions

Neutralization Target T	IE[T] (eV)	CH_5^+ in NRMS	Observations	Ref.
K	4.3	No	Kinetic Energy Release (KER), $CH_5^{\cdot} \rightarrow CH_4 + H^{\cdot}$; 2-3 eV	16,18[a]
Na	5.1	No	KER $CH_5^{\cdot} \rightarrow CH_4 + H^{\cdot}$; 2-3 eV	16,20[a]
$NO^{\cdot}$	9.3	No	KER $CH_5^{\cdot} \rightarrow CH_4 + H^{\cdot}$; 2-3 eV	21[a]
Zn	9.4	Yes	KER $CH_5^{\cdot}$ (High Rydberg) $\rightarrow CH_4 + H^{\cdot}$; ca 8 eV[b]	20[a]
Xe	12.1	No	KER $CH_5^{\cdot} \rightarrow CH_4 + H^{\cdot}$; 2-3 eV	21[a]
		Yes	No report.	19[a]
Kr	14.0	No	KER $CH_5^{\cdot} \rightarrow CH_4 + H^{\cdot}$; 2-3 eV	21
		Yes	No report.	19
N_2	15.6	No	KER $CH_5^{\cdot} \rightarrow CH_4 + H^{\cdot}$; 2-3 eV	21
		Yes	No report.	

(a)   Includes observations on CD_5^+.

(b)   This KER is close to the energy, ΔH_r for $CH_4 + H^{\cdot} \rightarrow CH_5^+$, 8.0 eV.

References

1. G.I. Gellene and R.F. Porter, Acc. Chem. Res. 16, 200 (1983).
2. C. Wesdemiotis and F.W. McLafferty, Chem. Rev. 87, 485 (1987).
3. J.K. Terlouw and H. Schwarz, Angew. Chem. Int. Ed. Engl. 26, 805 (1987).
4. J.L. Holmes, A.A. Mommers, J.K. Terlouw and C.E.C.A. Hop, Int. J. Mass Spectrom. Ion Processes 68, 249 (1986).
5. R.P. Morgan, J.H. Beynon, R.H. Bateman and B.N. Green, Int. J. Mass Spectrom. Ion Phys. 28, 171 (1978).
6. J. Bordas-Nagy, J.L. Holmes and A.A. Mommers, Org. Mass Spectrom. 21, 629 (1986); J. Bordas-Nagy and J.L. Holmes, Chem. Phys. Letters 132, 200 (1986).
7. J.K. Terlouw, C.B. Lebrilla and H. Schwarz, Angew. Chem. Int. Ed. Engl. 26, 354 (1987).
8. J.L. Holmes, F.P. Lossing, J.K. Terlouw and P.C. Burgers, Can. J. Chem. 61, 2305 (1983).
9. P.O. Danis, C. Wesdemiotis and F.W. McLafferty, J. Am. Chem. Soc. 105, 7454 (1983).
10. J.K. Terlouw, W.M. Kieskamp, J.L. Holmes, A.A. Mommers and P.C. Burgers, Int. J. Mass Spectrom. Ion Processes 64, 245 (1985).
11. C. Wesdemiotis, R. Feng, P.O. Danis, E.R. Williams and F.W. McLafferty, J. Am. Chem. Soc. 108, 5847 (1986).
12. C.E.C.A. Hop, J. Bordas-Nagy, J.L. Holmes and J.K. Terlouw, Org. Mass Spectrom. 23 (3) (1988).
13. C. Wesdemiotis, R. Feng, M.A. Baldwin and F.W. McLafferty, Org. Mass Spectrom. 23 (3) (1988).
14. B.F. Yates, W.J. Bouma and L. Radom, J. Am. Chem. Soc. 106, 5805 (1984); idem, ibid, 109, 2250 (1987).
15. W.A. Latham, W.J. Hehre, L.A. Curtis and J.A. Pople, J. Am. Chem. Soc. 93, 6377 (1971).
16. B.W. Williams and R.F. Porter, J. Chem. Phys. 73, 5598 (1980).
17. G.I. Gellene and R.F. Porter, J. Phys. Chem. 88, 6680 (1984).
18. G.I. Gellene and R.F. Porter, Int. J. Mass Spectrom. Ion Processes 64 55 (1985).

19. W.J. Griffiths, F.M. Harris, A.G. Brenton and J.H. Beynon, Int. J. Mass Spectrom. Ion Processes 74, 317 (1986).
20. S.F. Selgren and G.I. Gellene, J. Chem. Phys. 87, 5804 (1987).
21. J. Bordas-Nagy and J.L. Holmes (submitted for publication).

NEW INFORMATION ON THE STRUCTURE AND DYNAMICS OF MOLECULAR CATIONS

FROM EXPERIMENTS ON THE SPECTROSCOPY OF POLYATOMIC RYDBERG STATES

J. W. Zwanziger, K. S. Haber, F. X. Campos and E. R. Grant

Department of Chemistry
Purdue University
West Lafayette, IN 47907

I. Introduction

Historically, electronic spectroscopy has played a central role in the effort to structurally characterize small molecules, and, particularly, reactive intermediates.[1] Increasingly, spectroscopic knowledge and analysis has also had a tremendous impact on our understanding of the radiationless dynamics of molecular excited states.[2] The theoretical apparatus for dealing with such problems is well developed, and has been profitably extended to considerations of intramolecular relaxation in vibrationally excited electronic ground states.[3]

Attention has also been moving higher in energy, to molecular levels with spectroscopic terms in the vacuum ultra-violet, near and above first ionization thresholds.[4] There exists the opportunity for a fascinating common ground, in which theories of radiationless processes are extrapolated upward in energy to meet cation-electron scattering formalisms.[5] High energy excited states, or in generic terms, molecular Rydberg states are particularly interesting because they lie in regions of energy just now accessible to developing high-resolution laser and synchrotron techniques in VUV and multiresonant multiphoton spectroscopies that are applicable to samples cooled in free-jet expansions. Thus, they constitute an inviting prospect for study because comparatively much remains to be learned. They are also inviting because, so far above thresholds for homolytic rearrangement and fragmentation, this class of states promises a rich short-time dynamics of competing decay processes. Yet, finally, these states are inviting, because, from a zeroth-order perspective, their spectroscopy and electronic structure can be viewed in terms of an extraordinarily simple separation.

This separation, of course, is that of an ideal Rydberg limit, in which a single diffuse excited electron moves hydrogenically in the central field of a distant cation. The implications of this simple picture are particularly important for molecular spectroscopy. In this limit, the rovibrational structure impressed upon the electronic spectrum of a Rydberg state is viewed as the vibration-rotation level structure of the core shifted down by the binding energy of the electron. This is given systematic expression in terms of the Rydberg formula:

$$T_{v,J} = IP_{v,J} - \frac{R}{(n-\delta)^2} \, ,$$

which predicts, in an ideal Rydberg limit, series in selected ν and J converging with constant quantum defect, δ, to corresponding internal states of the ion.

This extremely simplified model for the electronic structure of high-lying molecular excited states is subject to obvious limitations. For a given system, at energies corresponding to low values of the principal quantum number, n, one may find important contributions from higher-lying excited valence configurations, even at internuclear displacements not far from equilibrium.[6] Such states can be dissociative, and interactions can be highly quantum state specific.[7] At intermediate values of n, where Rydberg term spacings are commensurate with core vibrational separations, rovibrationally mediated Rydberg-Rydberg interactions can be expected, and even without other vibronic effects, very near ionization thresholds, Rydberg-electron core angular momentum uncoupling effects can be expected to scramble core-structural regularity in Rydberg spectra.[5]

Despite these anticipated complexities, this simple separation has been of great general utility.[8] In particular, it has had extremely important implications for the assignment of polarized absorption spectra of polyatomic Rydberg states converging to electronically degenerate cations.[9-18] In those cases it has proven possible to interpret spectroscopic evidence for distortion and fluxionality in terms Jahn-Teller effects in a strong vibronic coupling limit of the cation core.[19] In this interpretation, seemingly paradoxical electronic transition symmetry assignments, associated with series in degenerate Rydberg orbitals, are seen to follow as a consequence of a comparatively weak interaction of the Rydberg electron with the core.

In numerous cases where perturbations have been observed and studied, cation-electron scattering formalisms, such as that of multichannel quantum defect theory (MQDT), have proven instructive.[4] Implications of such interpretation for the physical description of short-time intramolecular relaxation, promise a rich new radiationless dynamics within a universal class of systematically related molecular electronic states. This particular aspect figures prominently in our own program of current research. Meaningful progress on dynamics, however, relies on well-systematized spectroscopic knowledge. Moreover, experience, such as that cited above, has shown that much can be learned about the structure and intramolecular coupling dynamics of polyatomic cations themselves from spectroscopic studies of molecular Rydberg states. In keeping with the theme of this workshop, it is thus appropriate to focus in this report on the simpler questions of: When can we have confidence in the simple Rydberg separation and, under such circumstances, what can Rydberg spectroscopy tell us about the structure and dynamics of molecular cations?

We begin with a brief survey of selected examples which supply spectroscopic grounds for the one-electron Rydberg separation. We shall mention the role of Frank-Condon factors in establishing isolated Rydberg character, as well as regions mixed by surface crossings or Rydberg-vibration interactions. We will then turn to a pair of specific examples, in which it is believed that the observed vibrational structures well represent respective cores. In both cases cores are degenerate, and insights born of vibrational structure illuminate first, in the case of sym-triazine, the effects of a simple conical intersection, and secondly, in the case of cyclopropane, the extraordinary intramolecular dynamics of a large-amplitude multimode unhindered pseudorotator.

II.   The Ideal Rydberg Limit

It has long been appreciated that electronically excited states for various diatomic molecules exist at high energy in atomic-like series.[20] Helium dimer was the first such system so assigned,[21] and has since become a prototype for the class of so-called Rydberg molecules.[22] Work following this has vastly expanded the range of diatomic systems for which high energy absorption spectra are available, adding information on vibrational/rotational structure. In cases where comparable data exists for corresponding ions, reasonable connections between core vibrational structure and ionic levels can be drawn. In the few cases where Rydberg spectra are sufficiently resolved, rotational assignments match equilibrium internuclear separations. Table I summarizes selected representative comparisons.

Table I. Vibrational frequencies and rotational constants of selected diatomic Rydberg excited states and cation ground states[20]

	ν_0	ν_R	ν_+	B_e^R	D_e^+
H_2	4401	2444(2pπ)	2322	31.363	30.21
He_2	——	1746(3sσ)	1698	7.365	7.211
N_2	2358	2202(3pσ)	2207	1.9612	1.9318
CO	2170	2176(3pσ)	2214	1.9533	1.9772
NO	1904	2374(3sσ)	2376	1.9965	1.9878

Rydberg entries above have been chosen broadly to typify fundamental frequencies found throughout the various neutral excited state manifolds. It must be added, however, that perturbations are common, and in some cases it has been possible to reconstruct complex underlying internuclear potentials.[23] These are usually found to be Rydberg in nature, characterized by the restoring forces of the ion at low internal energies and small internuclear distances (comparable to that of the free ion), but, at higher energies, show avoided crossings producing second minima at larger internuclear distances, in which the electronic structure is enriched by mixing with valence or charge-transfer configurations.

It can be anticipated that a similar picture extends to polyatomics, although fewer measurements exist, and in no case is a set of frequencies sufficiently complete to establish fully comparative force fields. Results which are available suggest a degree of correspondence not unlike that found for diatomics. Table II below summarizes selected comparisons.

Again, at least rough conformance between Rydberg and cationic vibrational level structure is typically the case. Particularly noteworthy is the match between the complex pattern of vibronically active bands found repeated throughout the Rydberg manifold of benzene, and the level structure observed for Jahn-Teller distorted $^2E_{1g}$ ground state cation. Moreover, as also tabulated above, in polyatomic systems such as HCO and H_3 for which high-resolution data exist on Rydberg states and cations, rotational fits favorably relate equilibrium internuclear distances. As with diatomics, however, exceptions to a picture which completely separates Rydberg states into ion-core and single electron components are also evident. Though their vibronic spacings are roughly invariant, the lower lying Rydberg levels of benzene exhibit homogenous widths that indicate a

Table II. Available vibrational frequencies for selected polyatomic Rydberg excited states and cation ground states.

		ν_R	ν^+	ref.
HCO	(ν_2)	822	830	
	(ν_3)	2177	2184	(25)
	$(B_o$	1.500)	(1.4875)	(24,25)
DCO	(ν_2)	657	655	
	(ν_3)	1900	1904	(24,25)
H_3	$(B_o$	44.19)	(43.56)	(22)
D_3	$(B_o$	21.98)	(21.82)	(22)
NO_2	(ν_1)	1404	1400	(26,27)
	(ν_2)	614	615	
CO_2	(ν_2)	498	519	(28)
C_6H_6	$6^1(3/2)$	313	347	
	$6^1(1/2)$	680	686	(14,29)
	1^1	969	992	
	$6^2(1/2)$	1253	1331	
C_2H_2	ν_2	1807	1829	(30)
	ν_4	597	670	

femtosecond decay of Rydberg character[14] (as confirmed by time-resolved pump-probe measurements[31]). Regions of the acetylene spectrum are too diffuse to register resonant structure at all.[30]

A more sensitive test of the appropriateness of this separation hypothesis, which matches the internuclear structure of Rydberg states with that of their core cations, is found in probes of vibrational wavefunctions, as manifested, for example, in observed Franck-Condon factors. Rydberg-Rydberg double resonance experiments on NO, for example, show a high propensity for vertical transitions, in accord with a picture that largely maintains a constant, well-defined core structure.[32] Electronic bound-continuum transitions, in which cation final- state distributions are determined by photoelectron velocity analysis, also probe core-ion Franck-Condon overlaps. There, for NO[33] as well as other diatomics, such as H_2,[34] N_2[23,35] and Cl_2,[36] and polyatomic C_2H_2,[30] one finds regions of substantially vertical Franck-Condon character, indicating that Rydberg vibrational states are diagonal or nearly so in a basis of ion internuclear wavefunctions. These same systems, however, all show regions of energy in which the vibrational character of excited electronic states is decidedly mixed in such bases. As with the spectra themselves, such observations serve to set limits beyond which the many-body aspects of the problem cannot be safely ignored. Within these limits it is clear that Rydberg separation constitutes a useful zeroth-order perspective from which to approach molecular spectroscopy near ionization thresholds. Beyond this exist outstanding challenges to quantum chemists and spectroscopists to experimentally and theoretically characterize regions of mixed nature, and relate these findings to fundamental questions of high-energy many-body electronic structure. Scattering formalisms will be useful in such contexts, and in the time domain itself these interactions promise a fascinating short-time dynamics. As a more immediate return, it is clear that, interpreted with care, high-resolution Rydberg spectra of polyatomic molecules cooled in free-jet expansions can provide useful information on the structure and bonding of otherwise difficult molecular cations. We amplify this point with two examples taken from work carried out in our laboratory: sym-triazine and cyclopropane.

III. Structure and dynamics of the simple conical intersection: $^2E'$ sym-$C_3H_3N_3^+$

We most often obtain information on the positions and intensities of transitions from levels of neutral electronic ground states to Rydberg states by two-photon resonant three-photon ionization spectroscopy: A level populated by two-photon absorption is detected by subsequent one-photon ionization.[9,10,14] Our experiments are conducted in a laser-crossed pulsed-molecular-beam quadrupole mass spectrometer system. Spectral congestion is reduced by the substantial cooling associated with the isentropic expansion of the sample seeded in He. Laser powers for two-plus-one experiments are maintained such that the ionization step is saturated, while two-photon absorption is rate limiting in the quadratic regime. In this way measured relative spectral intensities accurately reflect relative two-photon absorption cross sections.

When sym-triazine is seeded in the beam and a single, focussed, excimer-pumped dye laser is scanned through the region from 360 to 340 nm, a complex two-photon absorption spectrum containing well over two-dozen vibronic transitions is observed.[11,18] The overall electronic transition is readily assigned as $X\ ^1A_1' \rightarrow 3s\sigma\ ^1E'$, in which, according to the simple separation discussed above, a single Rydberg electron occupies a diffuse totally symmetric hydrogenic orbital, and the molecular core has the structure of the $^2E'$ cation. Indeed the observed spectrum conforms well with the lower-resolution picture of the vibrational states of the cation painted by the VUV excited photoelectron spectrum.[37] In the absence of vibronic coupling effects, one would expect, on the basis of the normal modes of the ground state, to find the first spectroscopically accessible vibrationally excited level of the upper state in a region 1000 cm^{-1} above the origin. Here one finds instead fully five bands active with frequencies lower than this. Moreover, their distribution is completely non-harmonic, while the hot-band spectrum is dominated by transitions from v=1 in the degenerate mode ν_6 to the origin and low lying vibrationally excited levels. Clearly, the electronic degeneracy of the core confers a special property on the vibrational positions and intensities of the $3s\sigma$ Rydberg state of sym-triazine.

A qualitative appreciation of the effects of the electronic degeneracy can be gained simply by examining the associated adiabatic potential surfaces in selected normal coordinates. These are found in the usual way, by diagonalizing the Hamiltonian exclusive of the nuclear kinetic energy operator. Our minimal basis necessarily contains the pair of degenerate electronic wavefunctions:

$$\psi = a\,\psi^+ + b\psi^-.$$

which immediately raise the possibility of off-diagonal elements in our matrix Schrodinger equation:

$$
\begin{bmatrix}
\langle\psi^+|H_e(Q)|\psi^+\rangle - E(Q) & \langle\psi^+|H_e(Q)|\psi^-\rangle \\
\langle\psi^-|H_e(Q)|\psi^+\rangle & \langle\psi^-H_e(Q)|\psi^-\rangle - E(Q)
\end{bmatrix}
\begin{bmatrix}
a(Q) \\
b(Q)
\end{bmatrix}
= 0
$$

In the limit where these are zero, the roots are equal. Jahn and Teller showed in 1935, however, that for all symmetric molecules there exists at least one normal coordinate for which off-diagonal matrix elements above are non-zero by symmetry.[38] In practice, matrix elements are expanded in Taylor's series retaining the lowest-order quadratic (harmonic) restoring force on the diagonal, while retaining linear and, when necessary, quadratic terms off the diagonal.[39]

For normal coordinates, Q, corresponding to the two-dimensional isotropic harmonic oscillator, the resulting surfaces form the familiar conical intersection. Even at the zero point, vibrational motion, which can be expressed in terms of an amplitude ρ and a phase ϕ, is not well-referenced to a single adiabatic sheet of this potential when coupling is weak. However, in a limit of strong linear coupling, levels well below the conical intersection can be characterized adiabatically in terms of a radial oscillation about a distorted equilibrium position ρ_o, and a fluxional adiabatic free pseudorotation in the phase angle. Only in the limit of strong linear-plus-quadratic coupling is this rotation quenched and distorted levels trapped.

Thus, in this overall picture we can recognize a basis for expecting low-frequency, irregular structure in the degenerate $3s\sigma$ $^1E'$ Rydberg state of sym-triazine. To obtain exact vibronic levels from the model for comparison with experiment we return to our matrix Hamiltonian, assume, in accordance with the hot band spectrum, a single active mode and now, including the nuclear kinetic energy operator, diagonalize in an appropriate basis of harmonic oscillator wavefunctions. In the linear limit the problem block factors so that only basis states of the same value of $\ell\pm1/2$ couple, where ℓ is the vibrational angular momentum quantum number: $\ell = \nu, \nu-2 \ldots -\nu+2, -\nu$. Thus, eigenstates in this limit can be characterized by a good half-integer quantum number $j = \ell\pm1/2$.[40]

Figure 1 below shows a correlation diagram for the single mode linear problem which gives the eigenenergies in units of the zeroth-order basis frequency, as a function of the off-diagonal expansion coefficient, k, related to the depth of the pseudorotation trough, by $D = k^2/2$. Note that for zero coupling we recover the levels of the isotropic two-dimensional harmonic oscillator. As we turn on coupling the levels split, preserving the original degeneracy, into combinations of states of E and A symmetry. The first level above the ground state, for example, splits into an E' ($j = 1/2$) radially excited state, which remains at high energy, and a pair of accidentally degenerate excited pseudorotation states ($j = 3/2$) of symmetry A_1' and A_2', which fall in energy with increasing distortion amplitude (moment of inertia). Quadratic coupling splits this degeneracy. Two-photon transitions from an A_1' ground state are allowed to E' levels but Franck-Condon forbidden to A' states. In the limit of linear coupling, however, only a subset of E' levels (those with $j = 1/2$) will find themselves in the right symmetry block to derive Franck-Condon intensity from the origin, and so, in this limit, only the radial fundamental and its overtones should appear. Note, that the complete pattern of levels which starts as quite complex for weak coupling, develops the greater regularity, of quadratically spaced free-pseudorotation states built on radial oscillator levels as distortion deepens.

Comparison of the relative vibronic level spacings observed in the two-photon absorption spectrum of the 3s $^1E'$ Rydberg state of sym-triazine with the range of possibilities suggested by the above correlation diagram produces reasonable correspondence with simple linear coupling for $k \approx 2.1$ ($D = 2.2$). However, though this single coupling parameter qualitatively accounts for the irregular low-frequency vibrational structure and hot-band activity, it fails to predict splittings observed in sequence band transitions to $j = 3/2$ A' states, as well as non-zero intensity in higher angular momentum E' states.

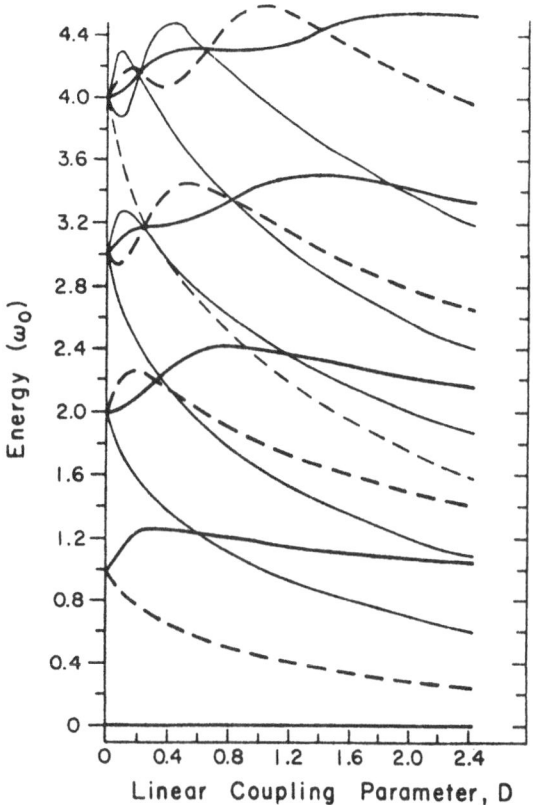

Figure 1. Correlation diagram for the levels of the linearly-coupled single-mode Jahn Teller problem as a function of coupling strength.

Both of these latter effects can be predicted to arise from higher order coupling. Quadratic terms modulate the adiabatic surfaces, tending to quench the free adiabatic pseudorotation. As a consequence of these higher-order terms in the Hamiltonian, the problem block factors as $j' = j$ mod 3. The result is a splitting of the accidentally degenerate $j = \pm 3/2, 9/2, 15/2, \ldots$ A' levels, and mixing, with resultant intensity borrowing, among states of the E' $j' = 1/2$ mod 3 block ($1/2, -5/2, 7/2 \ldots$).

Inclusion of a small quadratic coupling coefficient, g = 0.05, has a dramatic effect on the theoretically predicted spectrum. Intensity redistribution increases more than two-fold the number of active bands forecast in the first 1000 cm^{-1}. With optimization, precise quantitative correspondence is found for positions and intensities of all vibronic transitions observable in ν_6 within 2200 cm^{-1} of the origin. Representative of the fit achieved is the figure below, showing schematically the predictions of the simple single mode linear plus quadratic coupling model.[18]

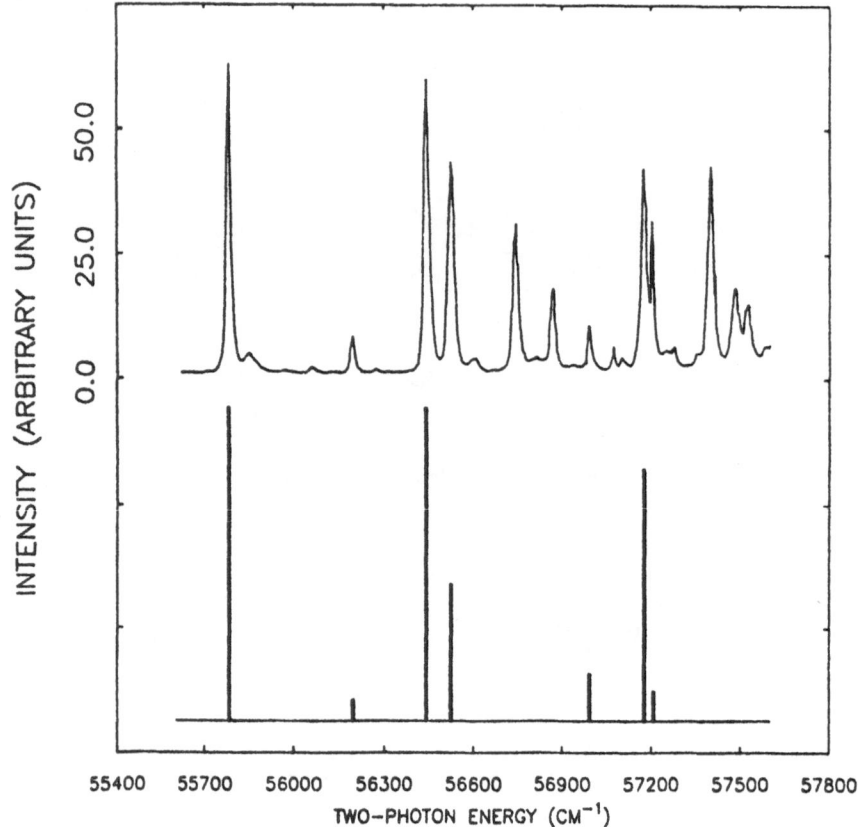

Figure 2. Two-photon absorption spectrum of sym-triazine in the region from 346 to 361 nm. The stick figure shows theoretical positions and intensities for linear-plus-quadratic coupling in ν_6 with k = 2.14 and g = 0.046 (symmetric modes and their combinations with ν_6 are suppressed).

The adiabatic potentials derivable from this calculation must be regarded as very nearly exact to an energy at least as high as the highest measured eigenvalue. It is interesting to compare the potential in the pseudorotation coordinate around the bottom of the trough with the internal rotor levels built on the first radial oscillator state. As shown above, the quadratic barriers to pseudorotation only weakly alter the shape of the potential surface.

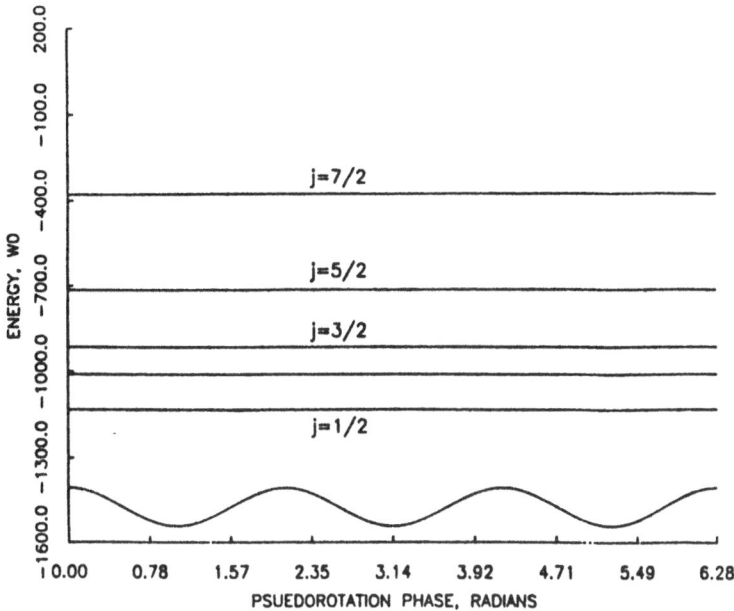

Figure 3. Lower surface adiabatic potential in the pseudorotation coordinate along the minimum energy path in (ρ,ϕ).

Despite this, and the reasonably close correspondence between linear and linear-plus-quadratic eigenvalues, evident splittings and transfers of intensity show that, by their presence in the vibronic hamiltonian, these quadratic terms have a significant effect on the wavefunctions.

These are conveniently visualized as nuclear probability densities. Figures below show plots of nuclear probability densities (for complex vibronic wavefunctions) for the first two levels in the $j = (\pm 1/2)$ mod 3 manifold.

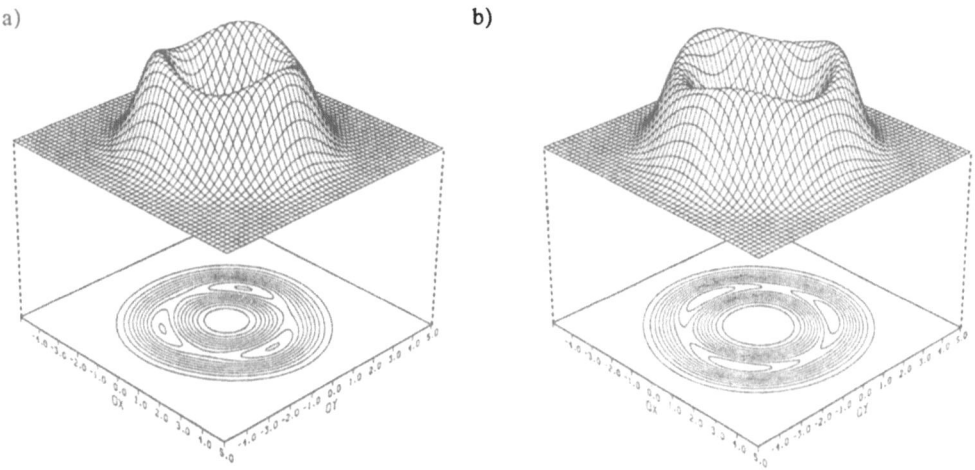

Figure 4. Contours of nuclear probability density for state 0° (left) and state $6^2(5/2)$.

301

For state (a), which corresponds to the vibrationless level, we see that quadratic coupling modulates nuclear probability density (which is cylindrically symmetric for linear coupling alone) by causing this density to accumulate over surface depressions. Even though the j quantum number from the linear nonadiabatic problem is no longer strictly good, sorting the eigenvectors by their basis state composition shows that, by and large, linear-j labels carry over, eigenlevel by eigenlevel, as indicators of predominant composition. Thus the vibrational ground state is dominated by the lowest basis function of the internal rotor ground state, $j = 1/2$, block. We find that other density distributions for similarly rotationless wavefunctions are mildly localized over surface wells. By contrast, density distributions for states with vibrational angular momentum such as (b) calculated for $6^2(5/2)$ localize over surface maxima. In classical language the pseudorotation slows as it passes over the barriers.

To summarize, sym-triazine offers remarkable simplicity in an isolated nonadiabatic system. With its single active mode and large linear stabilization, it is possible to easily visualize the dynamics of its vibronic pseudorotation in the adiabatic limit. From the precise correspondence between high resolution jet spectroscopy and theory, when carried to second order in coupling terms, emerges a quantitative picture of the role of quadratic effects and the loss of cylindrical symmetry in splitting and mixing vibronic levels.

IV.   The extraordinary intramolecular dynamics of the Rydberg-core $^2E'$ cyclopropane cation

In its electronic ground state cyclopropane is readily acknowledged as among the most rigid of polyatomic hydrocarbons. In its point group, D_{3h}, electron ejection yields a degenerate $^2E'$ ground state cation. Indeed, to the extent that its hydrogens can be neglected, cyclopropane constitutes a triangular molecule, and its degenerate excited states (all Rydberg) have been considered ideal test cases for parameterized vibronic coupling theories.[41]

There exist, however, portents of complexity in the making of a cation (or Rydberg state) from cyclopropane. The highest occupied molecular orbital is $(3e')^4$. Ionization to $(3e')^3$ yields the $^2E'$ cation, but, in the process, removes a σ-bonding electron from the C-C network. Thus, we might expect an impact on vibrational structure beyond that associated with the electronic degeneracy.

Several calculations have been carried out on the electronic structure of cyclopropane cation.[41,43] The most thorough of these notes a Jahn-Teller distortion along the ring bending normal coordinate, ν_{10}, and implicates as well another e' mode, ν_{11}, the in-plane degenerate CH_2-wag.[43] Other calculations have found frequency lowering as a result of the reduction in C-C bond order on ionization.[44]

These promises of complexity are fulfilled by the absorption spectrum of the 3s $^1E'$ Rydberg state, which here, as with sym-triazine, fits well with the crude vibronic band contours available for the ion from the photoelectron spectrum.[45] Hot bands, weakly observed in the jet, definitively place the origin at 52,964 cm^{-1}. To the blue, one finds an irregular progression of intense low-frequency bands, which form a pattern unfamiliar on the basis of the linearly-coupled single-mode correlation diagram for any coupling parameter.

Closer examination of the hot band spectrum shows why. Figure 5 displays the two-photon absorption spectrum to the red of 52,964 cm^{-1}. Evident with comparable intensity are transitions to the zero-point and first cold-band vibrationally excited state from both of the low-frequency e' modes, ν_{10} (1029 cm^{-1}) and ν_{11} (866 cm^{-1}). The next higher e' hot band, which would originate with v = 1 in ν_9 (1476 cm^{-1}) is undetectable.

The Franck-Condon intensities from ground state e′ vibrational levels to the origin of the excited state gauge the degree to which Jahn-Teller distortion mixes the ground state's two-dimensional isotropic harmonic oscillator basis in the vibronic wavefunctions of the excited states. Thus the hot band intensities tell us immediately that two modes are active to an approximately equal degree in coupling with the electronic degeneracy. Cyclopropane clearly presents a much more complex Jahn-Teller problem than sym-triazine.

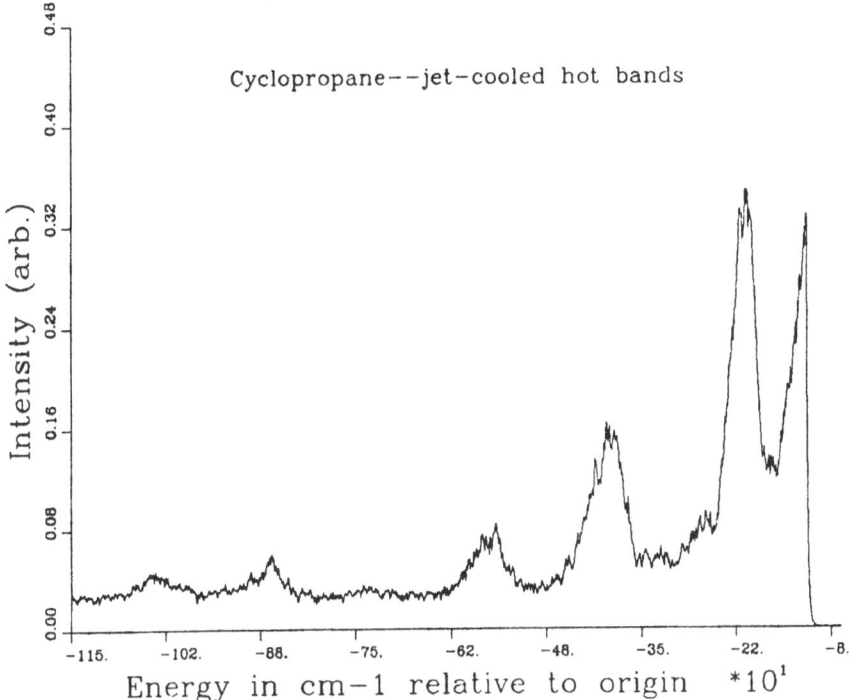

Figure 5. Two-photon absorption spectrum of the hot-bands of the 3s ^{1}E′ Rydberg state of cyclopropane. The band immediately to the red of the origin is the fully Franck-Condon allowed sequence transition, 14^1_1.

The upper frame of Figure 6 below shows the cold-band spectrum. Among the most distinctive features are: (1) the number of bands which have frequencies lower than any ground state fundamental, and (2) the high intensities of these same bands. Significant transfer of intensity to vibronically active bands signals strong linear coupling. Low frequencies could indicate either: (1) large quadratic terms as well, which, as we saw with sym-triazine, further redistribute intensity to higher angular momentum states of the j′ = 1/2 mod 3 block (which fall in energy as the pseudorotation amplitude increases), or (2) in the absence of higher-order coupling, lowered basis frequencies (reduced harmonic restoring forces in the excited state).

303

Preliminary calculations easily show that coupling strengths sufficient to transfer the intensity observed in the low-frequency cold bands produce a density of Franck-Condon active transitions which rises far too fast with energy to fit the comparatively sparse structure found experimentally. The other alternative, strictly linear coupling in two modes with adjustable frequencies (as manifested in the chosen harmonic basis), produces a much more acceptable qualitative fit to the density of observed transitions. Optimization yields the best quantitative fit shown as the lower frame of Figure 6. The parameters are: $\nu_{10} = 441$ cm^{-1}, k = 1.59, $\nu_{11} = 726$ cm^{-1}, $\ell = 2.87$.

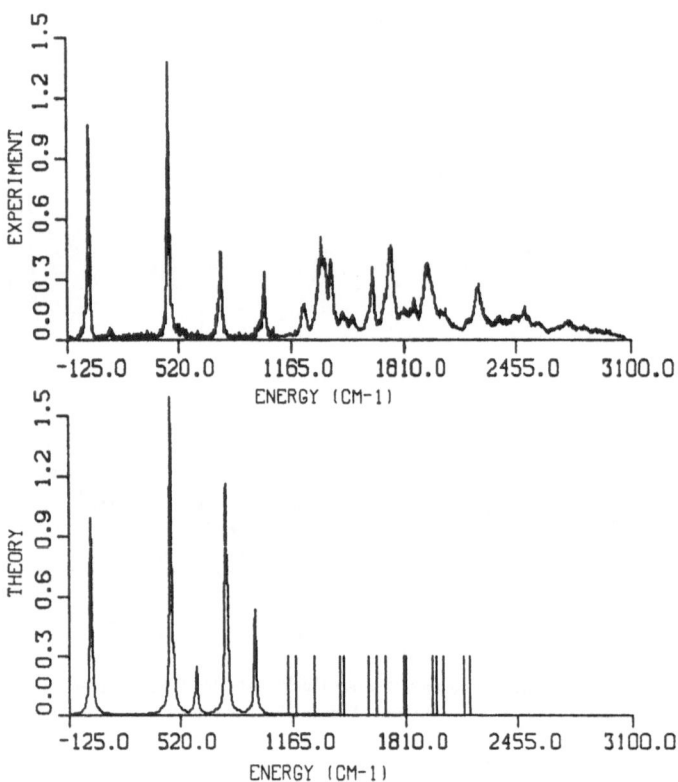

Figure 6. Two-photon absorption spectrum of the cold bands of the 3s ^{1}E′ Rydberg state (top), with theoretical simulation of positions and intensities for linear-coupling parameters and excited-state frequencies, k = 1.59, $\ell = 2.87$, $\nu_{10} = 441$ cm^{-1}, $\nu_{11} = 726$ cm^{-1}. Eigenvalues above 1000 cm^{-1}, which are to be taken as approximate, are marked by sticks of arbitrary height because intensities fail to meet a 0.1 percent convergence criterion.

While its line-for-line correspondence with experiment is imperfect above 1000 cm^{-1}, the model accounts well for the number of transitions observed, it predicts the positions and intensities of the first few bands with quantitative accuracy, and it also correctly returns the observed hot-band intensities (as measured at room temperature). Thus, the spectrum observed is explained in terms of a moderate degree of Jahn-Teller distortion in the two normal coordinates directly implicated by the hot band spectrum, together with a degree of frequency lowering well supported by ab initio calculations.

This same model additionally places a far greater number of states of higher angular momentum (j > 1/2) throughout the region of this spectrum. The inclusion of even very small higher-order off-diagonal terms would transfer significant intensity to transitions involving these states. That such transitions are apparently absent in the observed spectrum has profound significance for the dynamics of the cyclopropane cation core.

We illustrate these extraordinary dynamics by first tracing a full cycle of the pseudorotation phase at the equilibrium amplitudes of the distortion coordinates, following the minimum energy on the lower adiabatic surface. Figure 7 shows this path as mapped on the symmetry coordinates for C_3 ring-bend and CH_2-wag. Here the relative phases are maintained at zero through the complete cycles in both normal coordinates. The significance of the spectrum, with its apparent absence of intensity redistribution, is the following: Without higher order coupling, potential barriers between any of these configurations are zero, that is, the vibronic pseudorotation is unhindered.

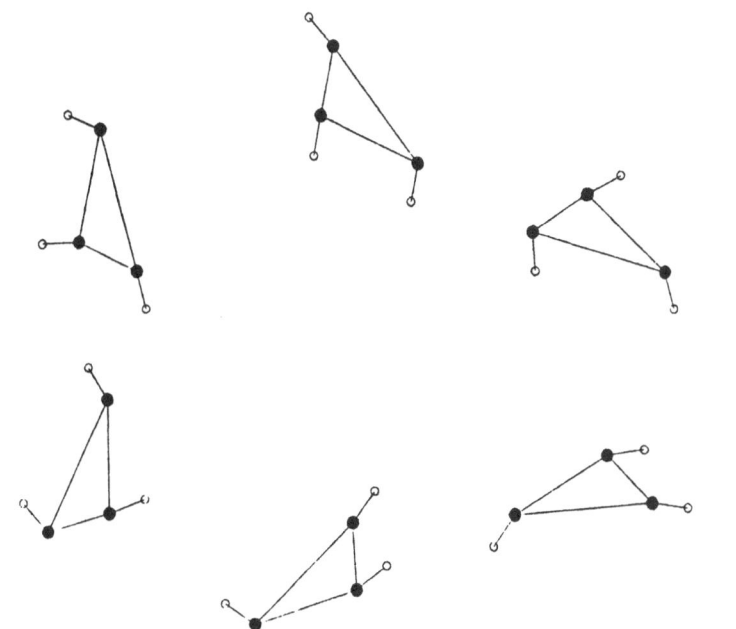

Figure 7. Cartesian representation of full cycles in the pseudorotation phase angles for cyclopropane normal coordinates ν_{10} and ν_{11} at distortion amplitudes corresponding to the minimum/energy path. Relative pseudorotation phase arbitrarily maintained at zero degrees.

Now, inspection of the minimum energy path on the lower adiabatic surface of this highly nonadiabatic system gives only the most primitive view of the vibrational dynamics even for the zero-point level. As the molecule undergoes real vibration, the pseudorotation phases in the two normal coordinates will not be locked, though our calculation shows a substantial barrier to the free oscillation of relative phase. Moreover, we can expect the dynamics of higher vibrational states to be increasingly enriched by coupling with the electronic degrees of freedom.[16,46]

Acknowledgement

This work was supported by the National Science Foundation under Grant CHE-8614703.

References

1. G. Herzberg, *Molecular Spectra and Molecular Structure I. Spectra of Diatomic Molecules* (Van Nostrand Reinhold, New York, 1950); G. Herzberg, *Molecular Spectra and Molecular Structure III. Electronic Spectra of Polyatomic Molecules* (Van Nostrand Reinhold, New York, 1966).
2. J. Kommandeur, W. A. Majewski, W. L. Meerts and D. W. Pratt, *Ann. Rev. Phys. Chem.* **38**, 433 (1987).
3. See for example papers in *Faraday Discuss. Chem. Soc.* **75**, (1983).
4. J. Jortner and S. Leach, *J. Chim. Phys.* **77**, 7 (1980).
5. C. H. Greene and Ch. Jungen, *Adv. At. Mol. Phys.* **21**, 51 (1985).
6. R. S. Mulliken, *Acc. Chem. Res.* **9**, 7 (1976).
7. A. Hodgson, J. P. Simons, M. N. R. Ashfold, J. M. Bayley and R. N. Dixon, *Mol. Phys.* **54**, 351 (1985).
8. M. B. Robin, *Higher Excited States of Polyatomic Molecules* (Academic, New York, 1974).
9. R. L. Whetten, K. J. Fu and E. R. Grant, *J. Chem. Phys.* **79**, 2626 (1983).
10. R. L. Whetten and E. R. Grant, *J. Chem. Phys.* **80**, 1711 (1984).
11. R. L. Whetten and E. R. Grant, *J. Chem. Phys.* **81**, 691 (1984).
12. S. G. Grubb, R. L. Whetten, A. C. Albrecht, E. R. Grant, *Chem. Phys. Lett.* **108**, 420 (1984).
13. R. L. Whetten, K. J. Fu and E. R. Grant, *Chem. Phys.* **90**, 155 (1984).
14. R. L. Whetten, S. Grubb, C. E. Otis, A. C. Albrecht and E. R. Grant, *J. Chem. Phys.* **82**, 1115 (1985).
15. S. G. Grubb, C. E. Otis, R. L. Whetten, E. R. Grant and A. C. Albrecht, *J. Chem. Phys.* **82**, 1135 (1985).
16. R. L. Whetten, G. S. Ezra and E. R. Grant, *Ann. Rev. Phys. Chem.* **36**, 277 (1985).
17. R. L. Whetten and E. R. Grant, *J. Chem. Phys.* **84**, 654 (1986).
18. R. L. Whetten, K. S. Haber and E. R. Grant, *J. Chem. Phys.* **84**, 1270 (1986).
19. L. Bigio and E. R. Grant, *J. Chem. Phys.* **83**, 5361 (1985).
20. For references pertaining to specific systems see: K. P. Huber and G. Herzberg, *Molecular Spectra and Molecular Structure IV. Constants of Diatomic Molecules* (Van Nostrand Reinhold, New York, 1979).
21. See: G. Herzberg and Ch. Jungen, *J. Chem. Phys.* **84**, 1181 (1986).
22. G. Herzberg, *Ann. Rev. Phys. Chem.* **38**, 27 (1987).
23. See for example: D. Stahel, M. Leoni and K. Dressler, *J. Chem. Phys.* **79**, 2541 (1983); S. T. Pratt, P. M. Dehmer and J. L. Dehmer, *J. Chem. Phys.* **81**, 3444 (1984).
24. P. J. H. Tjossem, T. A. Cool, D. A. Webb and E. R. Grant, *J. Chem. Phys.* **88**, xxx (1988).
25. C. S. Gudeman and R. J. Saykally, *Ann. Rev. Phys. Chem.* **35**, 387 (1984).
26. R. S. Tapper, R. L. Whetten, G. S. Ezra and E. R. Grant, *J. Phys. Chem.* **88**, 1273 (1984).
27. K. S. Haber, J. W. Zwanziger, F. X. Campos, R. T. Wiedmann and E. R. Grant, *Chem. Phys. Lett.*, in press.
28. C. Cossart-Magos, S. Leach, M. Eidelsberg, F. Launay and F. Rostas, *J. Chem. Soc. Faraday Trans II*, **78**, 1477 (1982).
29. S. R. Long, J. T. Meek and J. P. Reilly, *J. Chem. Phys.*, **79**, 3206 (1983).
30. T. M. Orlando, S. L. Anderson, J. R. Appling and M. G. White, *J. Chem. Phys.* **87**, 852 (1987).
31. J. M. Wiesenfeld and B. I. Greene, *Phys. Rev. Lett.* **51**, 1745 (1983).
32. W. Y. Cheung, W. A. Chupka, S. D. Colson, D. Gauyacq, P. Avouris and J. J. Wynne, *J. Chem. Phys.* **78**, 3625 (1983).
33. K. S. Viswanathan, E. Sekreta, and J. P. Reilly, *J. Phys. Chem.* **90**, 5658 (1986), and references therein.

34. S. T. Pratt, P. M. Dehmer and J. L. Dehmer, *J. Chem. Phys.* **85**, 3379 (1986).
35. S. T. Pratt, P. M. Dehmer and J. L. Dehmer, *J. Chem. Phys.* **80**, 1706 (1984).
36. B. G. Koenders, D. M. Wieringa, K. E. Drabe and C. A. DeLange, *Chem. Phys.* **118**, 113 (1987).
37. C. Fridh, L. Asbrink, B. O. Jonsson and E. Lindholm, *Int. J. Mass. Spectrom. Ion. Phys.* **8**, 85 (1972).
38. H. A. Jahn and E. Teller, *Proc. R. Soc. London Ser. A* **161**, 220 (1937).
39. H. C. Longuet-Higgins, *Adv. Spec.* **2**, 429 (1961); A. D. Liehr, *J. Phys. Chem.* 67, 389, 471 (1963).
40. See: J. W. Zwanziger and E. R. Grant, *J. Chem. Phys.* **87**, 2954 (1987), and references therein.
41. W. Duch and G. A. Segal, *J. Chem. Phys.* **76**, 2951 (1983).
42. K. Ohta, H. Nakatsiji, H. Kubodera and T. Shida, *Chem. Phys.* **76**, 271 (1983).
43. J. R. Collins and G. A. Gallup, *J. Am. Chem. Soc.* **104**, 1530 (1982).
44. J. D. Shao, T. Baer, J. C. Morrow and M. L. Fraser-Monterio, *J. Chem. Phys.* **87**, 5242 (1987).
45. H. Basch, M. B. Robin, N. A. Kuebler, C. Baker and D. W. Turner, *J. Chem. Phys.* **51**, 52 (1969).
46. J. W. Zwanziger, E. R. Grant and G. S. Ezra, *J. Chem. Phys.* **85**, 2089 (1986).

PRODUCTION AND RELAXATION OF NEGATIVE CLUSTER IONS
BY USE OF HIGH-RYDBERG RARE GAS ATOMS

Tamotsu Kondow and Kozo Kuchitsu

INTRODUCTION

Electron attachment to a van der Waals cluster leading to formation of a cluster anion has attracted much attention, since it provides valuable information on the dynamical processes characteristic of cluster systems, which are related closely to those encountered in particle collisions in the gas phase and in relaxation processes in the condensed phase. Studies of electron attachment phenomena have recently made remarkable progress, partly because of the advancement in techniques for efficient production of cluster anions.[1-5] We have developed a novel method of gentle and efficient production of cluster anions by transfer of the outermost electron (Rydberg electron) of a high-Rydberg rare gas atom to a van der Waals cluster and applied this method to a variety of van der Waals cluster systems.[1] In the present paper, we describe the essential features of the experimental techniques and recent results.

BASIC PRINCIPLE AND IONIZATION PROCESS

The outermost electron (Rydberg electron) of a high-Rydberg atom, Rg**, has a very small kinetic energy, ranging from 10 to 20 meV which corresponds to the principal quantum number of 20-40; it is collisionally transferred to a van der Waals cluster, $(M)_m$, if its vertical electron affinity is positive. This process can be expressed as

$$(M)_m + Rg^{**} \longrightarrow (M)_m^{-*} + Rg^+ \text{ (electron transfer)} \qquad (1)$$

$$(M)_m^{-*} \longrightarrow (M)_n^- + (m-n)M \text{ (evaporation)} \qquad (2)$$

In the first step of the electron transfer, the Rydberg electron is supposed to be accommodated in an extended affinity state, i.e., a state with an electron not localized in any of the single component molecules. Such an excited cluster ion is represented by $(M)_m^{-*}$, where $(M)_m^{-*}$ retains the geometry of $(M)_m$, since the electronic motion is much faster than the nuclear motion causing geometrical rearrangement. The cluster system tends to relax toward its stable state by localization of the captured electron accompanied by nuclear rearrangement. The excess energy associated with the relaxation is transmitted to intracluster modes; i.e., the cluster is "heated". When the excess energy shared by each intermolecular van der Waals bond is sufficiently low, it is retained in the cluster system and no evaporation is expected (non-evaporative). Otherwise, evaporation occurs, and a part of the excess energy is released out of the cluster system by evaporation (evaporative).

Since the Rydberg electron can be regarded as a free electron whose momentum and kinetic energy are identical with those of the Rydberg electron (essentially free electron model),[6] the electron transfer process (process(1)) can be treated as capture of the free electron by a cluster particle. The electron capture mechanism has been elucidated theoretically by Tsukada et al. on the basis of the electronic structure theory and a strong-coupled electron-phonon picture.[7] Suppose that a free electron having a wave vector, k, and a kinetic energy, $\varepsilon(k)$, is captured in an affinity state i having an energy of ε_i. Then the cross section, σ, for process (1) is given by,[7]

$$\sigma \simeq \frac{2\pi}{\hbar v} \sum_i |V^i(k)|^2 \frac{e^{-\varepsilon_{ki}^2/4\Gamma_i^2}}{2\sqrt{\pi}\Gamma_i} \exp\left[-\frac{2\pi}{\hbar}<|V^i(k)|^2>\rho(\varepsilon_i)\tau_c^i\right]. \qquad (3)$$

where v is the velocity of the incoming electron, $\rho(\varepsilon)$ is the density of states of a free electron and τ_c^i is the escape time from the reemission zone defined in reference 7, $<\ >$ represents the average over k contributing significantly to the summation, and Γ_i is the width of the affinity level arising from interaction of the electron with the cluster vibration involved. The coupling constant, $V^i(k)$, can be approximated by

$$|V_i(k)|^2 \sim \frac{m_s}{m}\left(\frac{z_B-z_s}{2z_B}\right)^2 W^2\Omega. \qquad (4)$$

Here, m is the total number of the component molecules (cluster size), m_s is the number of the surface molecules, Z_B and Z_s represent the coordination numbers of the bulk site and the surface site, respectively, W is the band width of the affinity level in the bulk, and Ω has an order of the volume of the component molecule. The width of the affinity level i, Γ_i, is given by the following approximate relation:

$$<\Gamma_i>^2 \simeq <\Delta E_i><\hbar\omega> \qquad (5)$$

In the preceding relation, $<\Delta E_1>$ represents the excess energy associated with the relaxation, and $<\hbar\omega>$ is the energy of the vibrational mode involved in the electron attachment. Furthermore, the escape time, τ_c^1, is approximately proportional to $<\omega>^{-1}$.

In the absence of significant evaporation, the theoretical cross section, σ, given by eq. (3) turns out to be the cross section for electron attachment to the cluster. This theoretical equation can predict the dependences of the cross section on (1) the cluster size, m, and on the kinetic energy of the incoming electron, (2) the energies of the affinity levels related to the vertical electron affinity, and (3) the vibrational modes involved in the electron attachment. As shown in the following section, the width of the affinity level, Γ, and the escape time, τ_c, can be deduced by a comparison of the m-dependence of the measured cross section with the theoretical expression given by eq. (3). These parameters provide important information on the vibrational modes and the negative ion sites involved in the electron attachment process. Ionization of ammonia clusters, $(NH_3)_m$, and carbon dioxide clusters, $(CO_2)_m$, are discussed in the following sections as typical examples of non-evaporative and evaporative electron attachment processes, respectively.

EXPERIMENTAL ASPECTS

The basic principle of the present ionization method is to transfer Rydberg electrons having kinetic energies of $\sim 10^{-1}$ meV to van der Waals clusters. This method of ionization has such advantages that the ions are produced efficiently and that disturbance due to the ionization is minimized because of the very small kinetic energies of the Rydberg electrons.

An ion source for formation of cluster anions has been designed: A high-Rydberg atom, Rg**, is produced in the ion source (see Fig.1) by electron impact on the rare gas, Rg. A cluster particle produced in a supersonic expansion is allowed to collide with Rg** inside G_A of the ion

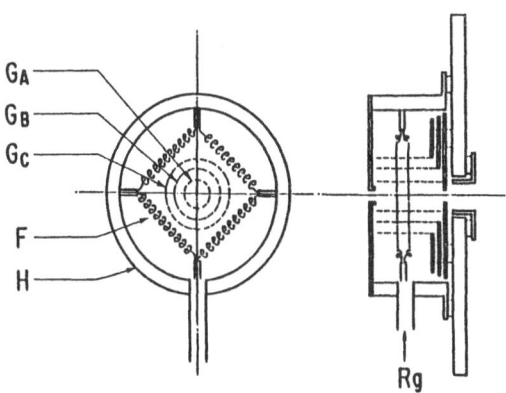

Fig.1 Schematic diagram of the triple grid ion source. Three cyclindrical grids, G_A, G_B and G_C are mounted concentrically. Filaments, F, are located outside of the grids. H denotes a housing. (Reprinted with permission from reference 1)

source and negative ions produced are extracted out of the collision region
and are detected by a conversion dynode and a ceratron electron multiplier
after mass-selection by a quadrupole mass spectrometer (Extrel, 162-8).
The grids, G_A, G_B and G_C, are made of stainless-steel mesh with a
transparency of 80%. The collision region surrounded by the inner grid G_A
has a length of 20 mm and a diameter of 10 mm. All the grids are insulated
by ruby balls of 3 mm diameter. The housing is isolated from the grids by
spacers made of talc porcelain. The four pieces of helical filaments made
of thoriated tungsten wire of 0.15 mm diameter form a rectangular square.

Argon or krypton gas with more than 95% purity is introduced to the
ion source through a stainless steel pipe of 4 mm diameter. The rare gas
atoms are excited in the exterior of G_C by 50 eV electrons. Ionic species
and electrons are retarded by application of appropriate potentials to the
three grids, and only neutral species including Rg** are allowed to enter
the central collision region. Typical potentials applied are -20, -150,
-50, and -100 V for G_A, G_B, G_C and the filaments, respectively.

The pushing pressure of the rare gases introduced to the ion source is
typically 0.05 - 0.2 torr. Test experiments have shown that single
collision conditions are found to be fulfilled in this pressure region and
that the observed negative ions originate from transfer of the Rydberg
electrons to the nertral clusters in collision with Rg**.

The principal quantum numbers, n_p, of the Rg** atoms are estimated to
be in the range of 25 - 35, since Rg** atoms having $n_p \sim 35$ are ionized by
the field of 430 V/cm between G_A and G_B, while those having $n_p \sim 25$ decay
radiatively before reaching the collision region.[8]

The transmission and detection efficiencies of the mass spectrometer
and the detector are calibrated by use of the known fragmentation patterns
of the positive and negative ions produced by electron impact on
perfluorokerosene (PFK),[9] The mass-to-charge ratios are calibrated by the
fragment ions of PFK and the cluster ions of CO_2.

NON-EVAPORATIVE ELECTRON ATTACHMENT

The ammonia cluster, $(NH_3)_m$, produced in a supersonic expansion is
collisionally ionized by a high-Rydberg krypton atom, Kr**, produced by
electron impact on Kr gas. Figure 2 shows a typical mass spectrum of the
ions thus produced: $(NH_3)_n^-$, is the only product ion. Its intensity starts
to rise at n=37, reaches a broad maximum at 50-60 and decreases with
increasing n.

The energetics of the electron attachment processes involving $(NH_3)_m$
indicates that no significant evaporation occurs for the following reason.
The excess energy is estimated from the energy of the solvated electron

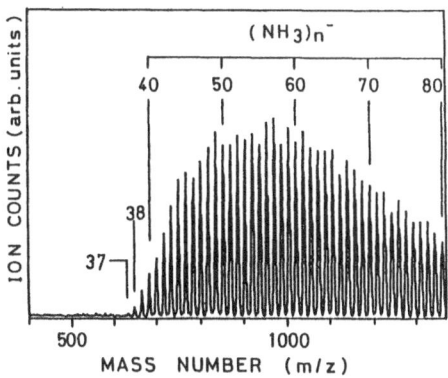

Fig.2 Mass spectrum of
$(NH_3)_n^-$ produced from
$(NH_3)_m$ in collision with
Kr**. The numbers in
the figure represent the
cluster size, n.

state in liquid ammonia, which is about 2 eV.[10] The solvated state in the
cluster system is not necessarily the state directly involved in the
ionization process, but the energy of this state can be used for the
present purpose, because this energy provides the upperbound of the excess
energy. In the electron attachment to a cluster having a threshold size,
m=37, an energy of about 0.01 eV is transmitted to one intermolecular mode,
if the excess energy is distributed statistically among (6m-6)
intermolecular vibrational modes. This energy is much smaller than the
energy, 0.24-0.2 eV, required to remove one NH_3 molecule from NH_3 liquid or
to dissociate a NH_3 dimer.[11,12] It is concluded, therefore, that the
evaporation does not occur significantly, that is,

$$Kr** + (NH_3)_m \longrightarrow Kr^+ + (NH_3)_m^- \qquad (6)$$

Because of the absence of significant evaporation, the size-dependence
of the electron attachment cross section, σ, can be deduced from the size-
dependence of the ion intensity and that of the neutral cluster. In the
size-range concerned, the size-distribution of the neutral cluster is a
smoothly decreasing function with its size and can be estimated from the
size-dependence of the positive cluster ion, $(NH_3)_{n-1}H^+$.[13] Figure 3 shows
the size-dependence of the relative cross section, σ, thus obtained. The
cross section, σ, starts to rise at n=37 and tends to level off with
increasing n. In the later discussion, n is replaced by m because of the
absence of significant evaporation.

The m-dependence, $\sigma(m)$, can be explained in terms of eq.(3) given by
Tsukada et al.[7]. As m increases, the energies of the affinity levels $\{\varepsilon_i\}$
which are originally negative turn to be positive at a critical size, m_c.
According to the increase in $\{\varepsilon_i\}$, the m-dependence of the cross section
changes as given in eq.(3). For $m < m_c$, i.e. $\{\varepsilon_i\} > \varepsilon(k)$, $\sum_i \exp(-\varepsilon_{ki}^2/4\Gamma_i^2)$ in

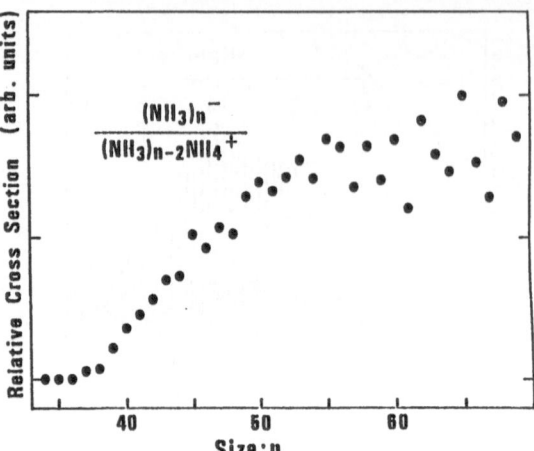

Fig.3 The relative cross section for the electron attachment to a NH$_3$ cluster is plotted as a function of the cluster size. Here, the cross section is obtained from the intensity of the negative ion (NH$_3$)$_n^-$ divided by that of the positive ion, (NH$_3$)$_{n-2}$NH$_4^+$ (see text).

eq.(3) can be approximated by $\exp(\varepsilon^2/4\Gamma^2)$ for a rough estimation of $\sigma(m)$. Here, ε is $\varepsilon(k)$-(the energy of the lowest affinity level). On the other hand, for $m \geq m_c$ there always exists an affinity level in $\{\varepsilon_i\}$ which satisfies $\varepsilon_{ki}=\varepsilon(k)-\varepsilon_i=0$ and contributes dominantly to the summation of eq.(3). Therefore, $\sum_i \exp(-\varepsilon_{ki}^2/4\Gamma_i^2)$ can be set equal to unity. In addition, ε may be linearly dependent on $m^{-1/3}$, as shown in the m-dependence of the adiabatic electron affinity of (NH$_3$)$_m$.[14] By taking these arguments into consideration, one obtains the following approximate size-dependence of the cross section:

$$\ln \sigma \simeq C_0 - \frac{1}{4\Gamma^2}(\varepsilon_0 + \varepsilon_1 m^{-1/3})^2 - C_1 \rho \tau_c m^{-1/3} \quad (m < m_c) \qquad (7)$$

$$\simeq C_0 - C_1 m^{-1/3} \qquad\qquad\qquad (m \geq m_c), \qquad (8)$$

where C_0 and C_1 are almost independent of m, and ε_0 and ε_1 are the parameters associated with the affinity energy of (NH$_3$)$_m$.[14]

The experimental cross sections follow eqs(7) and (8), as shown in Fig.4, where $\ln \sigma$ is plotted as a function of $m^{-1/3}$, and hence the

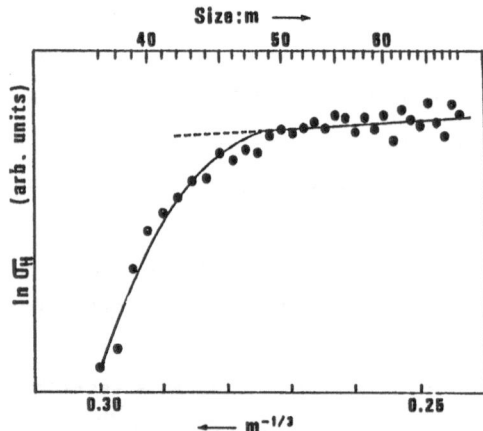

Fig.4 $\ln \sigma$ is plotted as a function of $m^{-1/3}$. The solid curve is obtained from eqs.(7) and (8) by adjusting the parameters C_0, C_1, ε_0 and ε_1. The linear portion gives τ_c, while the remaining part of the curve gives Γ.

parameters proportional to Γ^2 and τ_c can be derived. Evidently, the vertical electron affinity turns to be positive at $m=m_c \sim 50$. The threshold size (m=37) at which the cross section rises sharply is slightly smaller than m_c: Namely, the vertical electron affinity of $(NH_3)_m$ is still negative by Γ at the threshold size.

Similar measurements have been performed by use of the isotope species, $(ND_3)_m$. In this case, the threshold is located approximately at m=45. The isotope effect in the $\sigma_{ND_3}/\sigma_{NH_3}$ - m curve is significant in the vicinity of the threshold and tends to disappear as m increases. The isotope effect appearing in the threshold size and the cross section originates from the isotope effect of the vibrational modes which interact with the electron. Namely, the Γ and τ_c values should have the isotope effect. A similar analysis by use of eqs.(7) and (8) gives the parameters related to Γ and τ_c. In comparison with the data for $(NH_3)_m$, the isotopic ratios Γ_H/Γ_D and τ_c^H/τ_c^D are determined to be 1.4 ± 0.2 and 0.3 ± 0.2, respectively, where only Γ and τ_c are postulated to be dependent on the isotopic change. The isotopic ratio of the vibrational frequencies, ω_H/ω_D is estimated to be less than 2. It is likely that vibrational modes associated with hydrogen, such as hydrogen bonds, are involved in the attachment process.

EVAPORATIVE ELECTRON ATTACHEMENT

Figure 5 shows the mass spectrum of $(CO_2)_n^-$ produced from $(CO_2)_m$ in collision with Kr**. This spectrum has the following characteristic features: (1) The threshold size is 3. (2) The peak intensities are weak in the region of $11 \le n \le 13$. (3) Certain sizes, such as n=9, 14 and 16, are magic numbers.

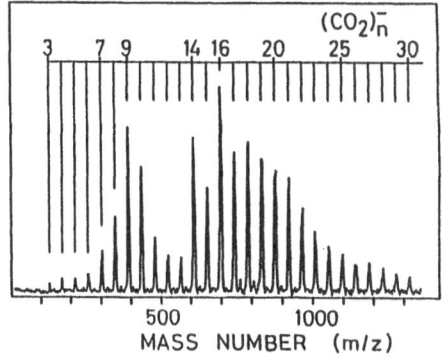

Fig.5 Mass spectrum of $(CO_2)_n^-$ from $(CO_2)_m$ in collision with Kr**. Significant evaporation is evidenced by the presence of the depletion region, $11 \le n \le 13$. (Reprinted with permission from reference 1).

These feature can be explained in terms of the evaporative electron attachment process,

$$Kr^{**} + (CO_2)_m \longrightarrow Kr^+ + (CO_2)_m^{-*} \qquad (9)$$
$$(CO_2)_m^{-*} \longrightarrow (CO_2)_n^- + (m - n)CO_2 \qquad (10)$$

In the cluster ion, $(CO_2)_n^-$, a dimeric cluster ion is likely to be solvated by the rest of the CO_2 component molecules, and hence, the excess energy amounts to 3.5 eV at maximum. A similar calculation cited in the previous section shows that the energy shared by one van der Waals mode is comparable to the evaporation energy from $(CO_2)_n^-$ (0.22 eV per one CO_2 molecule).[15] Therefore, evaporation of CO_2 molecules from $(CO_2)_m^{-*}$ should occur. Feature 2 is ascribable to evaporation of at least 4 CO_2 molecules from $(CO_2)_m^{-*}$. This conclusion is supported from the threshold size for $(CO_2)_n^-$ and $(CO_2)_m^{-*}$. The theory of Tsukada et al.[7] predicts that the threshold size for $(CO_2)_m^{-*}$ is 7. On the other hand, the threshold size for the observed cluster ion is 3, and hence, about 4 CO_2 molecules should be evaporated by electron transfer.

As the stagnation pressure of the nozzle increases, the signal for $(CO_2)_n^-$ starts to rise at a critical pressure, P_A. The P_A value is plotted as a function of the size of $(CO_2)_n^-$ in Fig.6. The P_A value for $(CO_2)_n^+$ is also plotted for comparison. These P_A-n curves indicate that $(CO_2)_n^-$ ($11 \leq n \leq 13$) are not very much stable, while $(CO_2)_n^-$ (n=9, 14, and 16) are stable, as expected from features 2 and 3.

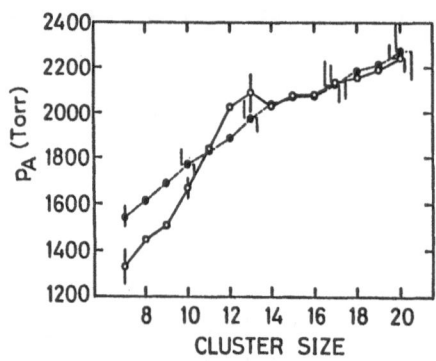

Fig.6 The critical value (—o—) of the stagnation pressure, P_A, is plotted as a function of the size of the cluster ion, $(CO_2)_n^-$. The P_A-value (--●--) for $(CO_2)_n^+$ is also plotted.

CONCLUSION

The method of gentle and efficient production of various cluster anions by use of Rg** impact has been developed and presented. This collisional ionization provides an opportunity to study the dynamics of electron attachment to weakly bound gas-phase clusters. The cluster systems investigated by us are summarized in Table I.

Table I. Negative Chuster Ions Observed by Mass Spectroscopy[a]

M	type	ion(RAI)	n_L	ion(EI)	n_L
CO_2	E	$(CO_2)_n^-$	3	$(CO_2)_n^-$	2.1[b]
OCS		$(OCS)_n^-$	2	no data	
CS_2	E	$(CS_2)_n^-$	1	$(CS_2)_n^-$	1
				$(CS_2)_nS^-$	0
H_2O		$(H_2O)_n^-$	~ 11	$(H_2O)_n^-$	11[c]
				$(H_2O)_nOH^-$	0[c]
D_2O		no data		$(D_2O)_n^-$	12[c]
				$(D_2O)_nOD^-$	0[c]
CD_3CN	N	$(CD_3CN)_n^-$	11	undetectable	
$CH_3CN + H_2O$	E	$(CH_3CN)_nH_2O^-$	~ 7	undetectable	
C_5H_5N	N	$(C_5H_5N)_n^-$	4	$(C_5H_5N)_n^-$	3
$C_5H_5N + H_2O$	E	$(C_5H_5N)_n(H_2O)_{n'}^-$	$n+n'=4$	$(C_5H_5N)_n(H_2O)_{n'}^-$	$n+n'=4$
SF_6	N	$(SF_6)_n^-$	1	$(SF_6)_n^-$	1
	R			SF_5^-	
N_2O	R	$(N_2O)_n^-$	~ 6	$(N_2O)_n^-$	1[d]
		$(N_2O)_n^-O$	1	$(N_2O)_nO^-$	0,0[d]
				$(N_2O)_nNO^-$	3,0[d]
$N_2O + H_2O$	R	$(H_2O)(N_2O)_nO^-$	3	$(H_2O)(N_2O)_nO^-$	1,0[d]
CCl_4	R	$(CCl_4)_n^-$	1	$(CCl_4)_nCl^-$	0
		$(CCl_4)_nCl^-$	0		
NH_3	N	$(NH_3)_n^-$	37	$(NH_3)_nNH_2^-$	
ND_3	N	$(ND_3)_n^-$	~ 45		

[a]RAI and EI stand for Rydberg atom and electron impact, respectively. N=nonevaporative electron attachment, E=evaporative electron attachment, and R=reactive electron attachment. [b]A. Stamatovic, K. Leiter, W. Ritter, K. Stephen, T. D. Märk, J. Chem. Phys. **83**, 2942 (1985). [c]M. Knapp, O. Echt, D. Kreisle, E. Recknagel, J. Chem. Phys. **85**, 636 (1986). [d]M. Knapp, O. Echt, D. Kreisle, T. D. Märk, E. Recknagel, Chem. Phys. Lett. **126**, 225 (1986).

ACKNOWLEDGEMENT

The authors are grateful to Professor M. Tsukada for his helpful discussion, and to Messrs F. Misaizu, H. Tada and S. Yamamoto for their collaboration. The present work has been supported by a Grant-in-Aid for Scientific Research by the Ministry of Education, Science and Culture of Japan and the Joint Studies Program of the Institute for Molecular Science (1985-1987).

References

(1) T. Kondow, J. Phys. Chem. **91**, 1307 (1987).

(2) M. Knapp, O. Echt, D. Kreisle, E. Recknagel, J. Chem. Phys. **85**, 636 (1986).

(3) A. Stamatovic, K. Leiter, W. Ritter, K. Stephan and T. D. Märk, J. Chem. Phys. **83**, 2942 (1985).

(4) H. Haberland, C. Ludewigt, H.-G. Schindler and D. R. Worsnop, Surf. Sci. **156**, 157 (1985).

(5) K. H. Bowen, G. W. Liesegang, R. A. Sanders and D. R. Herschback, J. Phys. Chem. **87**, 557 (1983).

(6) M. Matsuzawa, in Rydberg States of Atoms and Molecules, eds. R. F. Stebbings and F. B. Dunning (Cambridge Univ. Press, Cambridge, U. K., 1983) p. 267.

(7) M. Tsukada, N. Shima, S. Tsuneyuki, H. Kageshima and T. Kondow, J. Chem. Phys. **87**, 3927 (1987).

(8) T. F. Gallagher, in Rydberg States of Atoms and Molecules, eds. R. F. Stebbings and F. B. Dunning (Cambridge Univ. Press, Cambridge, U. K., 1983) p. 165.

(9) R. S. Gohlke and L. H. Thompson, Anal. Chem. **40**, 1004 (1968).

(10) J. Jortner, J. Chem. Phys. **30**, 839 (1959).

(11) JANAF, Thermochemical Tables, Dow Chemical Co. (1965 -)

(12) P. Kollman, J. Mckelvey, A. Johansson, and S. Rothenberg, J. Am. Chem. Soc. **97**, 955 (1975).

(13) O. Echt, P. D. Dao, S. Morgan and A. W. Castleman, Jr., J. Chem. Phys. **82**, 4076 (1985).

(14) P. Stampfli and K. H. Bennemann, Phys. Rev. Lett. **58**, 2635 (1987).

(15) M. L. Alexander, M. A. Johnson, N. E. Levinger and W. C. Lineberger, Phys. Rev. Lett. **57**, 976 (1986).

EXCITATION AND IONIZATION OF HIGH n H-ATOMS BY

MONOCHROMATIC AND BICHROMATIC MICROWAVE FIELDS

R. Blümel and U. Smilansky

Max-Planck-Institute for Quantum Optics, 8046 Garching, FRG
and
Department of Nuclear Physics
The Weizmann Institute of Science, 76100 Rehovot, Israel

ABSTRACT

We present recent theoretical results on the response of high n Rydberg atoms to strong microwave fields. For monochromatic driving we compare the measured ionization data with the classical and quantum mechanical descriptions of the ionization process, and identify features of genuine quantal origin. Extending the formalism to allow for fields with two non-commensurate frequencies, we show that many of the features which characterize the ionization by monochromatic fields persist in the regime of quasi-periodic driving. The theoretical ionization probabilities and thresholds agree very well with the experimental findings.

INTRODUCTION

The ionization of high n H-atoms by microwave fields is relevant to central issues in two seemingly unrelated fields of physics – it is a study case for the "quantum chaos" community, and it is the most transparent example for the study of multi-photon processes in free atoms. Because of the relative simplicity of this problem, and because of the availability of accurate and comprehensive experimental data,[1-5] this subject attracted much attention[1-15] and received the status of a paradigme in both these fields.

In the present talk, we would like to report on the recent progress in our understanding of microwave induced ionization of Rydberg H-atoms. We shall divide the discussion into two parts. In the first, we shall briefly discuss the response of the Rydberg atom to a monochromatic microwave field. We shall explain the sudden onset of ionization as the field strength reaches a threshold value, and discuss the dependence of the threshold field on the other relevant parameters. A comparison of the classical and quantum mechanical descriptions[13] shows that in the regime of field parameters studied experimentally, the two approaches reproduce the data equally well[4,7,8] (but for special features of quantum origin[10,13,15] which are of course lacking in the classical theory).

The report on the response of the atom to bichromatic (quasi-periodic) fields is of more pioneering nature. Experiments are being conducted now[16-18] and their results have been reported only very recently. From the theoretical point of view, this system bears on a fundamental issue concerning the nature of the response of a quantum system to a quasi-periodic perturbation. When the second frequency is switched on, the system ceases to be invariant under discrete time translations and hence, Floquet's theorem or any generalization thereof does not necessarily hold. This may have severe consequences for the way the system evolves, and may change drastically such observables as e.g. the rate of energy transferred by the field.[19] Several numerical and analytical studies of this problem were recently published,[19-22] and interesting new features were observed. Still, the issue is far from being settled theoretically and it deserves further study. We shall present our results on the bichromatic microwave ionization of the Rydberg H-atoms in this context and try to show how our results add to the understanding of the general problem.

Before plunging into the theoretical discussion we shall introduce some of the basic concepts by describing a typical experiment.[2-5]

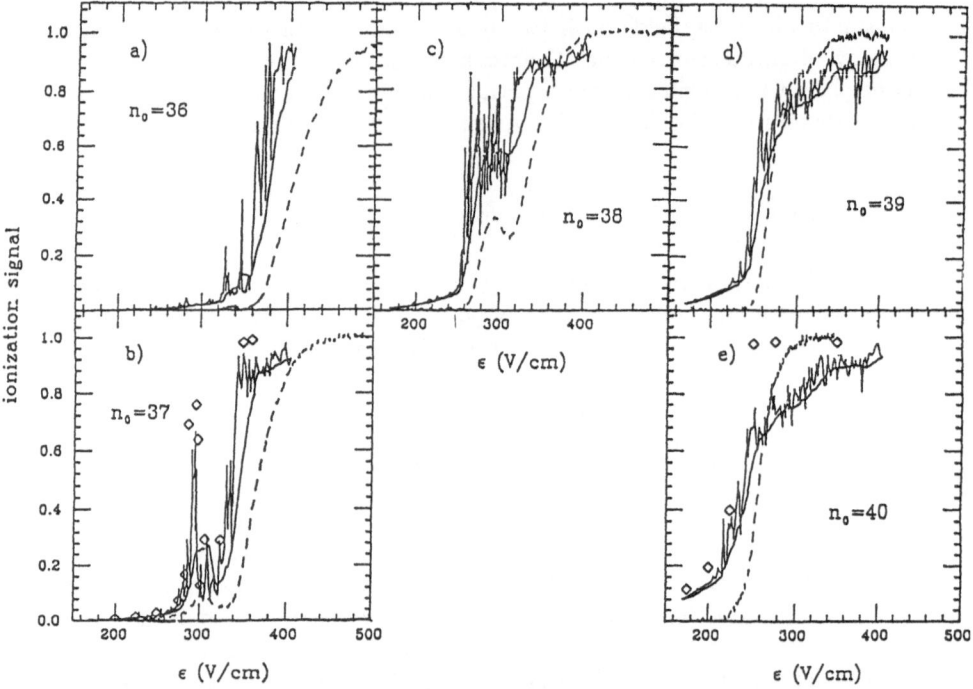

Fig. 1 . Theoretical (spiky full line) and experimental[3-5] (dashed lines) ionization curves for $36 < n_0 < 40$. The full line interpolating the spiky one is the theoretical results smoothed over the experimental field distribution. The diamonds in (b) and (e) are results of large basis calculations.

A beam of neutral H-atoms is excited to a specific high n state and then collimated to a microwave cavity, where it is exposed to a time dependent linearly polarized electric field. But for the regions of the entrance and exit holes, the field is spatially homogeneous and is characterized by its frequency (or two frequencies) and corresponding peak amplitude(s). In the present talk we shall use atomic units throughout and denote the frequencies and field strengths by ω_i and ϵ_i, respectively ($i = 1, 2$). The passage time of the atoms through the cavity is typically of the order of a few hundred field cycles.

One measures the ionization probability $P_I(\epsilon_1, \epsilon_2, \omega_1, \omega_2, n_0)$ keeping the frequencies fixed and varying the field strengths, for a range of initial quantum numbers n_0. At low fields there is no ionization ($P_I \approx 0$), while for high fields one achieves complete ionization ($P_I \approx 1$.) The transition between these extremes occurs rather abruptly, and we characterize the onset of ionization by the field values for which P_I reaches the value of 10% for the first time. (Note that P_I is not necessarily a monotonic function). The threshold values depend on the frequencies and the initial state. Some typical measured ionization curves are shown in Fig. 1, where both monotonic and structured ionization curves are shown. In the following theoretical discussion we shall present a simple physical framework in which the experimental findings can be easily interpreted. The picture is corroborated by detailed quantum and classical calculations. The description of the methods used to solve the quantum model is given in ref. 10). For the present paper it will suffice to say that we rely on the assumption that the extremal Stark states determine the ionization near threshold. We therefore replace the 3-dim. atom by a 1-dim. model governed by the hamiltonian:

$$H(z, t) = H_0 + z(\epsilon_1 \sin(\omega_1 t + \phi) + \epsilon_2 \sin(\omega_2 t)) \qquad (1.1)$$

$$H_0 = \begin{cases} \frac{p^2}{2} - \frac{1}{z} & z > 0 \\ \infty & z \leq 0 \end{cases} \qquad (1.2)$$

Here p and z are momentum and position of the atomic electron in field direction and ϕ is the relative phase of the two fields.

The numerical solution of the dynamics (classical or quantum mechanical) provides us with the theoretical ionization curves and thresholds to be compared with the experimental data.

MONOCHROMATIC DRIVING

In spite of the apparent simple form of the hamiltonian (1.1), the dynamics which it induces is very complicated. The excitation and further ionization depends crucially on the initial quantum number n_0, and in particular on the value ωn_0^3. Classically this is the ratio between the field- and the orbital (Kepler)-frequency. Quantum mechanically it expresses the ratio between the photon energy and the transition energy between the $(n+1)$'th and the n'th levels. Most of the experiments reported so far were performed in the regime $\omega n_0^3 < 1$, and our discussion will concentrate on this regime of parameters.

The atom absorbs energy from the field by first being excited (virtually) to discrete high n states, from which further direct transitions to the continuum are possible. We found that in order to enable ionization, states with $n > n_w = (3\epsilon)^{-1/4}$ must be excited.[10] The susceptibility to ionization changes abruptly at

n_w, and therefore the states with $n > n_w$ can be considered as doorway states for the ionization process. Of the two steps in this description of ionization, the excitation of the doorway states is the slow one, and therefore it is the step which determines the threshold and the rate of ionization. For a qualitative picture of the ionization at threshold, it is sufficient to consider the transitions induced by the field among the bound states of the atom. The threshold field will be determined as the field strength for which states with $n > n_w$ are effectively populated.

The fact that the perturbation is *periodic* is of tremendous help. By Floquet's theorem we know that at any moment

$$|\Psi(t)\rangle = \sum_\alpha e^{i\chi_\alpha t}|\phi_\alpha(t)\rangle\langle\phi_\alpha(0)|n_0\rangle \tag{2.1}$$

where time is expressed in units of the field period and χ_α are the "quasi-energies". $|\phi_\alpha(t)\rangle$ are periodic and we abbreviate the notation by denoting $|\phi_\alpha(0)\rangle$ by $|\alpha\rangle$. If the quasi-energy (qe) spectrum is discrete the wave function $|\Psi(t)\rangle$ evolves in a quasi-periodic way. The qe spectrum is always discrete if the Hilbert space in which the system evolves is of finite dimension. In more general situations this will happen if the qe states $|\alpha\rangle$ are normalizable (localized).

After N cycles of the field the transition amplitude is given by:

$$\langle n|\Psi(N)\rangle = \sum_\alpha \langle n|\alpha\rangle e^{i\chi_\alpha N}\langle\alpha|n_0\rangle \tag{2.2}$$

A transition to a high n states is possible if and only if there exists at least one $|\alpha>$-state which overlaps appreciably with *both* the n_0 and the high n state. This is why the localization properties of the qe states $|\alpha\rangle$ in the $|n\rangle$ basis are of such importance.

Close inspection of the qe states reveals that they fall into two categories (Fig. 2). The "narrow" states are sharply localized on the unperturbed states of the H-atom, and this happens for all n values lower than a critical value n_t. The "broad" qe states overlap with $n > n_t$ states. The expansion amplitudes $\langle n|\alpha\rangle$ vanish at low n, rise abruptly at some n ($> n_t$) from whence they fall off rather slowly (algebraically).[10,11] The Hilbert space spanned by the atomic bound states is of infinite dimension but the results of ref. 10,11) imply that the qe spectrum is always discrete.

If the H-atom is prepared initially in a state with $n_0 < n_t < n_w$ no excitation or ionization would occur. If however $n_0 > n_t$ (but still $n_0 < n_w$), ionization will be allowed since all the qe states of the "broad" class overlap with the doorway states for ionization. The appearance of an ionization threshold field is interpreted in the following way. At low field values n_t exceeds n_0 appreciably, and ionization is prohibited. As the field increases, n_t decreases till it assumes the value n_0. At this field value ionization starts. Detailed numerical results support this picture and reproduce the experimental values (Fig. 3).

The quantal mechanism described above can also be stated in classical terms.[13] Suffices it to say here that in the region of interest, the classical and the quantal descriptions do not differ by much, and account equally well for most of the observed features. There is, however, an exception to this good agreement. Several measured ionization curves show structures which deviate appreciably from the monotonic classical ionization curves. The quantum theory reproduces these "irregularities" (Fig. 1). They can be explained on the following grounds.[11] The qe spectrum and eigen-vectors are obtained by diagonalizing the one-period propagator, for all values of the parameter ϵ. The quasi energies $\chi_\alpha(\epsilon)$ show a typical pattern of avoided crossings (Fig. 4). Consider two qe states $|\alpha\rangle$ and $|\beta\rangle$ which belong to the "narrow" and the "broad" classes respectively. If the initial state of the atom overlaps appreciably with $|\alpha\rangle$, no ionization will occur. Suppose now that by changing the field strength, an accidental avoided crossing between χ_α and χ_β occurs. At field values just below or just above the crossing, the qe states do not change their structure by much, and ionization cannot be observed. At exactly the crossing point and in its narrow vicinity, the qe states are formed from the symmetric and anti-symmetric combinations of $|\alpha\rangle$ and $|\beta\rangle$. Therefore at the point

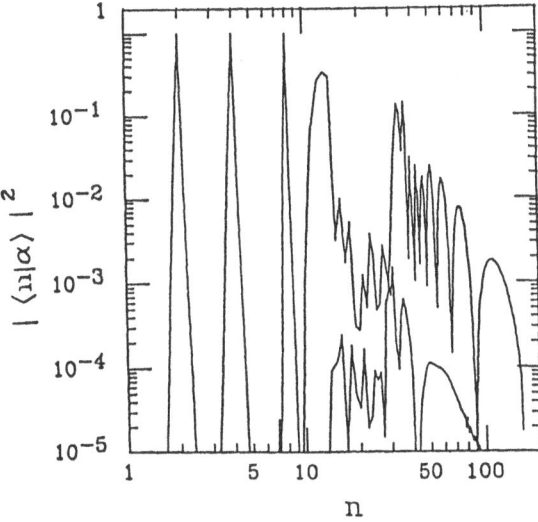

Fig. 2. Five typical quasi-energy states. Here $\epsilon = 10^{-5}$ and $\omega = 5 \cdot 10^{-5}$ (a.u.)

of avoided crossing, states with $n > n_w$ will be populated even though $n_0 < n_t$. This mechanism will cause narrow spikes in the ionization curve, as demonstrated in Fig. 1. The rather broad structures which appear in the experimental ionization curves are due to the clustering of a large number of avoided crossings. The experimental determination of the field is not sharp enough to resolve these fine details. Averaging the theoretical ionization curves to account for the experimental uncertainty, gives rise to structures which occur at the same field values and have the same width and relative heights as observed experimentally (Fig. 1).

To summarize we can say that the only genuine quantal effect in the ionization data is the appearance of the non-regular structures in the ionization curves. Our explanation of this phenomenon is strongly based on the properties of the qe spectrum, whose existence is a direct consequence of the periodic nature of the driving field. In the next chapter we shall see to what extent the Floquet theorem can be generalized to the domain of quasi-periodic driving, and in which ways this will affect the ionization curves.

As an illustration of the remarkable phenomena which may be encountered when a quantum system is exposed to a bichromatic perturbation, consider a two-level atom in an external field. If the field were periodic, the qe spectrum consist of two points and the wave function is quasi-periodic. Switching on a second field with a non-commensurate frequency may change the quasi-periodic nature of the wave function or leave it as it is, depending on the specific properties of the driving force. The exact conditions under which the transition from one mode of response to another occurs are not yet known.

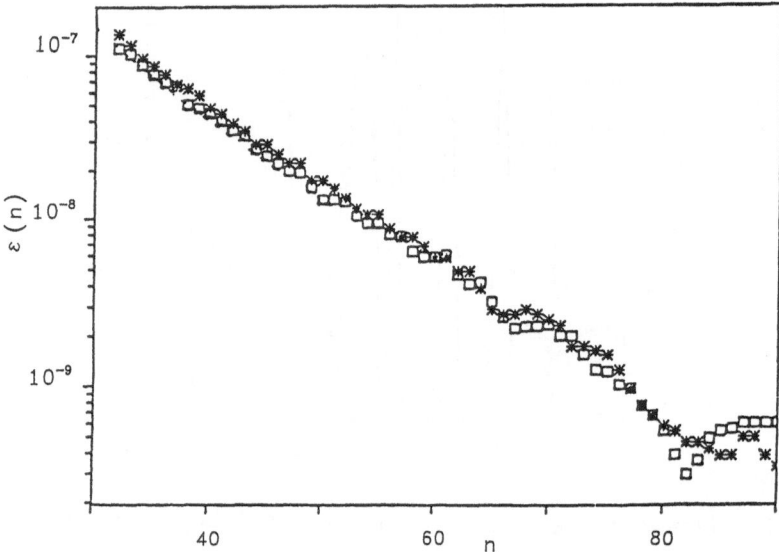

Fig. 3. Experimental (squares) and theoretical (stars) dependence of the ionization threshold field on the initial quantum number n_0 . The experimental values are taken from refs. 3-5). The Quantum theory from refs. 10,11).

It is convenient to consider the Fourier transform of the wave-function in order to characterize the time evolution. If the evolution is quasi-periodic, the Fourier transform vanishes everywhere but for a discrete (denumerable) set of frequencies. If the support of the Fourier transform is a denser set (e.g. a section of the real axis or a Cantor set), the time evolution is complicated and may have some attributes which warrant the title chaotic.

In practice, it is very difficult to assess the nature of the response numerically.[22] The only problem which was treated analytically is a schematic model [21] of a 2-level system driven by a quasi-periodic train of δ–"kicks". It was shown that in this case the support of the Fourier transform is not a discrete denumerable set for any non-zero value of the fields. Hence, the wave function is

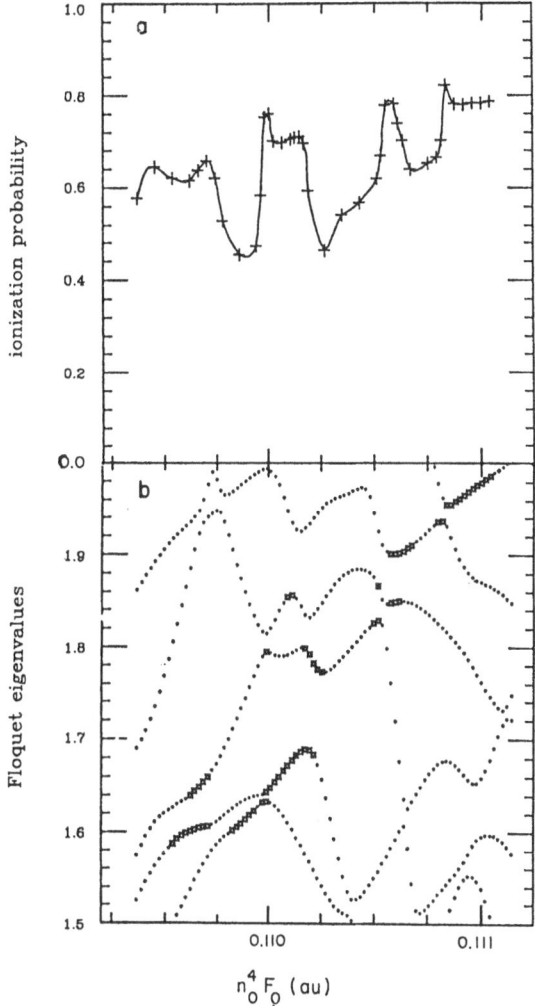

Fig. 4. a) A small section of the ionization curve of Fig 1 c) magnified to show individual spikes. b) qe spectrum as a function of ϵ. The avoided crossings which produce the spike structures are marked with crosses.

not quasi-periodic. Further studies on its exact characterization are in progress.

Much less is known for situations where the Hilbert space of the free system is of infinite dimension. The quasi-periodically kicked rotor is such a problem which was studied recently.[19,20] There, it was shown that an increase of the strength of the driving beyond a certain value was accompanied by a drastic change in the support of the Fourier transform. It changed from discrete to dense, although the exact characterization of the set is not given. This system, when kicked periodically, responds quasi-periodically and has a localized qe spectrum. Once kicked with the same strength but quasi-periodically, it may respond in an a-periodic fashion and since the localization property is lost, it can absorb an unlimited amount of energy from the field. It is not clear what determines the transition from one mode to another.

As was mentioned above, a system driven by bichromatic perturbation may respond quasi-periodically. Indeed if the series (3.2) converges uniformly, a simple generalization of the Floquet theorem does exist[23,24] and the wave function can be written as

$$|\Psi(t)\rangle = \sum_\alpha e^{i\eta_\alpha t}|\phi_\alpha(t)\rangle\langle\phi_\alpha(0)|n_0\rangle \qquad (3.1)$$

where η_α play the role of the "quasi-energies" and $|\phi_\alpha(t)\rangle$ are quasi-periodic vectors of the form

$$|\phi_\alpha(t)\rangle = \sum_{n,m} |a_\alpha^{(n,m)}\rangle e^{i(n\omega_1 + m\omega_2)t} \qquad (3.2)$$

Here, $|a_\alpha^{(n,m)}\rangle$ are constant vectors and α is a discrete index. Under these conditions it is tempting to propose the following (as yet unproved) argument. The frequency ratio is irrational, but it can be approximated to any desired accuracy by rationals p_n/q_n such that $|\omega_2/\omega_1 - p_n/q_n| < const./q_n^2$. Replacing in (1.1) the irrational frequency ratio by its rational approximants, we get problems with periodic driving, which, by Floquet's theorem, possess well-defined quasi-energies and eigenvalues. It seems reasonable to suggest that by increasing the order of the approximants one obtains in the limit the corresponding quantities for the case of the irrational frequency ratio. We do not know of a rigorous proof of the validity of this argument.

Microwave ionization of H-atoms is actually not the most convenient testing ground to check and study the problems stated above. From the theoretical point of view, the presence of the ionization channel introduces an imaginary part to the quasi-energies. Thus, all questions related to the recurrences of the wave functions, can be studied only within the finite lifetime of the neutral atom. From the experimental point of view, the finite exposure time, and the non-vanishing $Q-$ value of the microwave cavity, introduce a finite width to the frequency spectrum, and the irrationality (or rationality) of the frequency ratio can be determined only up to this uncertainty. The moral of this discussion is that the most relevant information which one can derive from the present study is to be found in the region of low fields where the lifetime against ionization is longer than the exposure time. The effects of the irrationality of the frequency ratio could thus be studied to the level of a fraction of a percent.

For this reason we shall concentrate on two observables which characterize the low ionization region – the ionization threshold fields and the subthreshold structures.

We performed classical and quantum mechanical calculations and deduced ionization curves and ionization thresholds. In both approaches we used the 1-dim. model for the H-atom, we neglected the slow turning-on and turning-off of the field experienced by the atom and did not average the results over the relative phases of the two fields. The calculations were performed for values of the field parameters which correspond to the experiments performed by the Stony-Brook group[16-18]. We avaraged ionization curve on an interval of ± 4 V/cm to allow for the distribution of field values experienced by different atoms in the beam.

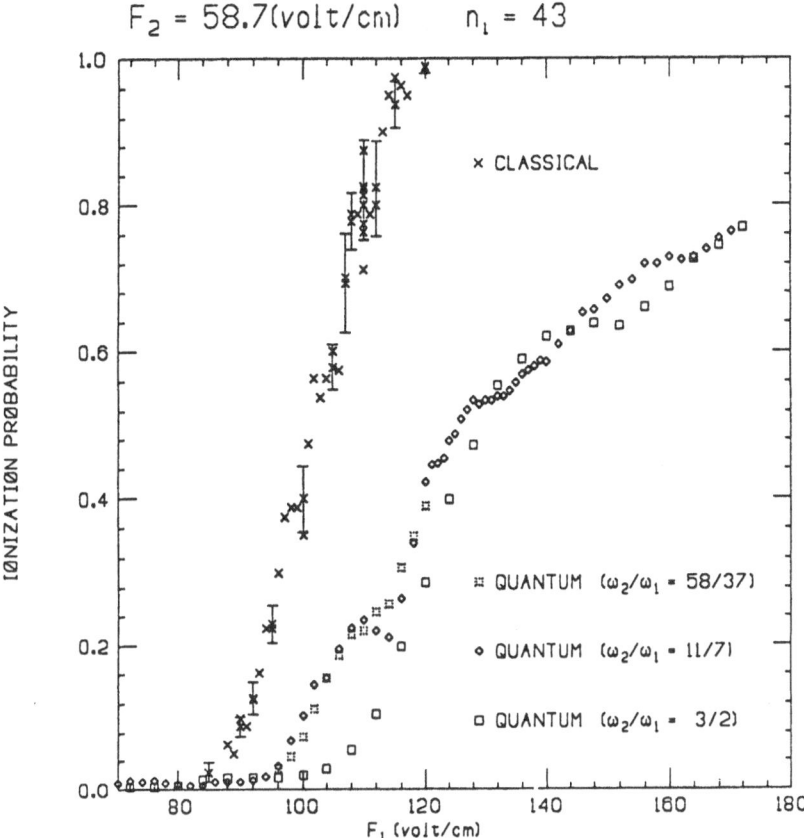

$$F_2 = 58.7(\text{volt/cm}) \qquad n_i = 43$$

Fig. 5. Classical and quantum mechanical ionization curves for bichromatic driving with $F_2 = 58.7$V/cm. The initial state is $n_0 = 43$. Three quantum mechanical calculations for the first rational approximants of the frequency ratio are shown.

A typical classical ionization curve is shown in Fig. 5. The error bars represent the estimated uncertainty due to finite sampling. At this level of accuracy we cannot distinguish any deviation from monotonicity, as was the case in the classical treatment of the ionization induced by monochromatic fields. The classical theory is invariant under a scale transformation, in which the fields are scaled by a factor n_0^4. In Fig. 6 we plot the locus of the scaled field values for which the ionization probability is 10%. The calculated points fall in a stripe which is centered about the line $\epsilon_1 n_0^4 + \epsilon_2 n_0^4 = 0.118$. This implies that the fields add together to induce the same level of ionization, and the most surprising observation is that they add linearly (and not quadratically) and with equal weights. This result is observed for the n_0 values studied here ($41 < n_0 < 45$) and the frequency ratio corresponding to the experimental conditions. The same correlation is observed experimentally and in the quantal theory. It is due to a simple and basic mechanism which will be discussed below.

Our codes which solve the quantum mechanical problem are much more efficient for a single frequency (periodic) driving. This fact adds a practical aspect to

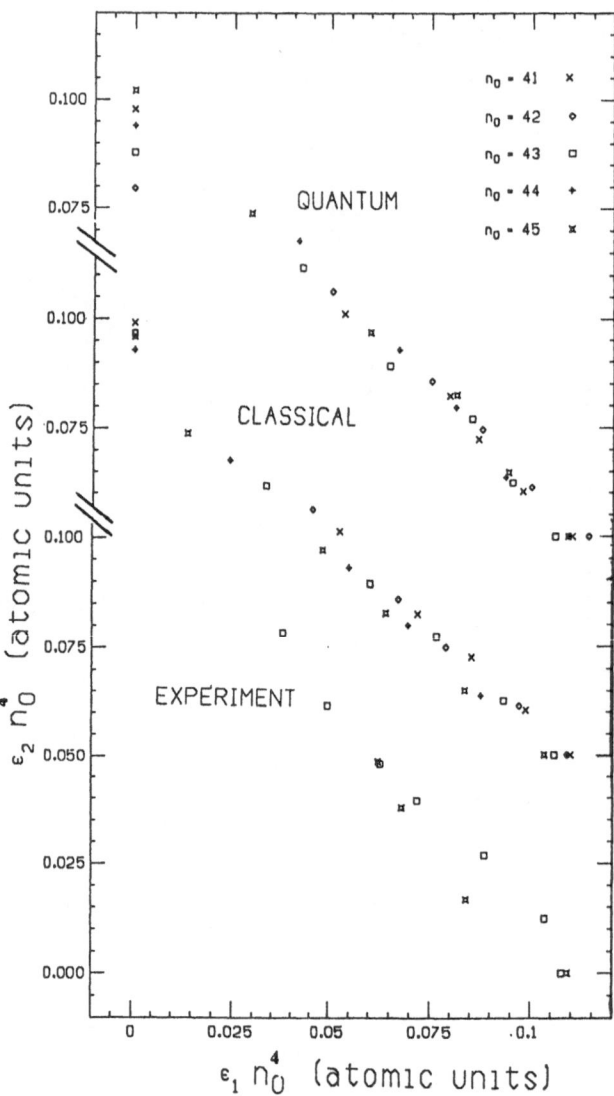

Fig. 6. The locus of the ionization threshold fields (scaled). From top to bottom: quantum mechanical, classical and experimental[16-18].

the general questions discussed above concerning the justification of using rational approximants instead of the exact frequency ratio. In Table 1 we show the first few rational approximants and the corresponding errors. In Fig. 5 we show ionization curves in which the first three approximants were used. Overall the three approximants give rather similar curves. Note however that the lowest approximant fails to produce the "bump" peaked at F_1=110 V/cm, while the next two sets reproduce each other to a rather high average accuracy. As was mentioned above, the data presented here is the average over a ±4 V/cm interval. The raw data shows spiky behaviour in this "bump" region, much the same as the avoided crossing structures which were discussed in detail in the previous section. This fact leads us to propose that in the present problem, and for the region of weak ionization, the generalized Floquet theorem applies. The sub-threshold structures are due to the clustering of avoided crossings, and an accurate description of these effects can be achieved by using the second rational approximant to the frequency ratio. Table 1 shows that the fraction 11/7 reproduces the experimental value to $2.1^{\circ}_{\circ\circ}$. Within a period of a few hundred cycles the system cannot resolve such a small frequency difference. On the basis of these observations we felt safe to carry on all the quantum calculations using for the frequency ratio the value 11/7.

Table 1

ω_2/ω_1	relative error (%)
$3/2 = 1.5$	4.35
$11/7 = 1.57143$	0.21
$58/37 = 1.56757$	0.038
$69/44 = 1.568182$	0.0014
experiment=1.568160	–

First four rational approximants to the experimental frequency ratio and their relative deviations (in percent).

The theoretical ionization curves reproduce rather well the experimental curves,[16−18] and in particular, they predict the location of subthreshold structures, their magnitude and width. We calculated the quantum mechanical locus of (scaled) threshold fields. The quantum results (Fig. 5) are consistent with the classical theory, and this agreement is reminiscent of the same behaviour in the case of monochromatic driving. This is probably due to the fact that for both frequencies used here, the classically scaled values satisfy $\omega_i n_0^3 < 1, i = 1, 2$. This is the regime of parameters where the perturbation is slow on the atomic time scale. In adiabatic situations the maximum value obtained by the field is the relevant parameter, and here it is the sum of the the two (scaled) fields. As was mentioned above, the linear correlation which we found theoretically is consistent with the threshold values observed experimentally.[16−18] We would like to stress that this correlation is linear on the *average*, and fluctuations which are due to specific resonance conditions, subthreshold structures etc. are definitely noticeable.

In conclusion we may say that our understanding of the microwave ionization of H-atoms in the domain $\omega n_0^3 < 1$ is rather complete. The study of bichromatic driving confirms the theoretical ideas and computational devices which were developed for the interpretation of the ionization induced by periodic fields. The appearance of subthreshold structures, and the fact that in the quantal theory, they are stable against increasing the rational approximant of the frequency ratio, is quite revealing. It is an indication that in this regime of the parameters, the simple generalization of Floquet's theorem does hold. In a sense this is disappointing, since we would have liked to see new physics emerging from these complicated and very expensive experiments and calculations.

The situation in the regime $\omega n_0^3 > 1$ may have some surprises in store. In this regime the theory for the periodically driven quantum problem predicts suppression of ionization due to localization of the qe states.[12] The localization phenomenon is sensitive to the quantum phases between matrix elements of the evolution operator, which may be drastically affected when a second field is switched on. This, and other related questions, are now under study.

ACKNOWLEDGEMENTS

We are grateful to Professor P.M. Koch for showing us his experimental data prior to its publications and for stimulating discussions. This research was supported in part by a grant from the United States – Israel Binational Science Fund (BSF), Jerusalem, Israel, and by the Israeli Basic Research Commission.

REFERENCES

1. J.E. Bayfield and P.M. Koch, Phys. Rev. Lett. 33:258 (1974); J.E. Bayfield, L.D. Gardner, and P.M. Koch, Phys. Rev. Lett. 39:76 (1977).

2. P.M. Koch, J. Physique 43:C2, 187 (1982).

3. P.M. Koch, Proc. NATO Advanced Research Workshop on Fundamental Aspects of Quantum Theory, Como, Italy (1985), V. Govini and A. Frigerio, ed., Plenum, London (1986).

4. K.A.H. van Leeuwen, G. v. Oppen, S. Renwick, J.B. Bowlin, P.M. Koch, R.V. Jensen, O. Rath, D. Richards, and J.G. Leopold, Phys. Rev. Lett. 55:2231 (1985).

5. P.M. Koch, Proc. First Int. Conf. on the Physics of the Phase Space, Univ. of Maryland, 20–23 May, 1986, to be published in "Physics of the Phase Space", Lecture Notes in Physics (Springer, 1987).

6. N.B. Delone, B.P. Krainov, and D.L. Shepelyansky, Usp. Fiz. Nauk. 140:355 (1983); Sov. Phys. Usp. 26:551 (1983), and references therein.

7. J.G. Leopold and I.C. Percival, J. Phys. B12:709 (1979).

8. R.V. Jensen, Phys. Rev. Lett. 49:1365 (1982); Phys. Rev. A30:386 (1984) and Proc. Workshop on "Photons and Continuum States of Atoms and Molecules", Cortona, Italy, June 1986, Springer, New York (1987).

9. J.G. Leopold and D. Richards, J. Phys. B18:3369 (1985); J. Phys. B19:1125 (1986) O. Rath and D. Richards, in preparation.

10. R. Blümel and U. Smilansky, Z.Phys. D6:83 (1987); Physica Scripta 35:15 (1987).

11. R. Blümel and U. Smilansky, Phys. Rev. Lett. 58:2531 (1987).

12. G. Casati, B.V. Chirikov, D.L. Shepelyansky, and I. Guarneri, Phys. Rev. Lett. 56:2437 (1986); Phys. Rev. Lett. 57:823 (1986); Physics Reports (in press) (1987).

13. R. Blümel and U. Smilansky, Proc. Conf. on Chaos and Related Non-Linear Phenomena, I. Procaccia, ed., Plenum, London (1987).

14. J.N. Bardsley, B. Sundaram, L.A. Pinnaduwage, and J.E. Bayfield, Phys. Rev. Lett. 56:1007 (1986).
    J.N. Bardsley and M.J Comella, J. Phys. B19:L565 (1986).

15. D. Richards, J. Phys. B20:2171 (1987).

16. L. Moorman, E.J. Galvez, B.E. Sauer, A. Mortazawi, K.A.H. van Leeuwen, G. v. Oppen and P.M. Koch, Bull. Am. Phys. Soc. 32:1264 (1987).

17. L. Moorman, E.J.Galvez, B.E. Sauer, A. Mortazawi, K.A.H. van Leeuwen, G. v. Oppen, and P.M. Koch, preprint (December 1987).

18. P.M.Koch, Proc. XV Intl. Conf. on the Physics of Electronic and Atomic Collisions, H.B. Gilbody, W.R. Newell, F.H. Read and A.C.H. Smith, ed. North-Holland Physics Publishing (1987).

19. D.L. Shepelyansky, Physica D8:208 (1983).

20. M.Samuelides, R.Fleckinger, L.Touziller, and J. Bellisard, Europhys. Lett. 1:203 (1986).

21. J.M. Luck, H. Orland, and U. Smilansky. In preparation (1987).

22. Y. Pomeau, B. Dorizzi, and B. Grammaticos, Phys. Rev. Lett. 56:581, (1986).
    See also comment by R. Badii and P.F. Meier, Phys. Rev. Lett. 58:1045 (1987).

23. T.S. Ho and S.I. Chu, J. Phys. B17:2101 (1984).

24. C.S. Chang and P. Stehle, Phy. Rev. A4:641 (1971).

A ZERO-ORDER ELECTRONIC STRUCTURE MODEL FOR IONIC DIATOMIC MOLECULES

Robert W. Field

Department of Chemistry
Massachusetts Institute of Technology
Cambridge, Massachusetts 02139

An atomic-ion-in-molecule electrostatic model for the valence state electronic structure of MX diatomic molecules (M=Ca-Cu, Ba-Hf, X=H,F,O) is proposed. The generalized Rittner[1], multiconfiguration crystal field model, which is expected to be valid in the $r \approx r_e$ region of internuclear separation, makes quantitative, predictive use of the Periodic Table via atomic spectral data and <u>ab initio</u> atomic-ion orbitals[1-4]. The key ideas are: (i) the couplings between core-like 3d or 4f electrons in compact $\langle r \rangle_{n\ell} \lesssim 0.3$ Å M-centered orbitals to form L-S-J terms are vastly more important than the molecular field due to a $\sim$2Å distant X^- ligand[4]; (ii) ligand-induced, M-centered $s \sim p$ and $p \sim d$ mixing causes the large, highly polarizable, and low-lying M-centered orbitals to "pooch" out the back side of the M, thereby justifying neglect of overlap (and all covalent bonding effects) between M and X centered orbitals[1,2]; (iii) rather than grouping molecular electronic states into the traditional Λ-S multiplets, larger groupings into super-multiplets and superconfigurations optimally display and utilize atomic multiplet and configurational structures[4,5]; (iv) the representation of molecular electronic structure by an effective Hamiltonian that includes atomic-ion structures via parametric reduction to F^k, G^k, $\zeta_{n\ell}$ electrostatic, exchange, and spin-orbit parameters and metal $\leftrightarrow$ ligand interactions via explicitly calculated crystal field multipole integrals (which depend only on the size, not the nodal structure of M-centered orbitals) contains all single-center correlation effects[2-4]; (v) some properties, such as spin-orbit coupling constants which are determined near one nucleus, are robust and therefore transferable atom $\rightarrow$ molecule and molecule $\rightarrow$ molecule, others are transferable only when the molecule displays the symptoms (unrecognizable without the supermultiplet model) of extreme separated-atomic-ion localization.

Features of this model have been tested against spectroscopic data generated by a variety of high resolution laser spectroscopic techniques[1-3,6]. All low lying states of all rare earth oxides are well represented by a zero-adjustable parameter $M^{+2}O^{-2}$ crystal field model[3,4]. All properties of the CaX (X=F, Cl, Br, I) $X^2\Sigma^+$, $A^2\Pi$, $B^2\Sigma^+$, and $A'^2\Delta$ states are reproduced by a generalized Rittner Ca^+X^- model which is a single-electron version of ligand-induced M-centered configuration interaction[2]. This multi-configuration crystal field model was extended to LaF, ThO, and HfO molecules which have two electrons outside closed shells[2]. CaO was found to display extreme localization onto Ca^+ and O^-, the two parts being transferable without modification from the structures of Ca^+F^- and Na^+O^- [1,7]. A supermultiplet treatment of NiH[8], modelled as $Ni^+(d^9)H^-$ despite the large size $[\langle r \rangle_{H^-}(1s) > r_e(NiH)]$ and polarizability of H^-, demonstrates the extreme reduction in number of parameters needed to fit the strongly interacting $^2\Delta$, $^2\Pi$, $^2\Sigma^+$ states arising from $Ni^+ d^9 \; ^2D$ even when Ni/H overlap cannot be neglected.

Acknowledgement

This work is supported by NSF grants PHY87-09759 and CHE86-14437.

References

1. S.F. Rice, H. Martin, and R.W. Field, "The Electronic Structure of the Calcium Halides. A Ligand Field Approach", J. Chem. Phys. 82:5023 (1985).
2. H. Schall, E. Murad, and R.F. Barrow, "The Electronic Structure of LaF: A Multi-Configuration Ligand Field Calculation", J. Chem. Phys. 87:2898 (1987).
3. M. Dulick, E. Murad, and R.F. Barrow, "Thermochemical Properties of the Rare Earth Monoxides", J. Chem. Phys. 85:385 (1986).
4. R.W. Field, "Diatomic Molecule Electronic Structure Beyond Simple Molecular Constants", Ber. Bunsenges. Phys. Chem. 86:771 (1982).
5. C. Linton, M. Dulick, R.W. Field, P. Carette, P.C. Leyland, and R.F. Barrow, "Electronic States of the CeO Molecule. Absorption, Emission and Laser Spectroscopy", J. Mol. Spectrosc. 102:441 (1983).
6. H. Schall, J. Gray, M. Dulick, and R.W. Field, "Sub-Doppler Zeeman Spectroscopy of the CeO Molecule", J. Chem. Phys. 85:751 (1986).
7. J.N. Allison, R.J. Cave, and W.A. Goddard, III, "Alkali Oxides. Analysis of Bonding and Explanation of the Reversed Ordering of the $^2\pi$ and $^2\Sigma^+$ States", J. Phys. Chem. 88:1262 (1984).
8. J.A. Gray, Laser Spectroscopy and A Supermultiplet Structure Model for Nickel Hydride. Ph.D. Thesis, M.I.T. (1987).

RECENT DEVELOPMENTS IN STUDIES OF OPEN-SHELL CATIONS

VIA THEIR ELECTRONIC TRANSITIONS

F.G. Celii, P.O. Danis, D. Forney, M. Rösslein,
T. Wyttenbach and J.P. Maier

Institut für Physikalische Chemie, Universität Basel
Klingelbergstrasse 80, CH-4056 Basel, Switzerland

Some of our recent developments aimed at spectral characterization of open-shell cations by means of their electronic transitions will be outlined in this contribution. These involve on the one hand the use of the well established methods - laser excitation and matrix absorption spectroscopy - to the study of the first fragment ion, C_2^+, and on the other hand the application of new approaches based on stimulated emission pumping or two-colour absorption spectroscopy on mass-selected ions.

Our studies in the past decade **have** focused on the spectral characterization of cations of stable, or semi-stable, molecules which decay radiatively from their excited electronic states.[1] The methods used involve the observation of emission spectra from effusive and supersonic free jets after electron beam excitation and of laser excitation of the fluorescence following Penning ionization. A further approach, the measurement of the absorption spectra of ions in 5 K neon matrices, produced by photoionization, has been applied to study some of the ions which do not fluoresce.[2] The present projects are aimed at spectral characterization of ions which are difficult to identify, such as fragment ions.

The first fragment ion we have studied is C_2^+. Although it is the simplest carbon molecular ion, and is involved in chemical schemes for formation of larger carbon species in interstellar space, plasmas and flames, its spectroscopic knowledge was based essentially on theoretical calculations.[3,4] Of the two earlier experimental studies, one was a low resolution measurement[5] whereas the assignment of the other[6] is in doubt.[3,4] In order to study fragment ions, such as C_2^+, our strategy is to characterize them first in a neon matrix, because the absorption measurement can be carried out with ease in the 220-1200 nm region to locate the electronic transition. With such data in hand, higher resolution gas-phase methods can be applied.

C_2^+ was generated by a two-step process in a 5 K neon matrix.[7] Either acetylene, or preferably chloroacetylene, was deposited with neon in a dilute concentration (~1:4000) and C_2 (monitored by its known absorption systems) was produced by photolysis with Xe I (147 nm) radiation. Subsequent photoionization of the matrix with Ne I (73.6 nm) radiation led to a new band system (fig. 1). The latter is identified to be the $\widetilde{B}\ ^4\Sigma_u^- - \widetilde{X}\ ^4\Sigma_g^-$ transition of C_2^+ on the basis of chemical evidence (different precursors), isotopic data (^{13}C shifts of bands), bleaching experiments (ionic species involved) and comparison with the calculated[4] transition energy and vibrational frequency of the $\widetilde{B}\ ^4\Sigma_u^-$ state. Laser excitation and dispersion of

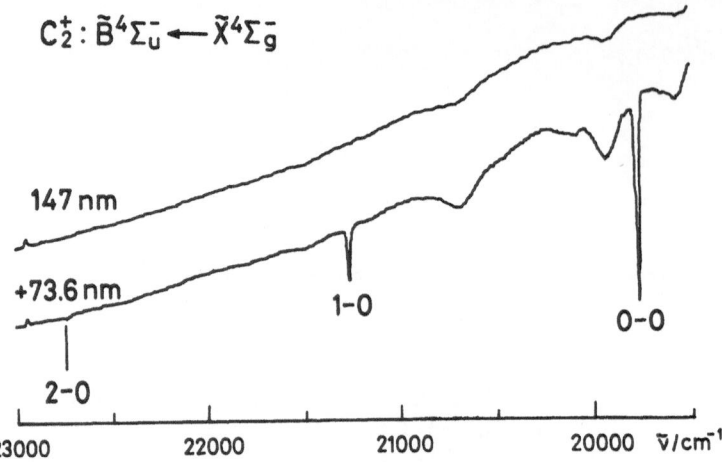

Fig. 1. The $\tilde{B}\,^4\Sigma_u^- \leftarrow \tilde{X}\,^4\Sigma_g^-$ v'-v" absorption spectrum of C_2^+
(0.1 nm resolution) in a 5 K neon matrix. ClCCH was de-
posited in neon ($\sim 1{:}4000$), C_2 was produced by Xe I (147
nm) photolysis and C_2^+ by subsequent Ne I (73.6 nm) photoionization.

the fluorescence in the matrix also enabled the frequency of C_2^+ in the
$\tilde{X}\,^4\Sigma_g^-$ state to be inferred.[7]

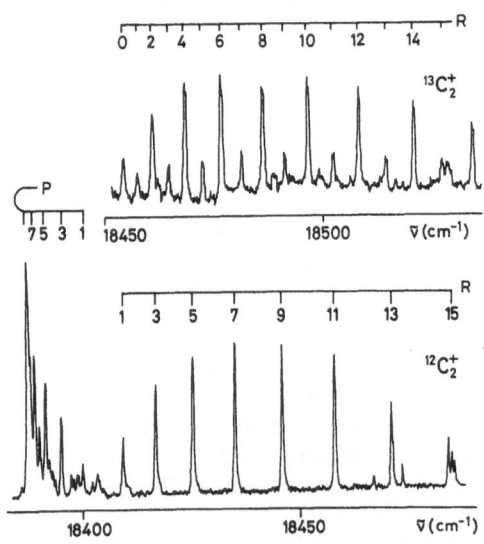

Fig. 2. The $\tilde{B}\,^4\Sigma_u^- - \tilde{X}\,^4\Sigma_g^-$ laser excitation spectra (0.3 cm^{-1}
fwhm) of the 0-1 band of $^{12}C_2^+$ and $^{13}C_2^+$.

With the knowledge of the location of the 0-0 band of the $\widetilde{B}$ $^4\Sigma_u^-$ -
$\widetilde{X}$ $^4\Sigma_g^-$ C_2^+ transition in the matrix (19765±4 cm^{-1}), it proved possible to
observe this band system in the gas phase (origin ~ 19731 cm^{-1}) by laser
excitation spectroscopy.[8] C_2^+ was produced by reacting He metastables with
acetylene, or cyanogen, under conditions such that collisions are minimised.
The rotationally resolved laser excitation spectrum (fig. 2) shows features
characteristic of a Σ-Σ transition of a homonuclear carbon species (even
lines are missing) and the spectrum on ^{13}C (nuclear spin of 1/2) substitution
shows a 3:1 intensity alternation (fig. 2), the even lines being more intense.

The rotational analysis is consistent with a $^4\Sigma_u^-$ - $^4\Sigma_g^-$ transition, and
the spin-splittings are partly resolved with increased resolution (0.04 cm^{-1}).
That $^{12}C_2^+$ is the species is unambiguously established from the ^{13}C isotopic
changes and relationships involving the rotational constants, the frequencies, and vibration-rotation interaction values. The lifetimes of selected
rotational levels within the $\widetilde{B}$ $^4\Sigma_u^-$ state were also measured and the oscillator strengths for the various vibronic transitions, $\widetilde{B}$ $^4\Sigma_u^-$ - $\widetilde{X}$ $^4\Sigma_g^-$ v'-v",
determined; for example, f_{0-0} = 0.022 (3). Thus C_2^+ can now be identified by
this electronic transition in environments such as of flames and plasmas
(laser excitation), comets (resonance fluorescence) and perhaps diffuse
interstellar clouds (starlight absorption) and its abundance can be evaluated.

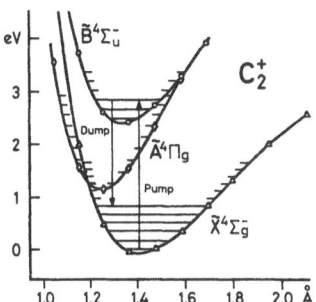

Fig. 3. Potential energy curves of the three lowest quartet states of C_2^+
redrawn from ref. [4], illustrating the scheme of the stimulated
pumping approach to study the $\widetilde{B}$ $^4\Sigma_u^-$ - $\widetilde{X}$ $^4\Sigma_g^-$ 2-5 transition.

The levels v"=0-3 in the $\widetilde{X}$ $^4\Sigma_g^-$ state were probed by the laser excitation method because these were sufficiently populated in the process leading
to C_2^+ formation. To obtain further information on the potential energy surface, higher lying vibrational levels have been studied by a stimulated
emission pumping approach. We recently demonstrated that this is a feasible
and attractive technique for the investigation of transient species such as
ions produced in a flow system[9]; previously it had been used for the study
of stable molecular species in static cells[10] and supersonic free jets[11].

337

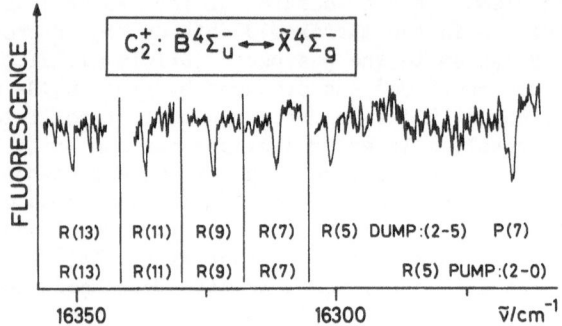

Fig. 4. Some rotational lines of the $\tilde{B}\ ^4\Sigma_u^-$ – $\tilde{X}\ ^4\Sigma_g^-$ 2-5
transition of C_2^+ measured by pumping a selected
line R(N) of the v' = 2 level and then observing
one R(N) and one P(N+2) line of the 2-5 band
system by scanning the dump laser (0.3 cm^{-1} fwhm)
while the fluorescence from the v' = 2 state is
monitored.

The scheme of the approach is shown in figure 3 using the calculated potential energy curves of the quartet states of C_2^+. The population is initially transferred by laser excitation (pump) from the $\tilde{X}\ ^4\Sigma_g^-$ state to a
level of the $\tilde{B}\ ^4\Sigma_u^-$ state (e.g. v' = 2), and transitions are then stimulated into the $\tilde{X}\ ^4\Sigma_g^-$ manifold by a second laser (dump). Thus highly excited
levels of the $\tilde{X}\ ^4\Sigma_g^-$ state manifold can be accessed (i.e. those not observable in normal emission due to the low Franck-Condon factors) and with resolution set by the laser band-width. In practice the fluorescence from the
v' = 2 level of the $\tilde{B}\ ^4\Sigma_u^-$ state is monitored while the wavelength of the
dump dye-laser is scanned. Such a spectrum is reproduced in figure 4 whereby the v" = 5 level of the $\tilde{X}\ ^4\Sigma_g^-$ state is rotationally characterized. A
chosen R(N) transition is pumped (v'=2 – v"=0) and one R(N) and one P(N+2)
line of the v'=2 – v"=5 system is observed in the stimulated emission pumping spectrum (cf. fig. 4). By pumping a series of R(N) lines, a rotational
spectrum is accumulated and an analysis can then be carried out. By this
means levels with v"=3 – 6 of the $\tilde{X}\ ^4\Sigma_g^-$ state of C_2^+ have been studied.[12]
In addition it should be possible to use this method to characterize the
$\tilde{A}\ ^4\Pi_g$ state (fig. 3) of C_2^+ by stimulating transitions to this from higher
excited vibrational levels of the $\tilde{B}\ ^4\Sigma_u^-$ state.

The final new approach to be outlined is based on two-photon absorption by mass-selected ions. The essential features of the optical and
experimental schemes are given in figure 5. The apparatus is a home-built
triple quadrupole mass spectrometer. Ions are produced by electron impact
in a high pressure source (0.01-0.1 mbar) so that they are collisionally
relaxed. The ion of interest is mass-selected (M^+) by the first quadrupole
(Q_1) and is then injected into the middle (Q_2) RF trapping field. The ions
there are electronically excited (say to state $\tilde{A}$ – fig. 5) by the first
laser photon ($h\nu_1$) and subsequently by a second laser photon ($h\nu_2$) to a
higher excited electronic state ($\tilde{J}$). The latter state is dissociative/pre-
dissociative leading to fragment ions (F^+), which are first constrained in
the RF field of Q_2 and then are mass-selected in the third quadrupole (Q_3).
Thus either $h\nu_1$ is scanned with $h\nu_2$ fixed or vice-versa, while F^+ are
detected.

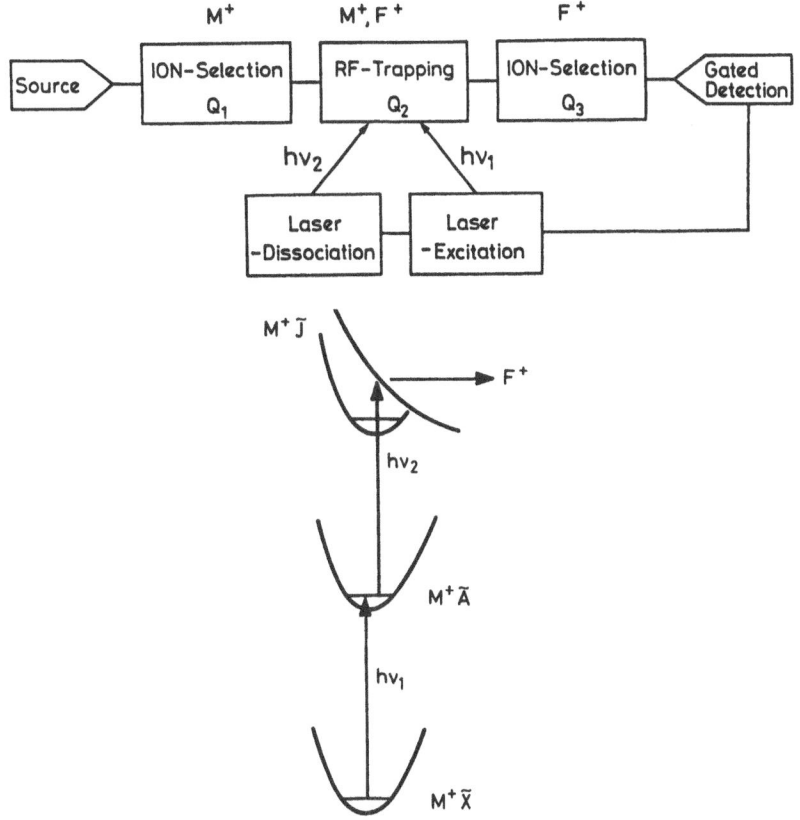

Fig. 5. Schematic diagram of the experimental arrangement and optical excitation used for two-photon spectroscopy of mass-selected ions.

The initial studies have been carried out on N_2O^+ and CS_2^+ [13] because their lowest lying electronic states are well known by emission spectroscopy. The first experiment demonstrates the approach to characterize stable electronic states of ions, e.g. $\tilde{B}\ ^2\Sigma_u^+$, $\tilde{X}\ ^2\Pi_g$ states of CS_2^+. The CS_2^+ species are excited in the Q_2 quadrupole by two photons: $h\nu_1$ is varied across the $\tilde{B}\ ^2\Sigma_u^+ \leftarrow \tilde{X}\ ^2\Pi_g$ transition while $h\nu_2$ is fixed on a $\tilde{C}\ ^2\Sigma_g^+ \leftarrow \tilde{B}\ ^2\Sigma_u^+$ band. The $\tilde{C}\ ^2\Sigma_g^+$ state of CS_2^+ falls apart to produce S^+ (or CS^+) ions which are monitored. In figure 6 is shown the rotational structure on the origin band of the $\tilde{B}\ ^2\Sigma_u^+ \leftarrow \tilde{X}\ ^2\Pi_g$ transition of CS_2^+ recorded by this technique. The second possibility is to scan $h\nu_2$ while $h\nu_1$ prepares state selected ions. However, the success in obtaining spectroscopic information on the predissociative state is limited to cases when its lifetime is long enough that the vibrational and rotational structure is not smeared out.

In the case of the $\tilde{B}\ ^2\Pi$ state of N_2O^+ the rotational structure is not resolved and only a complex vibronic pattern is observed when the $\tilde{B}\ ^2\Pi \leftarrow \tilde{A}\ ^2\Sigma^+$ transition is excited. A lifetime of 3×10^{-14} s is deduced for the upper electronic state.[13] This short lifetime is due to intramolecular coupling of electronic states[14] rather than to direct predissociation. On the other hand scanning through the $\tilde{C}\ ^2\Sigma_g^+ \leftarrow \tilde{B}\ ^2\Sigma_u^+$ transition of CS_2^+, while $h\nu_1$ prepares the $\tilde{B}\ ^2\Sigma_u^+$ state, shows a well defined vibrational pattern (fig. 7). The rotational structure can be resolved as well when individual rotational lines in the $\tilde{B}\ ^2\Sigma_u^+ \leftarrow \tilde{X}\ ^2\Pi_g$ transition are selected. Thus the rotational and vibrational constants of CS_2^+ in the $\tilde{C}\ ^2\Sigma_g^+$ state could be obtained.[13]

339

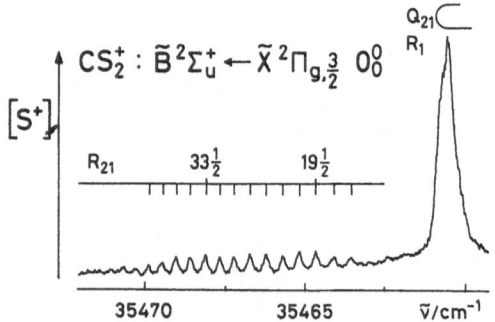

Fig. 6. Part of the $\tilde{B}\ ^2\Sigma_u^+ - \tilde{X}\ ^2\Pi_g\ 0_0^0$ band of CS_2^+ recorded by two colour photon absorption on a mass-selected ion beam. The first dye laser (0.04 cm^{-1} fwhm) scans the known transition while the second dye laser is set on the $\tilde{C}\ ^2\Sigma_g^+ \leftarrow \tilde{B}\ ^2\Sigma_u^+\ 0_0^0$ transition which yields S^+ ions which are monitored.

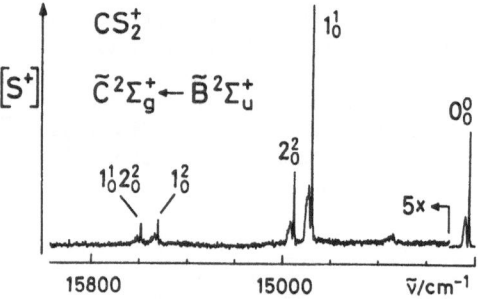

Fig. 7. The vibronic structure of the $\tilde{C}\ ^2\Sigma_g^+ \leftarrow \tilde{B}\ ^2\Sigma_u^+$ transition of CS_2^+ recorded by scanning a dye laser (0.6 cm^{-1} fwhm) while monitoring S^+ ions formed from the upper state. The CS_2^+ ions are prepared electronically excited by pumping the $\tilde{B}\ ^2\Sigma_u^+ \leftarrow \tilde{X}\ ^2\Pi_g\ 0_0^0$ transition.

For the future, it is hoped that the described approaches and results pave the way for the spectral characterization of a variety of interesting ions – fragment, isomer and small cluster ions - which have hitherto only been identified by mass-spectrometry.

Acknowledgements
     The financial support for the outlined research has been provided by the "Schweizerischer Nationalfonds zur Förderung der wissenschaftlichen Forschung".

REFERENCES

1. For a recent review see, J.P. Maier, J. El. Spectrosc. Rel. Phenom., 40:203 (1986).
2. J. Fulara, S. Leutwyler, J.P. Maier and U. Spittel, J. Phys. Chem. 89:3190 (1985).
3. C. Petrongolo, P.J. Bruna, S.D. Peyerimhoff and R.J. Buenker, J. Chem. Phys., 74:4594 (1981).
4. P. Rosmus, H.-J. Werner, E.-A. Reinsch and M. Larsson, J. El. Spectrosc. 41:289 (1986).
5. A. O'Keefe, R. Derai and M.T. Bowers, Chem. Phys., 91:161 (1984).
6. H. Meinel, Can. J. Phys., 50:158 (1972).
7. D. Forney, H. Althaus and J.P. Maier, J. Phys. Chem., 91: Dec. issue, 1987.
8. M. Rösslein, M. Wyttenbach and J.P. Maier, J. Chem. Phys., 87: Dec. issue, 1987.
9. F.G. Celii, J.P. Maier and M. Ochsner, J. Chem. Phys., 85:6230 (1986).
10. C.E. Hamilton, J.L. Kinsey and R.W. Field, Annu. Rev. Phys. Chem., 37: 493 (1986) and references therein.
11. S.H. Kable and A.E.W. Knight, J. Chem. Phys., 86:4709 (1987).
12. F.G. Celii and J.P. Maier, unpublished data.
13. P.O. Danis, T. Wyttenbach and J.P. Maier, J. Chem. Phys., 88, 1988 in press.
14. H. Köppel, L.-S. Cederbaum and W. Domcke, Chem. Phys., 69:175 (1982).

INTERNATIONAL WORKSHOP ON
"THE STRUCTURE OF SMALL MOLECULES AND IONS"
Held in the memory of Prof. Itzhak Plesser
Neve Ilan, Israel
December 13 - 18, 1987

PARTICIPANTS

AGRANAT, I.
Dept. of Organic Chemistry
The Hebrew Univ. of Jerusalem
91 904 Jerusalem
Israel

APELOIG, Y.
Dept. of Chemistry
Technion
Israel Institute of Technology
32 000 Haifa, Israel

BERRY, S.
Dept. of Chemistry
University of Chicago
Chicago, IL 60637
U.S.A.

BOWEN, K.
Dept. of Chemistry
John Hopkins University
Baltimore, MD 21218
U.S.A.

BUELOW, S.
Mail Stop J567
Los Alamos National Lab.
Los Alamos, NM 87545
U.S.A.

CASTLEMAN Jr., A.W.
Dept. of Chemistry
The Pennsylvania State Univ.
152 Davey Lab., University Park
PA 16802, U.S.A.

CHECHNOWSKY, O.
Chemistry Dept.
Tel Aviv University
Ramat Aviv
Israel

COHEN, S.
Dept. of Isotope Res.
Weizmann Institute of Science
76 100 Rehovot
Israel

DAVIDSON, E.R.
Dept. of Chemistry
University of Indiana
Bloomington, IN 47405
U.S.A.

FAIBIS, A.
Nuclear Physics Dept.
Weizmann Institute of Science
76 100 Rehovot
Israel

FIELD, R.W.
Dept. of Chemistry
MIT, Cambridge, MA 02139
U.S.A.

GERBER, B.
Dept. of Physical Chemistry
Hebrew University of Jerusalem
Jerusalem
Israel

GOLDRING, G.
Nuclear Physics Dept.
Weizmann Institute of Science
76 100 Rehovot
Israel

GOODSON, D.
Chemical Physics Dept.
Weizmann Institute of Science
76 100 Rehovot
Israel

GRANT, E.
Dept. of Chemistry
Purdue University
West Lafayette, IN 47907
U.S.A.

HOLMES, J.L.
Chemistry Dept.
University of Ottawa
Ottawa, KIN 6N5
Ontario, Canada

HORN, T.
Dept. of Physical Chemistry
The Hebrew Univ. of Jerusalem
91 904 Jerusalem
Israel

HOWARD, B.
Dept. of Chemistry
University of Oxford
Oxford OX1 3NP
U.K.

IRAQI, M.
Dept. of Physical Chemistry
The Hebrew Univ. of Jerusalem
91 904 Jerusalem
Israel

JELLINEK, J.
Chemistry Division
Argonne National Laboratory
9700 South Cass. Ave.
Argonne, IL 60439
U.S.A.

JORTNER, J.
School of Chemistry
Tel Aviv University
69 978 Ramat Aviv
Israel

KABABIA, S.
Dept. of Physical Chemistry
The Hebrew Univ. of Jerusalem
91 904 Jerusalem
Israel

KALDOR, U.
Dept. of Chemistry
Tel Aviv University
Ramat Aviv
Israel

KANTER, E.
Argonne National Lab. (Bldg. 203)
9700 South Cass. Ave.
Argonne, IL 60439
U.S.A.

KIMURA, K.
Dept. of Molecular Assemblies
Inst. for Molecular Science
Myodaiji, Okazaki 444
Japan

KLEMPERER, W.
Dept. of Chemistry
Harvard University
12 Oxford St., Cambridge,
Mass 02138, U.S.A.

KONDOW, T.
Dept. of Chemistry
University of Tokyo
Bunkyo-ku, Tokyo 113
Japan

KOOT, W.
FOM - Institute for Atomic
   and Molecular Physics
1098 SJ Amsterdam
The Netherlands

KOSLOF, R.
The Fritz Haber Res. Center
   for Molecular Dynamics
Hebrew University of Jerusalem
Givat Ram
91 904 Jerusalem, Israel

KOVNER, H.
Nuclear Physics Dept.
Weizmann Institute of Science
76 100 Rehovot
Israel

LEVINE, R.D.
Institute of Chemistry
The Hebrew Univ. of Jerusalem
91 904 Jerusalem
Israel

LEVINGER, A.
Chemical Physics Dept.
Weizmann Institute of Science
76 100 Rehovot
Israel

LEVY, D.
The James Franck Institute
University of Chicago
5640 S. Ellis Ave.
Chicago, IL 60637, U.S.A.

LIFSHITZ, C.
Dept. of Physical Chemistry
The Hebrew Univ. of Jerusalem
91 904 Jerusalem
Israel

LINEBERGER, W.C.
Univ. of Colorado and
Joint Inst. for Lab. Astrophysics
Boulder, CO 80309
U.S.A.

LLORENTE, J.G.
Dept. of Chemical Physics
Weizmann Institute of Science
76 100 Rehovot
Israel

LORQUET, J.C.
Dept. de Chimie Physique
Universite de Liege
Sart-Tilman, B-4000 Liege 1
Belgium

MAIER, J.P.
Inst. für Physikalische Chemie
der Universitat Basel
Klingelbergstrasse 80
CH-4056, Basel, Switzerland

MILLER, R.E.
Dept. of Chemistry
University of North Carolina
Chapel Hill, NC 27514
U.S.A.

MOISEYEV, N.
Technion
Israel Institute of Technology
32 000 Haifa
Israel

MORRISON, J.D.
Dept. of Chemistry
La Trobe University
Bundoora, Victoria
3083 Australia

NAAMAN, R.
Dept. of Isotope Research
Weizmann Institute of Science
76 100 Rehovot
Israel

NESBITT, D.
JILA
University of Colorado
Boulder, CO 80309
U.S.A.

NG, C.Y.
Dept. of Chemistry
  and Ames Laboratory-USDOE
Iowa State University
Ames, Iowa 50011, U.S.A.

OKA, T.
Dept. of Chemistry
University of Chicago
5735 S. Ellis Ave.
Chicago, IL 60637, U.S.A.

PAUNCZ, R.
Dept. of Chemistry
Technion
Israel Inst. of Technology
32 000 Haifa, Israel

PLESSER, R.
High Energy Physics Lab.
Harvard University
Cambridge, MA 02139
U.S.A.

POLLAK, E.
Chemical Physics Dept.
Weizmann Institute of Science
76 100 Rehovot
Israel

RADOM, L.
Research School of Chemistry
The Australian National Univ.
G.P.O.Box 4, Canberra, A.C.T. 2601
Australia

RAJWAN, M.
Dept. of Physical Chemistry
The Hebrew University
Jerusalem
Israel

RATNER, M.
Dept. of Chemistry
Northwestern University
2145 Sheridan Rd.
Evanston, IL 60201 - U.S.A.

REILLY, J.P.
Dept. of Chemistry
Indiana University
Bloomington, Indiana 47405
U.S.A.

RUDIK, I.
Dept. of Isotope Research
Weizmann Institute of Science
76 100 Rehovot
Israel

RUSCIC, B.
Rudjer Boskovic Inst.
P.O. Box 1016
41001 Zagreb
Yugoslavia

RUTTINK, P.J.A.
Theoretical Chemistry Group
Utrecht University
Padualaan 8, 3508 TB Utrecht
The Netherlands

SCHWARZ, H.
Dept. of Chemistry
Technical University
Strasse des 17 Juni 135
D-1000 Berlin 12, W. Germany

SEIDEMAN, T.
Chemical Physics Dept.
Weizmann Institute of Science
76 100 Rehovot
Israel

SMILANSKY, U.
Nuclear Physics Dept.
Weizmann Institute of Science
76 100 Rehovot
Israel

SOFER, I.
Dept. of Isotope Research
Weizmann Institute of Science
76 100 Rehovot
Israel

STAHL, D.
Institut de Chimie Physique
Ecole Polyt. Fed. de Lausanne
CH-1015 Lausanne
Switzerland

TERLOUW, J.K.
Analytisch Chemisch Laboratorium
Rijksuniversiteit te Utrecht
Croesestraat 77A, 3522 AD Utrecht
Holland

VAGER, Z.
Nuclear Physics Dept.
Weizmann Institute of Science
76 100 Rehovot
Israel

WEINHOLD, F.
Theoretical Chemistry Institute
Univ. of Wisconsin-Madison
Madison, WI 53706
U.S.A.

WOODS, R.C.
Dept. of Chemistry
Univ. of Wisconsin-Madison
Madison, WI 53706
U.S.A.

ZAJFMAN, J.
Nuclear Physics Dept.
Weizmann Institute of Science
76 100 Rehovot
Israel

INDEX